VOLUME ONE HUNDRED AND THREE

ADVANCES IN BOTANICAL RESEARCH

Stomata Regulation and Water Use Efficiency in Plants Under Saline Soil Conditions

ADVANCES IN BOTANICAL RESEARCH

Series Editor

Jean-Pierre Jacquot
Membre de l'Institut Universitaire de France, Unité Mixte de Recherche INRA, UHP 1136 "Interaction Arbres Microorganismes", Université de Lorraine, Faculté des Sciences, Vandoeuvre, France

VOLUME ONE HUNDRED AND THREE

ADVANCES IN BOTANICAL RESEARCH

Stomata Regulation and Water Use Efficiency in Plants Under Saline Soil Conditions

Volume Editor

SERGEY SHABALA

Professor and Head, Stress Physiology Laboratory, University of Tasmania, Australia; Distinguished Professor, Foshan University, China

ELSEVIER

ACADEMIC PRESS

An imprint of Elsevier

Academic Press is an imprint of Elsevier
125 London Wall, London, EC2Y 5AS, United Kingdom
The Boulevard, Langford Lane, Kidlington, Oxford OX5 1GB, United Kingdom
50 Hampshire Street, 5th Floor, Cambridge, MA 02139, United States
525 B Street, Suite 1650, San Diego, CA 92101, United States

First edition 2022

ISBN: 978-0-323-91217-4
ISSN: 0065-2296

For information on all Academic Press publications
visit our website at https://www.elsevier.com/books-and-journals

Publisher: Zoe Kruze
Acquisitions Editor: Sam Mahfoudh
Developmental Editor: Jhon Michael Peñano
Production Project Manager: James Selvam
Cover Designer: Vicky Pearson

Typeset by STRAIVE, India

Contents

Contributors *ix*
Preface *xiii*

1. Stomatal regulation and adaptation to salinity in glycophytes and halophytes **1**

Guang Chen, Hanna Amoanimaa-Dede, Fanrong Zeng, Fenglin Deng, Shengchun Xu, and Zhong-Hua Chen

1. Introduction 2
2. Stomatal physiological responses to salinity differ in glycophytes and halophytes 4
3. Membrane transporters regulate ion-homeostasis in stomata under salinity 9
4. Signaling pathways help plants regulate stomatal movement under salinity 22
5. Conclusions and future perspective 26

Acknowledgments 27
References 27

2. Water use efficiency in mangroves: Conservation of water use efficiency determined by stomatal behavior across leaves, plants, and forests **43**

Jie Liang, Graham D. Farquhar, and Marilyn C. Ball

1. Introduction 44
2. The definitions of water use efficiency 45
3. Long-term vs. short-term water use efficiency 47
4. Instantaneous WUE vs. intrinsic WUE 49
5. Compromise between WUE and photosynthetic nitrogen use efficiency 50
6. Into the future: Response of water-use efficiency to elevated CO_2 51
7. Ecosystem-level water use efficiency of mangrove forests is comparably high as the leaf level 52
8. Unresolved complications and future directions 54
9. Conclusion 56

Acknowledgments 56
References 56

3. Salty or sweet? Guard cell signaling and osmotic control under saline conditions 61
Yotam Zait and Sarah M. Assmann

1. Whole plant Na^+ responses 62
2. Overview of Na^+ transporters and their regulation 63
3. Guard cell operation in salt-stressed leaves 64
4. Conclusions 79

Acknowledgments 80
References 80

4. Modulation of abscisic acid signaling for stomatal operation under salt stress conditions 89
Jong-Joo Cheong

1. Introduction 90
2. Modulation of ABA metabolic pathways 92
3. Modulation of the ABA signaling network 97
4. Modulation of ABA signaling by protein degradation 105
5. Summary and concluding remarks 110

Acknowledgment 111
References 111

5. Impact of nutritional imbalance on guard cell metabolism and stomata regulation under saline hyperosmotic conditions 123
Christian Zörb, Bastian L. Franzisky, Patrick P. Lehr, Robin Kosch, Michael Altenbuchinger, and Christoph-Martin Geilfus

1. Introduction 124
2. The possible role of magnesium in stomatal regulation under sodium-salinity 125
3. The possible role of salt stress-induced GABA production as metabolic signal involved in reduction of stomatal aperture 129
4. The possible role of sulfur in guard cell metabolism under salt stress 132
5. Conclusion 134

References 135

6. Stomata under salt stress—What can mechanistic modeling tell us? 139
Nguyen B.A. Thu, Anna Amtmann, Michael R. Blatt, and Thanh-Hao Nguyen

1. Introduction 140

Acknowledgments 156
References 156

7. Signaling molecules and transcriptional reprogramming for stomata operation under salt stress **163**

Jahanzeb Jafar, Heba Hassan, Sergey Shabala, and Bo Ouyang

1. Introduction 164
2. Signaling molecules involved in stomatal closure 165
3. Transcription factors involved in the regulation of stomatal structure under salinity stress 172
4. Concluding remarks 182

Acknowledgments 183

References 183

8. Root-borne signals and their control of guard cell operation under saline conditions: The role of root signals in stomata regulation **195**

Mengliang Niu, Zhilong Bie, and Yuan Huang

1. Introduction 196
2. How do plants perceive stress and trigger signals in the root? 198
3. Root to shoot signaling 199
4. Local control of stomatal opening and closing under salinity 210
5. Conclusions and prospects 212

Acknowledgments 215

References 215

9. Salinity and night-time transpiration under current climate scenarios **221**

Wieland Fricke

1. Introduction 222
2. Transport processes: Getting things from A to B 224
3. Aquaporins 228
4. Night-time transpiration and growth 233
5. Night-time transpiration and energy 238
6. Outlook: Night-time transpiration in a changing climate 240

Acknowledgments 243

References 243

10. Application of omics technologies in single-type guard cell studies for understanding the mechanistic basis of plant adaptation to saline conditions **249**

Fatemeh Rasouli, Ali Kiani-Pouya, Yuan Wang, Ali Movahedi, Min Yu, Ayesha Tahir, and Sergey Shabala

1. Stomata and salinity 250

2. Application of omics approaches in single-cell-type studies in plants 252
3. Guard cell proteomics 254
4. Guard cell transcriptome 256
5. Guard cell metabolomics 257
6. Single cell isolation methods 258
7. Optimization of GC-enriched epidermal fragments for omics studies under saline conditions 259
8. Case studies of application of GC-enriched epidermal fragments method for proteomics analysis under saline conditions 261
9. Future prospects 264
References 266

Contributors

Michael Altenbuchinger
Department of Medical Bioinformatics, University Medical Center Göttingen, Göttingen, Germany

Hanna Amoanimaa-Dede
Hubei Collaborative Innovation Center for Grain Industry, College of Agriculture, Yangtze University, Jingzhou, China

Anna Amtmann
Laboratory of Plant Physiology and Biophysics, Bower Building, University of Glasgow, Glasgow, United Kingdom

Sarah M. Assmann
Department of Biology, Penn State University, University Park, PA, United States

Marilyn C. Ball
Plant Science Division, Research School of Biology, Australian National University, Acton, ACT, Australia

Zhilong Bie
Key Laboratory of Horticultural Plant Biology, Ministry of Education/College of Horticulture and Forestry Sciences, Huazhong Agricultural University, Wuhan, PR China

Michael R. Blatt
Laboratory of Plant Physiology and Biophysics, Bower Building, University of Glasgow, Glasgow, United Kingdom

Guang Chen
Central Laboratory, State Key Laboratory for Managing Biotic and Chemical Threats to the Quality and Safety of Agro-products, Zhejiang Academy of Agricultural Sciences, Hangzhou, PR China

Zhong-Hua Chen
School of Science; Hawkesbury Institute for the Environment, Western Sydney University, Penrith, NSW, Australia

Jong-Joo Cheong
Center for Food and Bioconvergence, Seoul National University, Seoul, Republic of Korea

Fenglin Deng
Hubei Collaborative Innovation Center for Grain Industry, College of Agriculture, Yangtze University, Jingzhou, China

Graham D. Farquhar
Plant Science Division, Research School of Biology, Australian National University, Acton, ACT, Australia

Bastian L. Franzisky
Department of Soil Science and Plant Nutrition, Hochschule Geisenheim University, Geisenheim, Germany

Wieland Fricke
School of Biology and Environmental Sciences, Science Centre West, University College Dublin (UCD), Belfield, Dublin, Republic of Ireland

Christoph-Martin Geilfus
Department of Soil Science and Plant Nutrition, Hochschule Geisenheim University, Geisenheim, Germany

Heba Hassan
Key Laboratory of Horticultural Plant Biology, Huazhong Agricultural University, Wuhan, China; Genetics Department, Faculty of Agriculture, Ain Shams University, Cairo, Egypt

Yuan Huang
Key Laboratory of Horticultural Plant Biology, Ministry of Education/College of Horticulture and Forestry Sciences, Huazhong Agricultural University, Wuhan, PR China

Jahanzeb Jafar
Key Laboratory of Horticultural Plant Biology, Huazhong Agricultural University, Wuhan, China

Ali Kiani-Pouya
State Key Laboratory of Molecular Plant Genetics, Shanghai Centre for Plant Stress Biology, Centre for Excellence in Molecular Plant Sciences, Chinese Academy of Sciences, Shanghai, China; Tasmanian Institute of Agriculture, College of Science and Engineering, University of Tasmania, Hobart, TAS, Australia

Robin Kosch
Department of Medical Bioinformatics, University Medical Center Göttingen, Göttingen, Germany

Patrick P. Lehr
Quality of Plant Products, Institute of Crop Sciences, University of Hohenheim, Stuttgart, Germany

Jie Liang
Plant Science Division, Research School of Biology, Australian National University, Acton, ACT, Australia

Ali Movahedi
Co-Innovation Centre for Sustainable Forestry in Southern China, Key Laboratory of Forest Genetics & Biotechnology, Ministry of Education, Nanjing Forestry University, Nanjing, China

Thanh-Hao Nguyen
Laboratory of Plant Physiology and Biophysics, Bower Building, University of Glasgow, Glasgow, United Kingdom

Mengliang Niu
School of Life Sciences, Jianghan University, Wuhan, PR China

Bo Ouyang
Key Laboratory of Horticultural Plant Biology, Huazhong Agricultural University, Wuhan, China

Fatemeh Rasouli
International Research Centre for Environmental Membrane Biology, Foshan University, Foshan, Guangdong; State Key Laboratory of Molecular Plant Genetics, Shanghai Centre for Plant Stress Biology, Centre for Excellence in Molecular Plant Sciences, Chinese Academy of Sciences, Shanghai, China; Tasmanian Institute of Agriculture, College of Science and Engineering, University of Tasmania, Hobart, TAS, Australia

Sergey Shabala
International Research Centre for Environmental Membrane Biology, Foshan University, Foshan, Guangdong, China; Tasmanian Institute of Agriculture, College of Science and Engineering, University of Tasmania, Hobart, TAS, Australia

Ayesha Tahir
Department of Biosciences, COMSATS University Islamabad, Islamabad, Pakistan

Nguyen B.A. Thu
Laboratory of Plant Physiology and Biophysics, Bower Building, University of Glasgow, Glasgow, United Kingdom

Yuan Wang
State Key Laboratory of Molecular Plant Genetics, Shanghai Centre for Plant Stress Biology, Centre for Excellence in Molecular Plant Sciences, Chinese Academy of Sciences, Shanghai, China

Shengchun Xu
Central Laboratory, State Key Laboratory for Managing Biotic and Chemical Threats to the Quality and Safety of Agro-products, Zhejiang Academy of Agricultural Sciences, Hangzhou, PR China

Min Yu
International Research Centre for Environmental Membrane Biology, Foshan University, Foshan, Guangdong, China

Yotam Zait
Department of Biology, Penn State University, University Park, PA, United States

Fanrong Zeng
Hubei Collaborative Innovation Center for Grain Industry, College of Agriculture, Yangtze University, Jingzhou, China

Christian Zörb
Quality of Plant Products, Institute of Crop Sciences, University of Hohenheim, Stuttgart, Germany

Preface

Global climate trends pose unprecedented and pressing demands to boost food production to match predicted population growth. With around 10 billion people expected to live on the planet in 2050, food production must be doubled to meet this demand (Razzaq et al., 2021). Achieving this goal is a highly challenging task, given the impact of climate on agricultural production systems. Over the last 50 years, the frequency of drought events in the world has already increased by about threefold (Liu, Pan, Allakhyerdiev, Yu, & Shabala, 2020), and current climate trends are expected to make droughts more frequent and more extensive. The risk of a 1-in-100 years extreme weather event is likely to increase to 1-in-30 by 2040 (Bailey et al., 2015), and major droughts are expected to occur every 5 years (Yu et al., 2018), affecting over 50% of productive lands. Drought already costs agriculture around US$60 billion in annual losses (Razzaq et al., 2021), and this number is only going to increase. Most agricultural production systems will be only profitable if they rely on irrigation, and this comes with a major issue of soil salinization. Soil salinization already takes out of production approximately 3 ha of arable land every minute (Shabala, Bose, & Hedrich, 2014) and results in over US$30 billion annual losses in crop and pasture production (Razzaq et al., 2021). To a large extent, this loss is a consequence of the salinity-induced restrictions in water availability and reduction in CO_2 assimilation due to compromised stomata operation. Stomata represent the microscopic sphincters inserted into the wax-covered epidermis that balance CO_2 intake for photosynthetic carbon gain and concomitant water loss. Upon salinity exposure, most plants close their stomata, in an attempt to conserve water. However, this stomata closure comes with a penalty of reduced CO_2 assimilation (hence, lower growth rate and yield).

Can the above dilemma be resolved? Can we make plants more efficient in their water use when grown under hostile saline conditions? The answer to this question comes from Mother Nature. A small group of plants termed *halophytes* are not only capable of surviving extreme salinity levels (e.g., seawater salt concentrations) but also significantly *benefit* from the presence of substantial amount of salt in the rhizosphere. How can they achieve this? What is special about their stomata operation? Is there anything unique about stomata in halophytes? Can these traits be transferred to traditional (salt-sensitive) crops?

This volume provides a timely and comprehensive update on the current knowledge of stomata development and operation in salt-grown plants. The range of topics covered includes evolution of stomata for adaptation to saline conditions; comparative analysis of proteomic, transcriptomic, and metabolomic profiles of stomata guard cells between halophyte and glycophyte species and their reprogramming under salt stress; molecular nature of the signals that control stomata aperture and their integration at the cellular and whole-plant level; and ecophysiological aspects of stomata operation in crop and halophyte species. The overarching aim of this work is to reveal the molecular identity of key genes mediating stress signaling and stomata operation under adverse soil conditions. Once this knowledge is in place, plant breeders will be able to devise breeding strategies to incorporate desirable features into traditional crops via molecular or traditional breeding tools thus expanding the sustainable limits for productive use of soil and water resources.

Sergey Shabala
Professor and Head, Stress Physiology Laboratory,
University of Tasmania, Australia; Distinguished Professor,
Foshan University, China

References

Bailey, R., et al. (2015). *Extreme weather and resilience of the global food system. Final Project Report from the UK-US Taskforce on Extreme Weather and Global Food System Resilience.* UK: The Global Food Security Programme.

Liu, M. M., Pan, T., Allakhyerdiev, S. I., Yu, M., & Shabala, S. (2020). Crop halophytism: An environmentally sustainable solution for global food security. *Trends in Plant Science, 25*, 630–634.

Razzaq, A., Wani, S. H., Saleem, F., Yu, M., Zhou, M. X., & Shabala, S. (2021). Rewilding crops for climate resilience: Economic analysis and *de novo* domestication strategies. *Journal of Experimental Botany, 72*, 6123–6139.

Shabala, S., Bose, J., & Hedrich, R. (2014). Salt bladders: Do they matter? *Trends in Plant Science, 19*, 687–691.

Yu, C., Huang, X., Chen, H., Huang, G., Ni, S., Wright, J. S., et al. (2018). Assessing the impacts of extreme agricultural droughts in China under climate and socioeconomic changes. *Earth's Future, 6*, 689–703.

CHAPTER ONE

Stomatal regulation and adaptation to salinity in glycophytes and halophytes

Guang Chen[a,*,†], Hanna Amoanimaa-Dede[b,†], Fanrong Zeng[b], Fenglin Deng[b], Shengchun Xu[a], and Zhong-Hua Chen[c,d,*]

[a]Central Laboratory, State Key Laboratory for Managing Biotic and Chemical Threats to the Quality and Safety of Agro-products, Zhejiang Academy of Agricultural Sciences, Hangzhou, PR China
[b]Hubei Collaborative Innovation Center for Grain Industry, College of Agriculture, Yangtze University, Jingzhou, China
[c]School of Science, Western Sydney University, Penrith, NSW, Australia
[d]Hawkesbury Institute for the Environment, Western Sydney University, Penrith, NSW, Australia
*Corresponding authors: e-mail address: g_chen0915@163.com; z.chen@westernsydney.edu.au

Contents

1. Introduction 2
2. Stomatal physiological responses to salinity differ in glycophytes and halophytes 4
3. Membrane transporters regulate ion-homeostasis in stomata under salinity 9
   3.1 Cation transporters 10
   3.2 Anion channels 16
   3.3 H^+ pumps and cotransporters 19
   3.4 The role of aquaporin in salt stress and stomatal regulation 21
4. Signaling pathways help plants regulate stomatal movement under salinity 22
   4.1 Abscisic acid (ABA) signaling 22
   4.2 Reactive oxygen species (ROS) signaling 24
5. Conclusions and future perspective 26
Acknowledgments 27
References 27

Abstract

Soil salinity is one of the major abiotic stresses, which negatively affects the productivity of plants in both cultivated and natural environments. Salinity stress increases osmotic stress and ROS accumulation and breaks the ionic balance in plant cells. In stomatal guard cells, these factors cause ABA biosynthesis and stomatal closure, resulting in the reduction of CO_2 assimilation and crop yield loss. Halophytes are the plants capable to thrive in an extremely saline environment, thus consider as one of the vital resources

† These authors contribute equally to this work.

Advances in Botanical Research, Volume 103
ISSN 0065-2296
https://doi.org/10.1016/bs.abr.2022.02.008

for breeding salt-tolerant crops. Understanding the common and specific adaptive mechanisms to salinity of halophytes and glycophytes (most crop species) is one of the effective approaches to mining the salt tolerance-associated genetic resources. Here, we first reviewed the physiological traits conferring salinity tolerance traits (stomatal density, stomatal aperture, photosynthetic characteristics, and ion content) between halophytes and glycophytes. Comparative genomic and transcriptomic analysis of key membrane transporters regulating stomatal opening and closure is then conducted to explore the adaptation mechanisms of halophytes and glycophytes to salinity stress. In summary, future research is suggested to focus more on halophytes to gain a better mechanistic understanding of salt tolerance before applications in glycophytic crop breeding.

1. Introduction

Soil salinity is a major abiotic stress limiting plant growth and productivity. Currently, over 20% of irrigated lands, as well as 6% of productive agricultural lands are salinized worldwide, which constraints sustainable agricultural production and global food supply (Song & Wang, 2015; Yuan, Leng, & Wang, 2016). Soil salinization results from the over-accumulation of salts in soil layers through natural or human activities (Deinlein et al., 2014; Zhu, 2001). Excessive accumulation of sodium ions (Na^+) in shoots significantly disrupts key morphological, physiological, biochemical, and molecular processes in plants (Guo, Huang, Li, & Hou, 2020; Shen et al., 2018). Generally, salinity causes hyperosmotic stress, ion imbalance, and oxidative stress which impedes plant growth and productivity (Deinlein et al., 2014; Flowers & Colmer, 2008). To combat the deleterious effects of salinity stress, plants have evolved several adaptive mechanisms such as selective ion influx and efflux, salt sequestration, activation of antioxidant defense system and compatible solute synthesis (Deinlein et al., 2014; Shabala, 2013). Understanding the key adaptive mechanisms and salinity tolerance is crucial for engineering salinity-tolerant plant species (Sanchez et al., 2011).

Identifying and utilizing naturally salt-tolerant species is a feasible strategy to cope with salinity-induced agricultural problems (Shabala, Hariadi, & Jacobsen, 2013). Halophytes (salt-tolerant plants), the native flora of saline soils employ various adaptive strategies to deal with salinity stress (Chen et al., 2018), but the growth and productivity of salt-sensitive plants (glycophytes) are adversely affected by salinity (Kiani-Pouya et al., 2020). Investigations of salt tolerance mechanisms of halophytes in the last 70 years demonstrated that the superior salinity tolerance of halophytes is achieved through several well-orchestrated strategies. These include a host of specialized physiological and morphological traits such as specialized salt bladders, extracellular Na^+ sequestration to salt glands, efficient Na^+ sequestration in vacuoles, effective control of xylem ion loading, high cytosolic K^+ retention, regulation of stomatal aperture, and enhanced water use efficiency (Kiani-Pouya et al., 2019; Munns et al., 2020; Shabala, 2013; Shabala, Chen, & Chen, 2020). Interestingly, stomata in halophytes possess some unique traits that improve plant water potential, leading to reduced transpiration and enhanced CO_2 uptake. For instance, efficient coordination of stomatal length and density effectively regulated stomatal movements in salt-tolerant varieties of a halophyte quinoa (*Chenopodium quinoa*) and a glycophyte barley (*Hordeum vulgare*), thereby optimizing transpirational water loss and CO_2 uptake under salinity stress (Kiani-Pouya et al., 2019; Liu et al., 2017; Shabala et al., 2013).

Plant adaptive response to salinity is determined by the microscopic valves—stomata—on the leaf epidermis formed by a pair of guard cells. Stomata are the major gateway for gaseous exchange between the plant and its atmosphere (Willmer & Fricker, 1996). The efficiency of stomatal operations depends on the balance between CO_2 assimilation and transpirational water loss (Hetherington & Woodward, 2003). Ion channels and transporters mainly guard cell signaling networks and their activity determines stomatal aperture (Kim, Böhmer, Hu, Nishimura, & Schroeder, 2010; Kollist, Nuhkat, & Roelfsema, 2014; Roelfsema, Hedrich, & Geiger, 2012; Song, Miao, & Song, 2014). Guard cells integrate these signals and adjust stomatal apertures appropriately to optimize growth performance (Hetherington & Woodward, 2003; Kim et al., 2010). Salinity stress significantly affects stomatal operations by inducing osmotic, oxidative, and ionic stresses (Liu et al., 2017). Under prolonged stress, plants require higher adaptations such as modification in stomatal size and density as well as activation of crucial mechanisms to deal with toxic ions in guard cells (Hedrich & Shabala, 2018; Kiani-Pouya et al., 2019, 2020). Recent studies showed that reduced stomatal density under stress improves water use efficiency without adversely affecting yield (Hughes et al., 2017; Wang & Chen, 2020). Under short-term salt stress, plants adjust the size of the stomata pore by regulating an equipoise between CO_2 uptake and water loss (Kiani-Pouya et al., 2019). Stomatal conductance is hypersensitive to both osmotic stress and ionic toxicity effects of salt stress, thus affecting the plant productivity under salt stress (Mohamed et al., 2020; Yuan et al., 2014). Moreover, stomatal speed has been found to improve water use efficiency, plant productivity, and stress tolerance among different plant species (Cai, Papanatsiou, Blatt, & Chen, 2017; Vialet-Chabrand et al., 2016). The speed of stomatal responses to fluctuating environments could enhance the balance between CO_2 uptake and water loss, thus maintaining optimal water retention and CO_2 assimilation under salinity (Lawson & Vialet-Chabrand, 2019).

Given the important roles of the stomatal movement in optimizing plant growth under salinity stress, this review discusses the regulation of stomatal movements and salinity tolerance mechanisms. We focus on physiology, gene expression and molecular evolution of important ion transporters, key signaling pathways and the major adaptive features by comparing halophytes and glycophytes under salinity stress. For reviews in plant salinity tolerance or stomatal regulation, the readers are directed to other excellent papers by Ozgur, Uzilday, Sekmen, and Turkan (2013), van Zelm, Zhang, and Testerink (2020), Bose, Rodrigo-Moreno, and Shabala (2014), Hedrich and Shabala (2018), Oprică and Vochița (2020).

2. Stomatal physiological responses to salinity differ in glycophytes and halophytes

Salt stress affects physiological activities of plants by osmotic and ionic stresses, influencing leaf gas exchange and associated photosynthetic processes (Hetherington & Woodward, 2003). Stomata as the key part in balancing CO_2 assimilation and water loss, their behavior in salt tolerant species play important roles in improving water status, leading to reduced transpiration and enhanced CO_2 uptake as well as a better water use efficiency. In short-term salt stress, halophytic plants keep the balance between water loss and CO_2 assimilation through optimizing the stomatal aperture (Rasouli, Kiani-Pouya, Tahir, et al., 2021). In glycophyte plant barley, salt-tolerance genotype showed larger stomatal aperture under 200 mM NaCl treatment, and the stomatal trait aperture width/length is highly correlate with grain yield under salinity (Liu, Mak, et al., 2014; Liu, Zhang, Dong, & Chu, 2014). In the glycophytic salt sensitive species rice (*Oryza sativa*), salt and osmotic stresses caused the rapid stomatal closure and damage to photosynthetic apparatus (Liu, Mao, Yuan, Chu, & Duan, 2021) (Table 1). Comparison between halophytic species quinoa and *Beta maritima* (sea beet) and their glycophytic relatives *Chenopodium album* (baconweed) and *Beta vulgaris* (sugar beet) provided interesting results in stomatal traits and photosynthetic responses to salinity differ between halophytes and glycophytes (Rasouli, Kiani-Pouya, Tahir, et al., 2021). Salinity reduced

Table 1 Physiological responses of selected glycophytes and halophytes to salinity.

Species	Parameters					References
Oryza sativa	NaCl	DW (g $plant^{-1}$)	FW (g $plant^{-1}$)	Chl a (μg g^{-1} FW)	Chl b (μg g^{-1} FW)	Tisarum et al. (2020)
	0	68.30 ± 1.05 a	260.33 ± 5.113 a	88.53 ± 1.48 a	53.96 ± 1.07 a	
	200 mM	49.73 ± 1.09 c	216.33 ± 4.50 b	28.91 ± 0.18 c	18.30 ± 0.42 c	
Hordeum vulgare		Shoot DW (g $plant^{-1}$)	Shoot height (cm)	Shoot FW (g $plant^{-1}$)		Zeeshan et al. (2020)
	0	0.21 a	40.1 b	1.8 a		
	200 mM	0.09 c	35.4 c	0.8 b		
Chenopodium quinoa cv. "Titicaca"	NaCl	Shoot DW (g $plant^{-1}$)	Shoot height (cm)	Chlorophylls	Carotenoids	
	0	2.04 ± 0.30 a	24.75 ± 2.32 a	23.42 ± 1.99 a	4.38 ± 0.40 a	Rodriguez-Hernandez et al. (2021)
	50 mM	1.72 ± 0.31 ab	23.25 ± 3.56 a	22.85 ± 0.97 a	4.31 ± 1.49 a	
	100 mM	1.38 ± 0.12 bc	25.36 ± 1.91 a	22.68 ± 1.45 a	3.27 ± 0.76 ab	
	200 mM	1.38 ± 0.12 bc	25.36 ± 1.91 a	15.45 ± 2.87 b	2.57 ± 0.47 b	
Chenopodium album	NaCl	A_N (μmol CO_2 m^{-2} s^{-1})	g_s (mol H_2O m^{-2} s^{-1})	T_r (mmol H_2O m^{-2} s^{-1})	C_i (μmol CO_2 mol^{-1})	
	0	14.4 ab	0.29 b	3.29 a	326 a	Rasouli, Kiani-Pouya, Shabala, et al. (2021)
	100 mM	11.7 bc	0.22 c	2.36 bc	287 bc	
	200 mM	12.2 bc	0.16 dz	1.71 de	255 cd	
	300 mM	9.6 cd	0.09 efg	1.06 fg	208 ef	
	400 mM	9.31 cd	0.06 fg	0.79 fg	148 h	
	500 mM	6.0 d	0.046 g	0.54 g	158 gh	
C. quinoa	0	15.4 ab	0.35 a	2.82 ab	293 ab	
	100 mM	16.4 a	0.28 b	2.63 b	285 bc	

Continued

Table 1 Physiological responses of selected glycophytes and halophytes to salinity.—cont'd

Species	Parameters					References
	200 mM	15.2 ab	0.14 c	1.90 cd	259 cd	
	300 mM	13.1 abc	0.11 de	1.23 ef	228 de	
	400 mM	11.5 bc	0.10 efg	1.04 fg	184 fg	
	500 mM	11.6 bc	0.091 efg	0.99 fg	178 fgh	
Beta vulgaris	0	13. bcd	0.33 c	2.93 b	302 a	
	100 mM	14.9 bc	0.29 c	2.67 b	288 ab	
	200 mM	11.9 cd	0.12 de	1.44 c	222 c	
	300 mM	10.0 de	0.09 de	0.99 c	199 cd	
	400 mM	8.31 ef	0.087 de	1.07 c	196 cd	
	500 mM	6.5 f	0.064 e	0.85 c	184 d	
Beta maritima	0	15.3 bc	0.064 e	3.85 a	317 a	
	100 mM	16.6 ab	0.41 b	3.16 ab	298 a	
	200 mM	18.7 a	0.30 c	2.68 b	261 b	
	300 mM	15.4 ab	0.13 d	1.40 c	216 cd	
	400 mM	13.7 bc	0.09 de	1.05 c	198 cd	
	500 mM	13.1 bcd	0.075 e	0.87 c	189 cd	
Atriplex halimus L.	NaCl	Photosynthetic dry mass (g)	A_N (μmoL m^{-2} s^{-1})	g_s (mmoL m^{-2} s^{-1})	Fv/Fm	Perez-Romero et al. (2020)
	171 mM	5.98 ± 0.44 a	4.35 ± 0.67 a	53.4 ± 7.47 a	0.70 ± 0.02 a	
	510 mM	4.55 ± 0.53 b	2.74 ± 0.32 b	46.6 ± 6.49 a	0.70 ± 0.02 a	

Note: DW, dry weight; FW, fresh weight; Chl a, chlorophyll *a*; Chl b, chlorophyll *b*; A_N, net photosynthesis; g_s, stomatal conductance; T_r, transpiration rate; C_i, intercellular carbon dioxide concentration.

stomatal conductance (g_s) in all species, but the halophytic species were able to retain water through reduced g_s without any penalty to the rate of CO_2 assimilation or fresh weight production (Table 1). Otherwise, halophytic species have higher maximum rates of Rubisco carboxylation and electron transport through photosystem II for regeneration of RuBP, smaller stomatal density, and faster stomatal movements (Rasouli, Kiani-Pouya, Tahir, et al., 2021). *Atriplex halimus*, a perennial shrub with C4 photosynthetic metabolic pathway, can survive in saline, sodic and drought conditions (Belkheiri & Mulas, 2013). Under 510mM NaCl treatment, *A. halimus* still keep the photosynthetic traits such as photochemistry (Fv/Fm), maximum quantum efficiency of photosystem II (Φ_{PSII}), stomatal conductance (g_s), intrinsic water use efficiency (iWUE) and intercellular CO_2 concentration (C_i) at the normal level, indicating its high capacity of photosynthesis and water use efficiency under salinity stress (Perez-Romero, Mateos-Naranjo, Lopez-Jurado, Redondo-Gomez, & Torress-Ruiz, 2020) (Table 1). In quinoa genotypes, there is a positive correlation between stomata density and salinity tolerance, all tested varieties adapted to saline environment through reduce the stomata density and increase leaf sap K^+ concentrations to deal with the NaCl-induced reduced water availability (Shabala et al., 2013). However, significant genetic variability in salinity stress tolerance was observed even in the same halophyte species. For example, quinoa genotypes split into two distinct clusters under saline condition: one is the exclusion strategy with very small amount of Na^+ accumulated in the shoot and another is the accumulation of substantial quantities of Na^+ in the shoot via vacuolar Na^+ sequestration (Shabala et al., 2013).

Osmotic stress is the first phase of NaCl-induced stress and ionic stress is usually mediated via Na^+ interference with K^+ homeostasis. Thus, maintaining a balanced cytosolic Na^+/K^+ ratio is a major salinity tolerance mechanism (Assaha, Ueda, Saneoka, Al-Yahyai, & Yaish, 2017; Shabala & Pottosin, 2014). Osmotic stress and ion homeostasis both directly affect stomatal movement. Glycophytes and halophytes have similar mechanisms of Na^+ and K^+ uptake and translocation, but glycophytes are more sensitive to ionic stress than halophytes (Assaha et al., 2017). In previous studies, it was found that there is no difference of the sodium content in shoot of glycophytes barley and maize and halophyte quinoa under normal condition (Ahmed et al., 2013; Akram, Malik, Ashraf, Saleem, & Hussain, 2007; Rodriguez-Hernandez, Morcillo, & Garmendia, 2021). When the plants treated by 100mM NaCl, Na content in quinoa was significantly higher than that in barley and maize (Fig. 1). Comparison of K content in these plants showed that shoot K content of quinoa is about 3–4 times higher than that

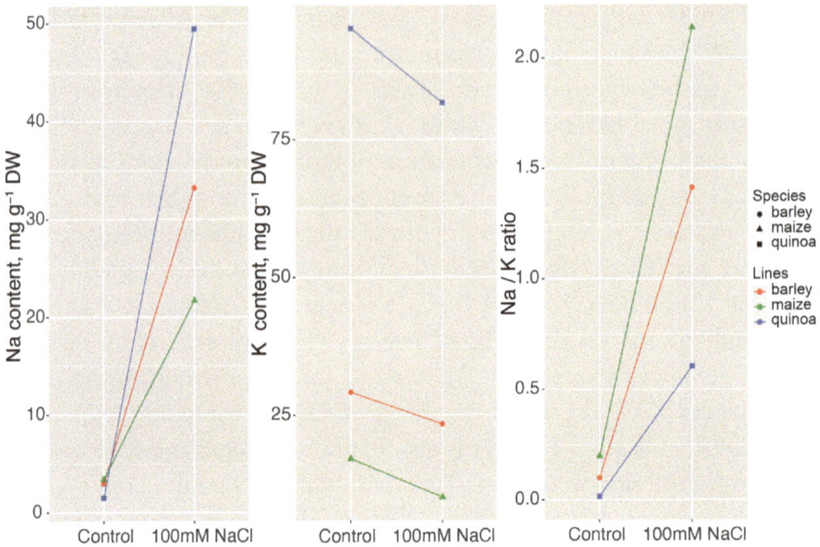

Fig. 1 Sodium and potassium contents in the shoot of glycophytic species barley, maize and halophytic species quinoa under control and salt treatment. *Data were summarised from Ahmed et al. (2013), Akram et al. (2007) and Rodriguez-Hernandez et al. (2021).*

barley and maize under normal condition and quinoa still maintains very high level of K content after NaCl treatment (Fig. 1). As a result, the Na^+/K^+ ratio in quinoa is much lower than maize and barley (Fig. 1). The efflux of K^+ induced by excess Na^+ influx leads to depolarization of membrane potential below resting potential, which activates the K^+ outward rectifier channel like GORK (guard cell outward rectifying K^+ channel) (Ache et al., 2000; Hosy et al., 2003). Thus, the ability of plant cells to enhance intracellular K^+ retention constitutes a vital mechanism for plant salinity tolerance (Adem, Chen, Shabala, Chen, & Shabala, 2020; Chen et al., 2005; Falhof, Pedersen, Fuglsang, & Palmgren, 2016).

The big difference between glycophytes and some halophytes is that halophytes have evolved specialized epidermal cells, such as bladder cells and salt glands, for sequestration or elimination of excess Na^+ (Bonales-Alatorre, Shabala, Chen, & Pottosin, 2013; Liphschitz & Waisel, 1982; Shabala & Pottosin, 2014; Yong, Solis, Amatoury, et al., 2022). The representative well-studied model halophytic species with epidermal bladder cells (EBCs) is *Mesembryanthemum crystallinum* (ice plant). EBCs are regarded as the storage reservoir for water and inorganic and organic compounds such as sodium, chloride, flavonoids, and betacyanins, thus regulating the salt sequestration and water relations (Adams et al., 1998; Adams, Thomas, Vernon, Bohnert, & Jensen, 1992). During CAM in mesophyll cells, EBCs may also serve to support the water relations of the mesophyll in conjunction with the osmotically important oscillations of malate levels (Rygol, Zimmermann, & Balling, 1989). Research on mutant ice plants lacked EBCs showed reduced leaf succulence and leaf and stem water contents, as well as dry weight gain under salt stress (Agarie et al., 2007). Otherwise, the ionic concentration in epidermal bladder cells and leaves including K^+, NO_3^-, and PO_4^{3-} were significantly reduced in mutant plants, and Na^+ and Cl^- were much higher than wild type (Agarie et al., 2007). These evidences suggested that the EBCs is one of the great advantages of halophytes to survive and thrive under saline conditions via the increase of toxic Na^+ in EBCs to reduce the accumulation of salt in mesophyll cells and guard cells for vital photosynthesis. However, it is still not clear how halophytes sense the salinity and genetically program higher differentiation of epidermal cells into the production of EBCs to store excessive NaCl and how stomatal guard cells and mesophyll cells are protected by EBCs to reduce the impact of NaCl-induced osmotic, ionic and oxidative stresses.

3. Membrane transporters regulate ion-homeostasis in stomata under salinity

Stomatal regulation is recognized as one of major components of salinity tolerance mechanism in plants (Kiani-Pouya et al., 2019). Although salinity-induced Na^+ toxicity and osmotic stress negatively influence stomatal parameters (Liu, Mak, et al., 2014; Liu, Zhang, et al., 2014), some plants especially halophytes can optimize their stomatal performance under saline conditions (Hedrich & Shabala, 2018; Shabala, 2013). Stomata rapidly reduce their size via ion efflux from guard cells into apoplast and plasma membrane- and tonoplast-localized ion channels, transporters, and pumps regulate the influx and efflux of solutes necessary for stomatal opening and closure (Roelfsema & Hedrich, 2005; Schroeder, Allen, Hugouvieux, Kwak, & Waner, 2001). However, the underlying tolerance mechanisms are unclear in stomatal guard cells as compared to our understanding of salinity tolerance in other cell types.

Benefiting from the dramatically increased pace of sequencing of land plants, a bulk of omics sequencing data can be used for comparative genetic and genomic analysis. This provides the possibility to explore the biological problems in a large evolutionary scale and comparative analysis among the land lineage plants. Here we use the data source of one thousand plant transcriptomes (OneKP) (One Thousand Plant Transcriptomes, 2019) to compare the origin of salt response stomatal regulation related gene families,

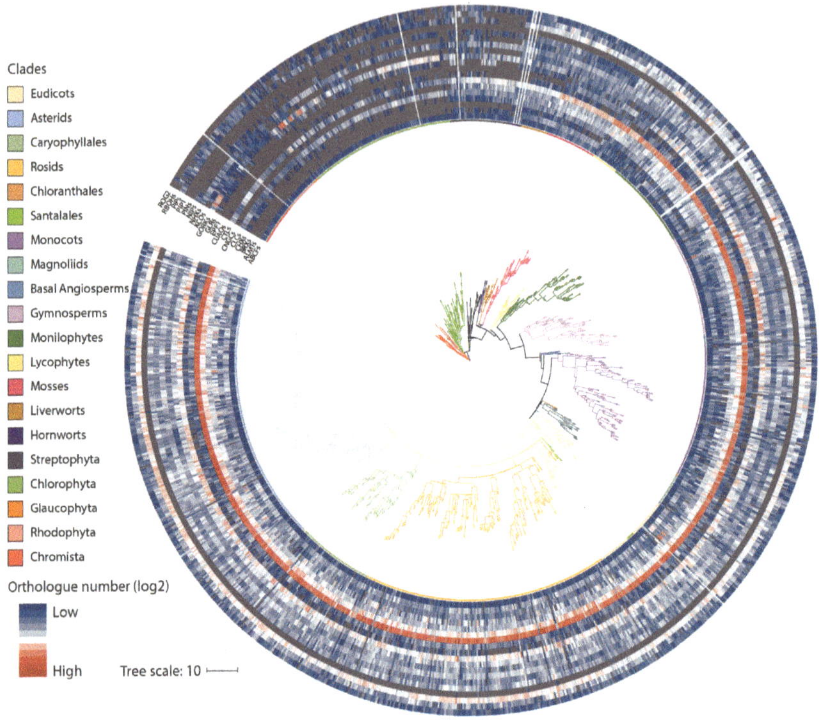

Fig. 2 Number of key membrane transporter gene family members in OneKP database sequenced green plants. Orthologue genes were extracted from OneKP Orthogroups Extractor (http://jlmwiki.plantbio.uga.edu/onekp/v2/). Gene numbers of key gene families for stomatal movement regulation under salinity stress in a large range of green plants were calculated. Phylogenetic species tree was got from https://doi.org/10.5281/zenodo.3255100, 1178 green plants were employed in this study, species list can be found in http://onekp.com/samples/list.php. The datasets of each predicted gene family orthologue numbers across these species were added on the outside of tree by iTOL (https://itol.embl.de). The color of the branches and the most inner color circles represent different clades in the phylogenetic tree. The color scale shows the number of gene family members from high (red) to low (blue), and gray represents a missing value. http://onekp.com/samples/list.php.

expansion and duplication and gene expression profiles in key green plant lineages (Figs. 2–4). Our analysis is likely to be important to expand our understanding in the evolutionary adaptation of plants to salinity stress.

3.1 Cation transporters

Maintaining a balanced cytosolic Na^+/K^+ ratio remains the key salinity tolerance mechanism in plants (Shabala & Pottosin, 2014). Plants employ two main mechanisms to control Na^+ accumulation by Na^+ sequestration into

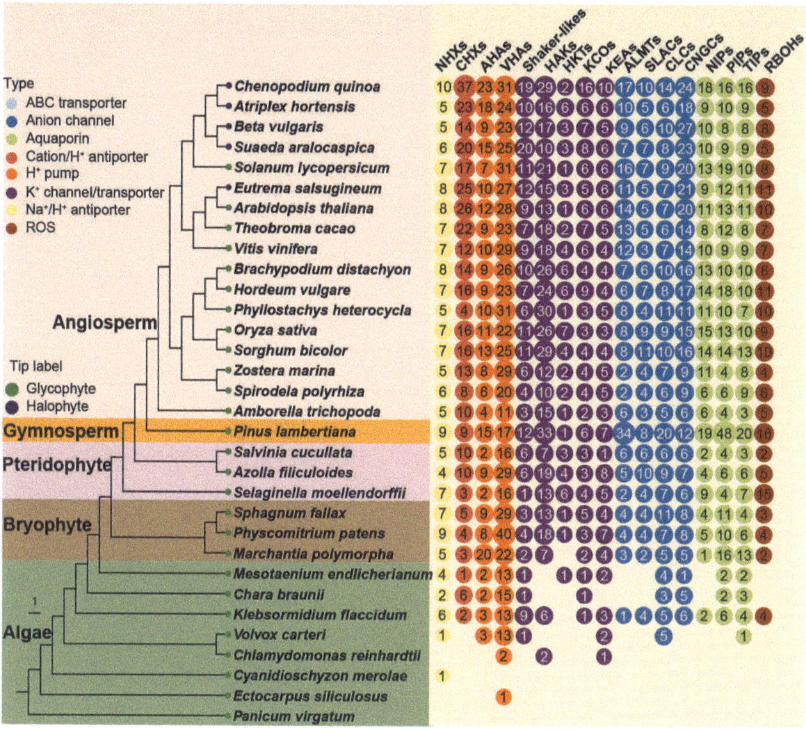

Fig. 3 Number of key membrane transporter gene family members in representative green plants. The whole genome sequenced representative plant species in each clade were selected for comparative genomic analysis. The genomes of plants were downloaded from Phytozome (https://phytozome-next.jgi.doe.gov) and Ensembl Plants (https://plants.ensembl.org/index.html). Phylogeny was constructed by OTL (Open Tree of Life, https://tree.opentreeoflife.org/about/open-tree-of-life). The blue tips indicate the halotyphic species, green tips were glycophytic species. Blastp with certification of E value $<10^{-5}$ was used to identify the homologue gene family members from selected species. The size of dots represents the number of gene family members, and the color of dots represents the different kind of gene families.

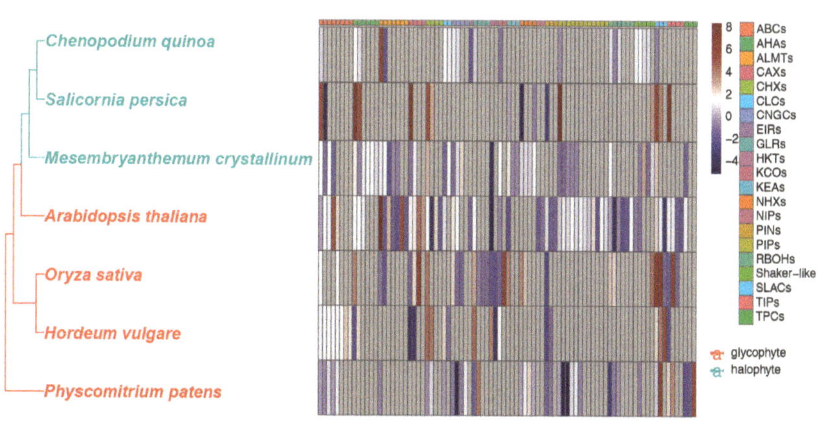

Fig. 4 Salinity-induced transcripts of transporter genes for stomatal-regulation in halophytes and glycophytes. Blue branches are halophytes, red branches are glycophytes. Combination of phylogeny and heatmap of expression levels was conducted by R package ggtreeExtra (Xu et al., 2021) and ggplot2 (Ito & Murphy, 2013).

the vacuole and minimizing Na^+ uptake by roots, in addition to the removal of Na^+ from xylem sap, control of xylem Na^+ loading, and Na^+ recirculation via the phloem in photosynthetically active mesophyll cells. Several studies suggest the crucial roles of Na^+/K^+ transporters and channels in regulating homeostatic balance in response to salinity (Yaish et al., 2017).

The high-affinity potassium ion transporters (HKTs) maintain K^+/Na^+ homeostasis and enhance salt tolerance in a large number of plants including many salinity-tolerant plant species to enhance the influx of K^+ and efflux of Na^+ in cells (Basu, Kumar, Benazir, & Kumar, 2020). HKT transporters, generally found in the xylem parenchyma cells are involved in ion homeostasis in plants (Munns et al., 2012). Class I HKTs are specific for Na^+ retrieval from the xylem sap in both dicots and monocots (Jabnoune et al., 2009) and Class II HKT transporters are involved in Na^+ uptake under K^+ starvation in the roots of monocots (Horie et al., 2001; Sunarpi, Motoda, Kubo, & Yang, 2005). HKT transporters balance the K^+/Na^+ ratio in the cytosol by simultaneously excluding Na^+ and allowing K^+ influx, thereby increasing tolerance to salinity (Horie et al., 2005). For example, several studies involving glycophytic rice (Plett et al., 2010), *Arabidopsis* (Baxter et al., 2010), tomato (Asins et al., 2013) and wheat (Munns et al., 2012) have demonstrated the role of HKTs in ion homeostasis (i.e., Na^+ transport) and salinity tolerance. For instance, the HKT transporter in *Arabidopsis*, *AtHKT1;1* and its orthologues in rice; *OsHKT1;4* and *OsHKT1;5* and wheat; *TmHKT1;5* and *TmHKT1;4* have been shown to reduce Na^+ transport and improve salinity tolerance (Cotsaftis, Plett, Shirley, Tester, & Hrmova, 2012; Horie et al., 2005; Horie, Karahara, & Katsuhara, 2012; James, Blake, Byrt, & Munns, 2011; Munns et al., 2012). Likewise, *OsHKT1;4* controlled Na^+ exclusion and improved tolerance to salt stress (Suzuki et al., 2016). James et al. (2011) and Zhu et al. (2016) showed the involvement of the wheat *Nax1* and *Nax2* genes encoding *HKT1;4* and *HKT1;5*, respectively in Na^+ retrieval from the xylem which enhanced salinity tolerance. The overexpression of *HvHKT2;1* gene in barley enhanced Na^+ xylem loading and subsequently increased Na^+ accumulation in shoots, which positively correlated with improved salinity tolerance in transgenic plants (Mian et al., 2011). Upregulation of SpHKT1;2 in stress tolerant wild tomato *Solanum pennellii* increased tolerance to salt stress compared to cultivated tomato *Solanum lycopersicum* (Albaladejo et al., 2017). Overexpression of HKT1 in a halophyte *Cochlearia pyrenaica* (scurvy grass) increased K^+ levels in roots and enhanced tolerance to salt (Nawaz, Iqbal, Bliek, & Schat, 2017). DNA and protein sequences of HKT1 in halophyte *Thellungiellla salsuginea* and glycophyte Arabidopsis showed high identities, but the *TsHKT1* was dramatically induced upon salt stress, whereas AtHKT1 was downregulated, and the *TsHKT1* acts as an efficient K^+ transporter under high Na^+ concentration by repression in yeast cells compared to *AtHKT1* (Ali, Cheol Park, Aman, Ali, & Yun, 2013). Transcriptome analysis revealed that salinity reduced *HKT* transcript abundance in shoots of three glycophytes (*Arabidopsis*, rice, and barley) and in two halophytes (ice plant and quinoa) (Fig. 4) (Fu, Shen, Kuang, Wu, & Zhang, 2019; Li et al., 2018). Interestingly, HKTs are not induced by NaCl treatment in the model moss *Physcomitrium patens* and a halophyte *Salicornia persica* (glasswort) (Fig. 4).

Sodium/proton exchangers (NHXs) are vital in maintaining intracellular Na^+/K^+ homeostasis as they function to counter excess accumulation Na^+ in the cytosol by mediating vacuolar sequestration of Na^+ and release of vacuolar K^+ under salt stress (Andrés et al., 2014; Huertas et al., 2013; Jiang, Leidi, & Pardo, 2010). Under low Na^+ concentration and control conditions, NHXs act as K^+/H^+ antiporter with no Na^+ transport activity (Jiang et al., 2010). For example, The upregulation of *AtNHX1* enhanced intracellular K^+ retention in the cytosol by mediating vacuolar K^+ but not Na^+ sequestration, thereby boosting salinity tolerance in tomato (Leidi et al., 2010; Yaish, 2015). In salt-stressed transgenic alfalfa, the overexpression of *TaNHX2* diminished K^+ efflux via activating K^+ outwardly rectifying (KOR) channels and minimizing plasma membrane depolarization to retain more intracellular K^+ (Zhang et al., 2012). The retention of cellular K^+ by NHX transporters was confirmed as no vacuolar K^+ pools were observed in the *Arabidopsis nhx1nhx2* double mutants (Barragán et al., 2012). Members of the NHX antiporter family also regulate stomatal functions through their $Na^+(K^+)/H^+$ exchange activity in plants (Andrés et al., 2014; Ashnest, Huynh, Dragwidge, Ford, & Gendall, 2015; Bassil et al., 2011; Huertas et al., 2013; Jiang et al., 2010; Wang et al., 2016; Wang, Wu, Liu, & Qiu, 2015; Wu, Ebine, Ueda, & Qiu, 2016). The expression of *AlNHX* from *Aeluropus littoralis* in soybean mediated a reduction in Na^+ sequestrated into the vacuole (Liu, Mak, et al., 2014; Liu, Zhang, et al., 2014). In addition, *PutNHX1* and *SeNHX1* cloned from halophytes *Puccinellia tenuiflora* and *Salicornia europaea* confer salt tolerance by altering cytosolic and vacuolar K^+ and Na^+ in Arabidopsis root cells (Liu, Cai, et al., 2017). The SOS signaling pathway, which comprises SOS1, SOS2, and SOS3, is crucial in the regulation of ion homeostasis in salt-stressed plants (Zhu, 2003). In *Arabidopsis*, the overexpression of glycophytic moss *Physcomitrium patens*

SOS1 (*PpSOS1*) increased K^+ uptake whereas plants lacking *PpSOS1* activity accumulated high Na^+ with impaired high-affinity K^+ uptake under combined salt stress and low K^+ nutrition, indicating the involvement of SOS1 in K^+ uptake in early divergent land plants (Fraile-Escanciano, Kamisugi, Cuming, Rodríguez-Navarro, & Benito, 2010). Likewise, electrophysiological studies revealed that SOS1 activity resulted in higher Na^+ efflux and enhanced salinity tolerance in Brassica species (Chakraborty, Bose, Shabala, & Shabala, 2016). We found there is no significant difference in numbers of NHX family member between glycophytes and halophytes, but all *NHXs* were significantly upregulated by NaCl in glycophytic rice, which is contrast to the downregulation of *NHXs* in halophytic *Salicornia persica* under salt stress, implying the different mechanisms of NHX regulation in different species (Figs. 2-4).

Many other K^+ transporters are involved in plant salinity tolerance. For instance, the K^+/H^+ antiporters KEAs are K^+ efflux antiporters localized on endomembrane to mediate H^+-linked K^+ transport (Detar et al., 2021; Sicilia, Romero, Rosales, & Venema, 2021). The *Arabidopsis* triple mutant *kea4 kea5 kea6* was found to be sensitive to salt stress (Zhu et al., 2018). The Ca^{2+} activated outward rectifying K^+ channel KCO, as known as TPK, is another important regulator of intracellular K^+/Na^+ ratio under salt stress (Assaha et al., 2017). Both transcriptome evidence and genome data suggest that these K^+ transporter gene families HAKs, HKTs, KCOs, KEAs and shaker-likes were originated from green algae and were identified in all tested land plants (Figs. 2 and 3). Interestingly, comparative genetic analysis showed that the Shaker-like family has evidence of gene duplication in a halophytic species *Suaeda aralocaspica*, while the halophyte *quinoa* has the largest gene numbers of KCOs and KEAs (Fig. 3). The expression of *KCO3* in Arabidopsis and rice was suppressed by salt stress as well as in quinoa (Fig. 4). While in barley and ice plant, the expression of KCO was significantly induced by threefold under salt stress compared to control (Fig. 4). All differentially expressed genes (DEGs) of shaker-like family were upregulated in comparison to the control in quinoa, which was different from those Arabidopsis, rice, barely, moss, ice plants and *Salicornia persica*, indicating the different regulation salt tolerance mechanisms in quinoa (Fig. 4) (Aliakbari, Razi, Alemzadeh, & Tavakol, 2020).

Cyclic nucleotide-gated cation channels (CNGCs) are non-selective cation channels, present in animals, plants, and some bacteria which have permeability for cations including Ca^{2+}, Na^+, and K^+ (Hanin, Ebel, Ngom, Laplaze, & Masmoudi, 2016). CNGC was first discovered in barley when screening for CaM-binding partners (Schuurink, Shartzer, Fath, & Jones, 1998). Subsequently, CNGCs were identified in *Arabidopsis* (Köhler, Merkle, & Neuhaus, 1999), rice (Nawaz, Kakar, Saand, & Shu, 2014), tomato (Saand, Xu, Li, Wang, & Cai, 2015), *Pyrus bretschneideri* (Chen et al., 2015) and *Brassica oleracea* (Kakar et al., 2017). CNGCs function as non-rectifying transporters that rely on an electrochemical gradient to mediate Na^+ fluxes (both influx and efflux) (Gobert, Park, Amtmann, Sanders, & Maathuis, 2006), suggesting their role in plant adaption to salt toxicity, possibly by mediating Na^+ transport (Kugler, Köhler, Palme, Wolff, & Dietrich, 2009). Electrophysiological studies involving the root epidermal cells of *Arabidopsis* revealed that CNGC is a Na^+ uptake channel that is non-selective to K^+ (Demidchik & Tester, 2002). CNGCs are implicated to regulate the influx and circulation of Na^+ ions, thereby improving tolerance to salt stress. Among the five *Arabidopsis* CNGC genes involved in salinity stress response, the expression of CNGC1, CNGC3 and CNGC13 were upregulated whereas CNGC7 and CNGC16 were downregulated upon salinity exposure (Maathuis et al., 2003). For instance, the *atcngc10* mutant accumulated less Na^+ and K^+ in shoots compared to those of the wild type and overexpression of *CNGC10* resulted in greater sensitivity to salt stress in *Arabidopsis* (Jin, Jing, Zhang, & Zhang, 2015). Besides, *atcngc3* mutants exhibited enhanced tolerance to high NaCl and KCl concentrations resulting from a possible restriction of ion influx in mutant plants (Gobert et al., 2006). The down-regulation of *OsCNGC1* in rice prevented Na^+ influx, which enhanced salinity tolerance of the salt-tolerant variety. In contrast, salinity stress up-regulated the expression of the gene in the salt-sensitive rice variety (Senadheera, Singh, & Maathuis, 2009). Likewise, Mekawy et al. (2015) partly linked the increased salinity tolerance in rice to the down-regulation of *OsCNGC1*, which inhibits excess Na^+ uptake in roots (Inada, Ueda, Shi, & Takabe, 2005) was simultaneously up-regulated. According to Massange-Sanchez et al. (2016), *AhCNGC5* was upregulated whereas *AhCNGC17* was downregulated in a halophyte *Amaranthus hypochondriacus* in response to salt stress. Evolutionary analysis presents the origin of CNGCs is from Streptophyta, and the gene family size increased gradually along with the plant evolution, then suddenly expanded after basal angiosperms (Figs. 2 and 3). Interestingly, the transcript abundance of *CNGCs* was slightly upregulated in two halophytic species (quinoa, ice plant) but downregulated in two glycophytes (Arabidopsis, rice) and the model moss under salt stress (Fig. 4). These evidences imply the important roles of CNGCs in plant adaptation to environmental stresses such as soil salinity.

In summary, a range of cation transporters (e.g., HKTs, NHXs, CNGCs) have been well established for their vital roles in plant salinity tolerance. However, there are a lack of investigations in the detailed function of guard cell specific cation transporters and their contribution to salinity tolerance. It is also important to understand the cell type specific expression and function of these cation transporters in relation to the stomatal regulation and high salt tolerance in halophytes from the physiological, molecular and evolutionary aspects.

3.2 Anion channels

Anion channels mediate intracellular signal transduction processes by coordinating the flow of anions across cell membranes (Kanwar et al., 2020; Roelfsema et al., 2012). Many plant's anion channels function in guard cells by regulating a wide range of cellular functions such as osmoregulation, signal transduction, anion transport, and stress tolerance (Barbier-Brygoo et al., 2011). For instance, patch-clamp experiment revealed two distinct types of anion channels in the guard cell plasma membrane; rapid-type (R-type) and slow-type (S-type). Many anion channels have been reported in the regulation of stomatal movements and/or salinity tolerance by mediating the influx and efflux of anions such as Cl^-, mal^{2-}, and NO_3^-. Here, we provide a brief summary of each of these anion channels in different plant species.

Slow anion channels (SLACs): Slow anion channel 1 (SLAC1) is ubiquitously expressed in the guard cells and highly permeable to NO_3^- and Cl^- (Negi et al., 2008; Vahisalu et al., 2008). Activities of SLAC1 are regulated through interactions with protein kinases (SNF1-related kinases (SnRKs), calcium-dependent kinases (CPKs), calcineurin B-like (CBL), and CBL-interacting protein kinase (CIPK) and deactivation of protein phosphatases 2C (PP2C) (Barbier-Brygoo et al., 2011; Li, Li, Wei, & Liu, 2020; Nejat & Mantri, 2017; Pantoja, 2021; Roelfsema et al., 2012; Saito & Uozumi, 2019). SLAC1 triggers varied signaling pathways in response to diverse environmental factors (Hedrich & Geiger, 2017; Pantoja, 2021; Pornsiriwong et al., 2017; Roelfsema et al., 2012; Zhao et al., 2019). Studies have shown that SLAC1 is involved in Cl^- exclusion, thereby improving tolerance to salinity. Transgenic overexpression of *AtSLAC1* in tobacco BY-2 cells enhanced cryptogein-induced Cl^- efflux (Kurusu et al., 2013). *slac1* mutants showed increased Cl^- content in guard cells (Negi et al., 2008). SLAH1, a homologue of SLAC1, modulates shoot Cl^- accumulation, and thus salt tolerance, in *Arabidopsis* (Cubero-Font et al., 2016; Qiu, Henderson, Tester, Roy, & Gilliham, 2016). SLAC/SLAH are the key regulators of stomatal closure in plants (Dreyer, Gomez-Porras, Riaño Pachón, Hedrich, & Geiger, 2012; Geiger et al., 2011; Hedrich, 2012; Pantoja, 2021; Vahisalu et al., 2008). SLAC1 mediate the efflux of anions (Cl^- and NO_3^-) from guard cells, depolarizing cell membranes to induce stomatal closure by reducing turgor pressure in guard cells (Hedrich, 2012). Lack of functional SLAC1 reduced stomatal closure (Kashtoh & Baek, 2021). SLAH3 are essential for efficient stomatal closure and opening in guard cells. SLAC1/SLAH have been reported to regulate stomatal closure and drought response (Zhang et al., 2016), but there is limited evidence for their direct involvement in plant salinity tolerance (Liu, Mak, et al., 2014; Liu, Zhang, et al., 2014).

Aluminum-activated malate transporters (*ALMTs*) are a unique family of membrane proteins found in land plants that mediate anion fluxes across cellular membranes (Barbier-Brygoo et al., 2011; De Angeli, Zhang, Meyer, & Martinoia, 2013). Plasma membrane-localized ALMTs mediate rapid-type currents, called quick-activating channel (QUAC). ALMTs have diverse biological functions and are distributed across multiple plant tissues and cell types including guard cells. For instance, ALMTs was first identified in *Arabidopsis* (Hoekenga et al., 2006) and wheat (Sasaki et al., 2004) with permeability to mal^{2-}, NO_3^-, and Cl^- (Sharma, Dreyer, Kochian, & Piñeros, 2016) and their activities might be enhanced by directly binding to Aluminum and Mal^{2-}, however the Arabidopsis guard cells ALMT channels are not aluminum-activated. The *Arabidopsis AtALMT6, AtALMT9, and AtALMT12* have been reported in the regulation of stomatal movements. *AtALMT6* is expressed specifically in guard cells and mediates Mal^{2-} and Fum^{2-} (fumarate) currents but not NO_3^- and Cl^- currents (Meyer et al., 2011). *AtALMT9*, a malate-activated Cl^- channel localizes in the tonoplast and mediates the vacuolar sequestration of Cl^- during stomatal opening (Baetz, Eisenach, Tohge, Martinoia, & De Angeli, 2016; De Angeli et al., 2013). Under salt stress, *AtALMT9* exhibited an up-regulated expression and *almt9* loss of function mutants have reduced accumulation of Cl^- in shoots (Baetz et al., 2016). *AtALMT12*, is located in the plasma membrane and is mainly permeable to Cl^- and NO_3^- (Meyer et al., 2010) and *ALMT12* interacts with *SRK2E/OST1* to regulate stomatal closure (Imes et al., 2013; Sasaki et al., 2004). In addition, the permeability ratio of *TaALMT1* and *ZmALMT2* toward malate and Cl^- at high external Cl^- suggest that these channels might mediate Cl^- efflux

(Ligaba, Maron, Shaff, Kochian, & Pineros, 2012; Piñeros, Cançado, & Kochian, 2008). *HvALMT1* is localized in the guard cells, seeds and roots for stomatal closure and plant development (Xu et al., 2015). Given the ability of *ALMT9* and *ALMT12* to mediate vacuolar sequestration of Cl^- in guard cells, we propose that the two genes and their orthologues in halophytes might be important for salt tolerance in plants. However, further studies are required to validate the prediction functions of ALMTs in salt tolerance.

Chloride channels (*CLCs*) family proteins are widespread in animals, bacteria, and plants (Jentsch & Pusch, 2018; Nedelyaeva, Shuvalov, & Balnokin, 2020) and function as anion channels/transporters in the sequestration of anions, including NO_3^- and Cl^- into the vacuole (Zifarelli & Pusch, 2010). For instance, combined electrophysiological and genetic studies revealed that *AtClCa* is a NO_3^-/H^+ exchanger involved in NO_3^- homeostasis (De Angeli et al., 2006; De Angeli et al., 2009; Zifarelli & Pusch, 2010) by mediating the influx and efflux of NO_3^- between cytoplasm and vacuole (De Angeli et al., 2006; Wege et al., 2014). Likewise, the CLCc protein was found to function in Cl^- homeostasis in guard cells (Harada, Kuromori, Hirayama, Shinozaki, & Leigh, 2004; Jossier et al., 2010). CLCs have been suggested to mediate stomatal response and salinity tolerance. For example, Wege et al. (2014) revealed that *AtCLCa* was required for ABA-induced stomatal closure or inhibition of stomatal opening. In another study, the phenotype of *clcc* mutants suggested its involvement in stomatal response to ABA and NaCl via modulating NO_3^-/Cl^- homeostasis (Jossier et al., 2010). Wei, Wang, Liu, Yu, and Lam (2016) showed that overexpression of *GmCLC1* in soybean hair roots increased Cl^- sequestration in roots and reduced Cl^- accumulation in shoots, leading to enhanced salt tolerance. In *Arabidopsis,* higher Cl^- accumulation in *AtCLCc* overexpression lines with increased overall salt tolerance (Hu et al., 2017; Nguyen et al., 2016). Moreover, CLCc increased Cl^- accumulation in vacuoles of roots and reduced Cl^- transport to the shoot under saline conditions, thus enhancing salt tolerance (Jossier et al., 2010). Given that CLC family is crucial in regulating stomatal responses and salt tolerance in plants, we speculate CLCs might mediate salinity-induced stomatal movement, but further research validations are required.

The origin of SLACs and ALMTs are from Streptophyta, but orthologues of CLCs exist in all clades of all tested green plant lineages including Chlorophyta (Figs. 2 and 3). *CLC-A* were induced by salt stress in Arabidopsis, barely and quinoa but suppressed in moss and *Salicornia persica* (Fig. 4). All differentially expressed *SLACs* were induced in barley, moss and ice plant, only *SpSLACs* were reduced after salt stress (Fig. 4). The expression of *ALMTs* are differ significantly among the halophytic and glycophytic species, showing the functional diversity of ALMTs under salt stress (Fig. 4). However, there are limited studies on SLAC/SLAH, ALMTs and CLCs in halophytes, which will be essential to advance our knowledge in the function of these anion channel families in the green plant evolution and adaptation to salinity stress.

3.3 H^+ pumps and cotransporters

H^+-ATPases are a member of the plant P-type ATPases superfamily that function as primary transporters that mediate the efflux of ions from the cytoplasm by activating various secondary transporters to mediate influx of metabolites and ions (Palmgren, 1998, 2001; Rengel, Bose, Chen, & Tripathi, 2015). H^+-ATPases are pumps that transport protons across plasma membrane via primary active transport, and showed a highly conservation along with the evolution of plants (Fig. 3) (Hasegawa, Bressan, Zhu, & Bohnert, 2000; Morsomme & Boutry, 2000). The plant H^+-ATPases are involved in diverse physiological processes such as growth, intracellular pH regulation, nutrient uptake, stomatal opening, and abiotic stress adaptations (Falhof et al., 2016; Janicka-Russak & Kabała, 2015; López-Pérez, del Carmen Martínez-Ballesta, Maurel, & Carvajal, 2009; Sondergaard, Schulz, & Palmgren, 2004; Yu, Kan, Zhang, Zeng, & Chen, 2016; Zhang et al., 2006). Many isoforms of H^+-ATPases have been identified in *Arabidopsis* (AHA1–AHA11) (Palmgren, 2001), rice (*OsA1–OsA10*) (Arango, Gévaudant, Oufattole, & Boutry, 2003), tobacco (*PMA1–PMA9*) (Oufattole, Arango, & Boutry, 2000), and fava bean (*VHA1–VHA5*) (Neuhaus, Geilfus, Zörb, & Mühling, 2013), all of which exhibiting functional and tissue specificity. For example, *VHA1* and *VHA2* are mainly expressed in guard cells of fava bean (Hentzen et al., 1996) whereas AHA1, AHA2, and AHA5 are highly expressed in guard cells of *Arabidopsis* (Ueno, Kinoshita, Inoue, Emi, & Shimazaki, 2005). The *Arabidopsis* AHA4 functions in salt tolerance (Vitart, Baxter, Doerner, & Harper, 2001) and the constitutive activation of AHA1 keeps stomata open (Merlot et al., 2002). Up-regulation of AHA2 in guard cells elevates light-induced stomatal opening, thus, improving plant growth (Wang et al., 2014). The Arabidopsis loss-of-function mutant *aha4* exhibited increased sensitivity to salinity stress (Vitart et al., 2001). H^+-ATPase are crucial in regulating ionic homeostasis and salinity adaptations in plants including halophytes by increasing the expression of H^+-ATPase encoding genes under salt stress (Janicka-Russak & Kabała, 2015; Sibole, Cabot, Michalke, Poschenrieder, & Barceló, 2005; Silva & Gerós, 2009). For instance, high salinity enhanced the expression of plasma membrane H^+-ATPase genes in a desert grass *Aeluropus littorali*, implicating their potential functions in salt stress adaptation (Olfatmiri, Alemzadeh, & Zakipour, 2014). Similar results observed in a halophyte *Populus euphratica* (desert poplar) was attributed to the enhanced vacuolar sequestration of salts (Ma, Liu, Li, & Zhang, 2002; Yang et al., 2007; Zhang et al., 2007). Under salt stress conditions, the ectopic expression of *PeHA1* significantly enhanced H^+-ATPase activity in *Populus euphratica* and salinity tolerance in transgenic *Arabidopsis* plants (Wang et al., 2013). Salinity inhibited H^+-ATPase activity in rice but was unrestrained in a seagrass *Zostera marina* (Muramatsu et al., 2002).

Activity of H^+-ATPase also provides a driving force for operation of Na^+/H^+ antiporter in the plasma membrane and tonoplast during salinity stress (Blumwald, Aharon, & Apse, 2000; Shabala, 2000; Shabala & Newman, 2000; Zhu, 2003). For example, the inhibition of H^+-ATPase activity by sodium orthovanadate in *P. euphratica* plants under long-term salinity exposure reduced Na^+ extrusion and H^+ uptake (Sun et al., 2009). Overexpression of the halophyte *Lobularia maritima LmVHA-E1* helped transgenic Arabidopsis to compartmentalize more Na^+ and enhanced osmotic adjustment in the leaves (Dabbous et al., 2017). Thus, the regulation H^+-ATPase activity may enhance the salinity tolerance mechanism in plants. H^+-ATPase activity in the guard cells is essential for stomatal movements and is modulated by both positive (blue light and auxins) and negative (Ca^{2+} and ABA) regulators (Merlot et al., 2002; Mishra, Singh, & Roy, 2022). During stomatal closure, H^+-ATPase is inhibited simultaneously with the activation of anion channels resulting in membrane depolarization. Both the R- and S-type anion channels facilitate the efflux of mal^{2-}, Cl^-, and NO_3^- (Cai et al., 2021; Chen, Hills, Lim, & Blatt, 2010; Roelfsema & Hedrich, 2005). Depolarization of the plasma membrane drives the efflux of K^+ via the K^+ outwardly rectifying channels such as GORK (guard cell outwardly rectifying K^+ channel) (Adem et al., 2020; Jeanguenin et al., 2008). Hosy et al. (2003) demonstrated that GORK is crucial for K^+ exclusion and stomatal closure as the *Arabidopsis gork* loss of function mutant showed impaired stomatal closure. Taken together, the efflux of solutes from the guard cells leads to reduced turgor and subsequently, stomatal closure.

On the other hand, the hyperpolarization of the plasma membrane via H^+-ATPase mediated efflux of H^+ from guard cells activates the inward K^+ rectifying channels (KAT1, KAT2, and AKT1) leading to K^+ intake, and thus, stomatal opening (Chen et al., 2017; Pilot et al., 2001; Szyroki et al., 2001). Furthermore, the release of H^+ from the guard cells acidifies the apoplast which activates the K^+ channels, resulting in K^+ influx. Transported anions (NO_3^- and Cl^-) and mal^{2-} from starch degradation accumulate in the guard cells to balance K^+ and together with water transported through aquaporins, the required turgor pressure is generated for stomatal opening. However, it is still elusive on how NaCl affects the well-studied membrane transport systems during stomatal opening and closure, which will be important for future investigation of stomatal regulation under salinity stress in halophytes.

3.4 The role of aquaporin in salt stress and stomatal regulation

Aquaporins are intrinsic membrane proteins that function as channels for water transport as well as small neutral molecules across intracellular membranes (Bienert & Chaumont, 2014; Kaldenhoff, 2012; Maurel et al., 2015). Plant aquaporins show a multiplicity of isoforms demystifying the high diversity in subcellular localization, function, transport selectivity, regulation effect and abundance. The plasma membrane intrinsic proteins (PIPs) and tonoplast intrinsic proteins (TIPs) are the most abundant constituents of the plasma membrane and tonoplast, respectively. Studies have shown that PIPs and TIPs are expressed in stomatal guard cells in diverse plant species (Chen et al., 2017; Hachez, Milhiet, Heinen, & Chaumont, 2017; Tyerman, McGaughey, Qiu, Yool, & Byrt, 2021). Many studies suggested the key roles of aquaporins in membrane water transport in guard cells, but recent reports point to their functions in stomatal movement and salinity tolerance (Heinen et al., 2014). The *Arabidopsis AtPIP2;1* partakes in ABA-mediated stomatal closure in guard cells by inducing H_2O_2 uptake and water efflux (Rodrigues et al., 2017). The *AtPIP2;1* lack of function mutants showed defects in ABA-triggered stomatal closure in epidermal peels (Cui, Hao, Chen, Chen, & Wang, 2008). Overexpression of *Vicia faba VfPIP1* in *Arabidopsis* enhanced darkness or ABA-induced stomatal closure (Cui et al., 2005). A diurnal peak expression of TIP isoform in guard cells was associated with stomatal closure in *Helianthus annuus* (sunflower) (Sarda et al., 1997). In tobacco, stomatal density increased in transgenic plants overexpressing *AtPIP1b* but was unaffected in transgenic *Arabidopsis* plants overexpressing *NtAQP1* (Cui et al., 2005). Considering that stomatal density is an essential determinant of stomatal conductance and photosynthesis in halophytic plants, these genes might exhibit similar functions. However, it is unclear how aquaporins regulate stomatal density.

Increasing evidence suggests the importance of aquaporins in salt-induced stress responses and tolerance mechanisms in plants. For instance, the *Arabidopsis* PIP2;1 and PIP2;2 genes were found to function as Na^+ channels, whereas PIP2;1 mediated the influx of Na^+ in roots, indicating their role in salinity tolerance (Byrt et al., 2017; Kourghi et al., 2017; Tyerman et al., 2021). Durand et al. (2020) demonstrated that high levels of Na^+ in guard cells positively correlated with the expression of some *PIP* and *TIP* aquaporins and *NHX1.13* in hybrid poplar, indicating that PIP and TIP could participate in salt-induced stomatal regulation. The constitutive expression of *OsPIP1;1* enhanced salt stress tolerance and subsequently, increased the growth of rice plants (Liu et al., 2013). Likewise, constitutive overexpression of *GmPIP1;6* significantly improved salt tolerance in soybean by enhancing Na^+ exclusion (Zhou et al., 2014). Overexpression of *TaNIP* from wheat enhanced salt tolerance in transgenic *Arabidopsis* plants (Gao et al., 2010) whereas *TsTIP1;2* in the halophytic plant *Thellungiella salsuginea* (salt cress) enhanced vacuolar accumulation of salt, thus improving salinity tolerance (Wang et al., 2014). In another study, the heterologous expression of *SpAQP1/SpPIP2* from the halophytic plant, *Sesuvium portulacastrum* (shoreline purslane) significantly enhanced salinity tolerance in tobacco plants (Chang, Liu, Zhu, Fan, & Zhang, 2016). We found that PIPs, TIPs and NIPs may have originated from Streptophyte algae, which are highly conserved in land plant species (Figs. 2 and 3). The expression of *PIPs*, *TIPs* and *NIPs* varies greatly among the halophytic and glycophytic species studied in this paper (Fig. 4). Nonetheless, it is unclear whether the expression of aquaporins in guard cells increases Na^+ permeability and directly contribute to stomatal movement. Given the expression guard cell localized aquaporins in stomatal and salinity stress responses, we propose that these proteins might be involved salt-induced stomatal regulation.

4. Signaling pathways help plants regulate stomatal movement under salinity

4.1 Abscisic acid (ABA) signaling

Many biotic and abiotic stress signals regulate stomatal movement. ABA is a major plant hormone with distinct roles in plant growth and developmental processes as well as plant responses to adverse environmental conditions (Merilo, Jalakas, Kollist, & Brosché, 2015). ABA signaling remains one of the primary signals in plants upon salinity exposure (Assmann, 1993; Hedrich & Shabala, 2018; Schroeder, Kwak, & Allen, 2001). To avoid the adverse effects of salt stress, ABA production is increased to activate enhanced ABA signaling (Golldack, Li, Mohan, & Probst, 2014). The main ABA signaling in response to stomatal regulation includes the ABA receptors; pyrabactin-resistance 1/pyrabactin-resistance like/regulatory component of ABA receptor (PYR1/PYL/RCAR), type 2C protein phosphatases (PP2Cs) and SNF1-related protein kinase (SnRK2) open stomata 1 (OST1) (Hedrich & Geiger, 2017; Imes et al., 2013; Roelfsema et al., 2012). Upon uptake through the ABC transporters in guard cells, ABA binds to the ABA receptors (PYR1/PYL/RCAR) and forms complexes such as HAB1 (hypersensitive to ABA1), ABI1 (ABA insensitive 1) and ABI2 (ABA insensitive 2) which act as negative regulators of ABA signaling (Ma et al., 2009; Nishimura et al., 2010; Park et al., 2009; Santiago et al., 2009). These complexes inactivate PP2C to stimulate the phosphorylation of downstream signaling components, which activate SnRK2 regulated ion channels (e.g., SLAC1, ALMTs) for membrane depolarization. It ultimately leads to the extrusion of guard cell anions and K^+ for stomatal closure (Fujii & Zhu, 2009; Fujita et al., 2009; Imes et al., 2013; Julkowska & Testerink, 2015; Kim et al., 2010; Pei, Kuchitsu, Ward, Schwarz, & Schroeder, 1997; Umezawa et al., 2009). Thus, ABA receptors, PP2Cs, and SnRKs form the core of the initial ABA signaling network, which are involved in both stomatal regulation and salinity stress response.

Under salt stress, ABA concentration in leaves increases which triggers stomatal closure (Merilo et al., 2015; Park et al., 2009; Pei et al., 1997), thus reducing transpirational water loss (Hedrich & Shabala, 2018; Hetherington & Woodward, 2003; Shabala, 2013). In guard cells, ABA inhibits H^+-ATPase activity in the plasma membrane, which restricts stomatal opening. In addition, H^+ can be released via the H^+-pump upon ABA degradation to accumulate water and ions (Cl^-, K^+, and mal^{2-}), which generates the required turgor for stomatal opening (Hayashi, Inoue, Takahashi, & Kinoshita, 2011). Notably, ABA signaling differs between halophytic and glycophytic guard cells. For instance, halophytes require much lower ABA concentrations to modulate stomatal opening than glycophytes (Hedrich & Shabala, 2018). In a study involving *Thellungiella* (halophyte) and *Arabidopsis* (glycophyte) under salinity stress, low ABA concentrations caused the downregulation of ABA receptors *PYL5, PYL6, PYL1, and*

PYL4 whereas *HAI1*, *SnRK2.9* and *CPK32* were upregulated in the guard cells of *Thellungiella* (Karimi et al., 2021). Under the same conditions, the upregulation of genes involved in ABA signaling including *PYL5, PYL6, SnRK2.9, CBL4/SOS3, AGH1, HAI1, HAI3, SLAH2, SLAH3* and some *CIPK/SnRK3* members in guard cells reduced stomatal conductance in *Arabidopsis* (Karimi et al., 2021). However, how these genes increase the sensitivity of guard cells to ABA and mediate stomatal movement under salt stress remain a mystery. Considering the possibility of decreased stomatal conductance via a decline in stomatal opening/stomatal density (Shabala, 2013), we speculate that the closure of the stomata in Arabidopsis guard cells reduced stomatal conductance. Taken together, salinity induces ABA accumulation which activates the ABA signaling network, the S-type and R-type anion channels mediated efflux of anions such as Cl^-, mal^{2-}, and NO^-, NO^{3-}, extrusion of K^+ via the GORK channel as well as the efflux of Ca^{2+} from the vacuole, all leading to stomatal closure.

4.2 Reactive oxygen species (ROS) signaling

ROS (e.g., superoxide anion radical $O_2{}^{\bullet -}$, hydroxyl radical $HO^{\bullet}$, singlet oxygen $^1O_2$, hydrogen peroxide H_2O_2) are toxic by-products of cellular metabolism and also function as key signaling molecules in diverse plant physiological processes (Baxter et al., 2010; Bose et al., 2014; Song et al., 2014). While low concentrations of ROS trigger cell signaling, high concentrations are extremely detrimental to plants (Neeraj et al., 2019). Functional disparity between ROS and antioxidants causes oxidative stress leading to cellular damage and eventually cell death (Ighodaro & Akinloye, 2018). Diverse environmental stresses including salinity rapidly induce ROS production and accumulation in plant cells (Nath et al., 2019; Niu et al., 2018; Ozgur et al., 2013), damaging proteins, lipids, and nucleic acids (Petrov, Hille, Mueller-Roeber, & Gechev, 2015). To prevent cellular damage, plants have evolved a complex anti-oxidative defense system consisting of enzymatic and non-enzymatic antioxidants that regulates excessive ROS production while maintaining optimal ROS levels for signaling (Hou et al., 2018; Wu et al., 2017).

ROS signaling is an essential component of plant adaptive response to salinity stress. Thus, maintaining ROS homeostasis may induce salinity stress tolerance in plants (Baxter, Mittler, & Suzuki, 2014; Choudhury, Rivero, Blumwald, & Mittler, 2017; Miller, Suzuki, Ciftci-Yilmaz, & Mittler, 2010; Suzuki, Koussevitzky, Mittler, & Miller, 2012). Salinity stress triggers the equipoise between ROS synthesis and detoxification. Increased antioxidant defense capacity, a common phenomenon during salt stress generally correlates with salt tolerance in plants. For instance, Rajaravindran and Natarajan (2012) reported that plants with high antioxidant activity are more tolerant to diverse environmental stresses than that with low anti-oxidative activity. High salinity tolerance of many halophytic plants in contrast to glycophytes is also attributed to their high antioxidant activity (Flowers & Colmer, 2008; Ozgur et al., 2013). H_2O_2 as a more stable ROS acts as regulatory signals during various cellular responses (Baxter et al., 2014; Camejo, Guzmán-Cedeño, & Moreno, 2016; Wrzaczek, Brosché, & Kangasjärvi, 2013). H_2O_2 has previously been found to be linked to mediate salinity tolerance (Xu et al., 2008; Li, Qiu, Zhang, & Wang, 2011; Tyerman, McGaughey, Qiu, Yool, & Byrt, 2021; Yong, Solis, Rabbi, et al., 2022). Although ROS detoxification strategies (in terms of antioxidant activity and isozymes expressed) in halophytes differ from glycophytes in response to salinity, their accumulation and toxicity mechanisms are very similar (Ozgur et al., 2013). Salinity tolerance mechanisms in many halophytic plants involve oxidative stress signaling and ROS detoxification. Perez and Brown (2014) reported a positive correlation between plants ability to combat oxidative stress and salt stress tolerance. Likewise, several studies demonstrated a correlation between antioxidant content and salt stress tolerance (Bor, Özdemir, & Türkan, 2003; Demiral & Türkan, 2005; Seckin, Turkan, Sekmen, & Ozfidan, 2010). Therefore, fine-tuning a balance between oxidative stress signaling and ROS scavenging may be vital for improving salinity tolerance in plants.

Upon salinity exposure, plasma membrane-localized NADPH oxidases mediate ROS production in the apoplast (Miller et al., 2010; Sierla, Rahikainen, Salojärvi, Kangasjärvi, & Kangasjärvi, 2013; Suzuki et al., 2011). ROS are important signals involved in the regulation of stomatal closure (Murata, Mori, & Munemasa, 2015; Song et al., 2014) and membrane transporter activity (Pottosin et al., 2014) in the guard cells. In the regulation of stomatal closure, apoplastic ROS mediates cell signaling leading to the activation of anion channels in the guard cell. However, ROS sensitivity of guard cell channels may differ among halophytes and glycophytes (Hedrich & Shabala, 2018). The activity of ion channels can be modified by H_2O_2, suggesting the involvement of these channels in H_2O_2 signaling. H_2O_2 inhibited both inward- and outward-rectifying K^+ channels in *V. faba* guard cells (Köhler, Hills, & Blatt, 2003; Zhang et al., 2001) and activated Ca^{2+}-permeable cation channels in *Arabidopsis* guard cells, resulting in stomatal closure (Pei et al., 2000). H_2O_2 has also been shown to directly activate the *Arabidopsis* plasma membrane K^+ channel, Stelar K^+ Outward Rectifier (SKOR) (Garcia-Mata et al., 2010). Effects of H_2O_2 on guard cell ion channels are obvious but whether their activities are directly fine-tuned by ROS remains unclear. Although the mechanisms of ROS production and signaling are somewhat understood, how ROS homeostasis is maintained in guard cells remains to be elucidated.

The *Arabidopsis* RBOHs (*AtRBOHF* and *AtRBOHD*) are involved in stomatal movements (Kwak et al., 2003). ABA-mediated stomatal closure was significantly impaired whereas ROS production ceased in *atrbohD atrbohF* double mutants. However, exogenous application of ROS rescued stomatal responses in *atrbohD atrbohF* double mutants (Chater & Gray, 2015). Under salt stress, both *AtRbohD* and *AtRbohF* were upregulated whereas concurrent mutations of these two genes resulted in hypersensitivity to salt stress (Ma et al., 2012). Salt-induced ROS production by *AtRbohD/AtRbohF* promoted K^+ influx into the cytosol, thereby reducing Na^+/K^+ ratios (Ma et al., 2012). *AtRbohF* restricted Na^+ circulation in the xylem sap, thereby decreasing Na^+ content in shoots (Jiang et al., 2012). During salt stress, *AtRbohD/F*-mediated H_2O_2 synthesis triggered an antioxidative response, which decreased oxidative damage to cells (Rejeb et al., 2015). Furthermore, H_2O_2 was shown to regulate the expression of the plasma membrane Na^+/H^+ antiporter, SOS1 in *Arabidopsis* under salt stress, thus maintaining K^+/Na^+ homeostasis (Chung, Zhu, Bressan, Hasegawa, & Shi, 2008). Halophytes exhibit more rapid H_2O_2 signaling than glycophytes (Bose et al., 2014), therefore, activation of salinity-induced stress signaling and adaptive response in halophytes may require higher H_2O_2 levels (Shabala, Wu, & Bose, 2015). Although RBOHs are so important in the regulation of salt stress in plants, we have not found the significant difference of gene family size between halophytes and glycophytes. This may imply that the key point of halophytes cope with salinity is the fine-tuned of the expression of *RBOHs*, but not through increase of the numbers of homologous genes (Figs. 3 and 4).

5. Conclusions and future perspective

Salinity stress induces ionic, osmotic, and oxidative stresses, which negatively influence stomatal operation and plant growth and productivity. The bulk majority of crops are glycophytes, which are sensitive to saline environment. In order to cope with the increased salinity problems in agriculture, one of the effective ways is utilizing the halophytes. Coordinated activation and/or inactivation of multiple ion channels, transporters, and pumps facilitate ion fluxes are required for stomatal movements. Stomatal function remains a crucial component of plant adaptive mechanisms to salinity, and this differs substantially between halophytes and glycophytes. Halophytic plants are capable of optimizing their stomatal performance under saline conditions by significantly increasing the speed of stomatal movements and adjusting stomatal size and density to enhance salinity tolerance. The EBCs of halophytes act as the storage reservoir for more effective stomatal regulation. However, understanding more about the key adaptive mechanisms of salinity-induced stomatal operations and salinity tolerance is crucial for engineering salinity-tolerant plant species. Salt response stomatal movement directly affect the water use efficiency and photosynthetic process. Halophytes deal with salinity problems not only through evolving the new specialize bladder cells, but also have the diversity and specific regulation network. It is interesting to note that though the regulation of various ion channels and transporters confer salinity stress tolerance and induce stomatal movements in plants, the underlying mechanisms of salinity-induced regulation of stomatal responses in both halophytes and glycophytes is unclear, presenting a significant knowledge gap that needs to be elucidated through further research endeavor. Learning from halophytes will help us to achieve crops with better water use efficiency for sustainable food production. It might also have chances to use genome editing and modification to create crops with EBCs to store excessive salt, which will ensure the high productivity of the modified elite crop cultivars on marginal or even saline land in the future. Understanding the differences between glycophytes and halophytes represents an important foundation of scientific advances for breeding salt tolerance crops.

Acknowledgments

G.C., F.Z. and Z.-H.C. conceived the review. G.C. analyzed the data and prepared all the figures together with H.A.-D. G.C., H.A.-D. and Z.-H.C. analyzed the results and wrote the manuscript with support from F.D., F.Z., S.X. Z.-H.C. conducted final editing of the manuscript. All authors read and approved the manuscript. We are grateful for the financial support from the Natural Science Foundation of China (32001456) and Australian Research Council (FT210100366).

References

Ache, P., Becker, D., Ivashikina, N., Dietrich, P., Roelfsema, M. R. G., & Hedrich, R. (2000). GORK, a delayed outward rectifier expressed in guard cells of *Arabidopsis thaliana*, is a K^+-selective, K^+-sensing ion channel. *FEBS Letters*, *486*(2), 93–98.

Adams, P., Nelson, D. E., Yamada, S., Chmara, W., Jensen, R. G., Bohnert, H. J., et al. (1998). Growth and development of *Mesembryanthemum crystallinum* (Aizoaceae). *New Phytologist, 138*(2), 171–190.

Adams, P., Thomas, J. C., Vernon, D. M., Bohnert, H. J., & Jensen, R. G. (1992). Distinct cellular and organismic responses to salt stress. *Plant and Cell Physiology, 33*(8), 1215–1223.

Adem, G. D., Chen, G., Shabala, L., Chen, Z. H., & Shabala, S. (2020). GORK channel: A master switch of plant metabolism? *Trends in Plant Science, 25*(5), 434–445.

Agarie, S., Shimoda, T., Shimizu, Y., Baumann, K., Sunagawa, H., Kondo, A., et al. (2007). Salt tolerance, salt accumulation, and ionic homeostasis in an epidermal bladder-cell-less mutant of the common ice plant *Mesembryanthemum crystallinum*. *Journal of Experimental Botany, 58*(8), 1957–1967.

Ahmed, I. M., Dai, H. X., Zheng, W. T., Cao, F. B., Zhang, G. P., Sun, D. F., et al. (2013). Genotypic differences in physiological characteristics in the tolerance to drought and salinity combined stress between Tibetan wild and cultivated barley. *Plant Physiology and Biochemistry, 63*, 49–60.

Akram, M., Malik, M. A., Ashraf, M. Y., Saleem, M. F., & Hussain, M. (2007). Competitive seedling growth and K^+/Na^+ ratio in different maize (*Zea mays* L.) hybrids under salinity stress. *Pakistan Journal of Botany, 39*(7), 2553–2563.

Albaladejo, I., Meco, V., Plasencia, F., Flores, F. B., Bolarin, M. C., & Egea, I. (2017). Unravelling the strategies used by the wild tomato species *Solanum pennellii* to confront salt stress: From leaf anatomical adaptations to molecular responses. *Environmental and Experimental Botany, 135*, 1–12.

Ali, A., Cheol Park, H., Aman, R., Ali, Z., & Yun, D. J. (2013). Role of HKT1 in *Thellungiella salsuginea*, a model extremophile plant. *Plant Signaling & Behavior, 8*(8).

Aliakbari, M., Razi, H., Alemzadeh, A., & Tavakol, E. (2020). RNA-seq transcriptome profiling of the halophyte *Salicornia persica* in response to salinity. *Journal of Plant Growth Regulation, 40*(2), 707–721.

Andrés, Z., Pérez-Hormaeche, J., Leidi, E. O., Schlücking, K., Steinhorst, L., McLachlan, D. H., et al. (2014). Control of vacuolar dynamics and regulation of stomatal aperture by tonoplast potassium uptake. *Proceedings of the National Academy of Sciences of the United States of America, 111*(17), E1806–E1814.

Arango, M., Gévaudant, F., Oufattole, M., & Boutry, M. (2003). The plasma membrane proton pump ATPase: The significance of gene subfamilies. *Planta, 216*(3), 355–365.

Ashnest, J. R., Huynh, D. L., Dragwidge, J. M., Ford, B. A., & Gendall, A. R. (2015). *Arabidopsis* intracellular NHX-type sodium-proton antiporters are required for seed storage protein processing. *Plant and Cell Physiology, 56*(11), 2220–2233.

Asins, M. J., Villalta, I., Aly, M. M., Olias, R., Álvarez De Morales, P., Huertas, R., et al. (2013). Two closely linked tomato HKT coding genes are positional candidates for the major tomato QTL involved in Na^+/K^+ homeostasis. *Plant, Cell & Environment, 36*(6), 1171–1191.

Assaha, D. V. M., Ueda, A., Saneoka, H., Al-Yahyai, R., & Yaish, M. W. (2017). The role of Na^+ and K^+ transporters in salt stress adaptation in glycophytes. *Frontiers in Physiology, 8*.

Assmann, S. M. (1993). Signal transduction in guard cells. *Annual Review of Cell Biology, 9*(1), 345–375.

Baetz, U., Eisenach, C., Tohge, T., Martinoia, E., & De Angeli, A. (2016). Vacuolar chloride fluxes impact ion content and distribution during early salinity stress. *Plant Physiology, 172*(2), 1167–1181.

Barbier-Brygoo, H., De Angeli, A., Filleur, S., Frachisse, J.-M., Gambale, F., Thomine, S., et al. (2011). Anion channels/transporters in plants: From molecular bases to regulatory networks. *Annual Review of Plant Biology, 62*, 25–51.

Barragán, V., Leidi, E. O., Andrés, Z., Rubio, L., De Luca, A., Fernández, J. A., et al. (2012). Ion exchangers NHX1 and NHX2 mediate active potassium uptake into vacuoles to regulate cell turgor and stomatal function in *Arabidopsis*. *The Plant Cell, 24*(3), 1127–1142.

Bassil, E., Tajima, H., Liang, Y.-C., Ohto, M.-A., Ushijima, K., Nakano, R., et al. (2011). The *Arabidopsis* Na^+/H^+ antiporters NHX1 and NHX2 control vacuolar pH and K^+ homeostasis to regulate growth, flower development, and reproduction. *The Plant Cell, 23*(9), 3482–3497.

Basu, S., Kumar, A., Benazir, I., & Kumar, G. (2020). Reassessing the role of ion homeostasis for improving salinity tolerance in crop plants. *Physiologia Plantarum, 171*(4), 502–519.

Baxter, I., Brazelton, J. N., Yu, D., Huang, Y. S., Lahner, B., Yakubova, E., et al. (2010). A coastal cline in sodium accumulation in *Arabidopsis thaliana* is driven by natural variation of the sodium transporter AtHKT1; 1. *PLoS Genetics, 6*(11), e1001193.

Baxter, A., Mittler, R., & Suzuki, N. (2014). ROS as key players in plant stress signalling. *Journal of Experimental Botany, 65*(5), 1229–1240.

Belkheiri, O., & Mulas, M. (2013). The effects of salt stress on growth, water relations and ion accumulation in two halophyte *Atriplex* species. *Environmental and Experimental Botany, 86*, 17–28.

Bienert, G. P., & Chaumont, F. (2014). Aquaporin-facilitated transmembrane diffusion of hydrogen peroxide. *Biochimica et Biophysica Acta (BBA)-General Subjects, 1840*(5), 1596–1604.

Blumwald, E., Aharon, G. S., & Apse, M. P. (2000). Sodium transport in plant cells. *Biochimica et Biophysica Acta (BBA)-Biomembranes, 1465*(1–2), 140–151.

Bonales-Alatorre, E., Shabala, S., Chen, Z. H., & Pottosin, I. (2013). Reduced tonoplast fast-activating and slow-activating channel activity is essential for conferring salinity tolerance in a facultative halophyte, quinoa. *Plant Physiology, 162*(2), 940–952.

Bor, M., Özdemir, F., & Türkan, I. (2003). The effect of salt stress on lipid peroxidation and antioxidants in leaves of sugar beet *Beta vulgaris* L. and wild beet *Beta maritima* L. *Plant Science, 164*(1), 77–84.

Bose, J., Rodrigo-Moreno, A., & Shabala, S. (2014). ROS homeostasis in halophytes in the context of salinity stress tolerance. *Journal of Experimental Botany, 65*(5), 1241–1257.

Byrt, C. S., Zhao, M., Kourghi, M., Bose, J., Henderson, S. W., Qiu, J., et al. (2017). Non-selective cation channel activity of aquaporin AtPIP2; 1 regulated by Ca^{2+} and pH. *Plant, Cell & Environment, 40*(6), 802–815.

Cai, S., Huang, Y., Chen, F., Zhang, X., Sessa, E., Zhao, C., et al. (2021). Evolution of rapid blue-light response linked to explosive diversification of ferns in angiosperm forests. *New Phytologist, 230*(3), 1201–1213.

Cai, S., Papanatsiou, M., Blatt, M. R., & Chen, Z. (2017). Speedy grass stomata: Emerging molecular and evolutionary features. *Molecular Plant*, 912–914.

Camejo, D., Guzmán-Cedeño, Á., & Moreno, A. (2016). Reactive oxygen species, essential molecules, during plant–pathogen interactions. *Plant Physiology and Biochemistry, 103*, 10–23.

Chakraborty, K., Bose, J., Shabala, L., & Shabala, S. (2016). Difference in root K^+ retention ability and reduced sensitivity of K^+-permeable channels to reactive oxygen species confer differential salt tolerance in three *Brassica* species. *Journal of Experimental Botany, 67*(15), 4611–4625.

Chang, W., Liu, X., Zhu, J., Fan, W., & Zhang, Z. (2016). An aquaporin gene from halophyte *Sesuvium portulacastrum*, *SpAQP1*, increases salt tolerance in transgenic tobacco. *Plant Cell Reports, 35*(2), 385–395.

Chater, C., & Gray, J. E. (2015). Stomatal closure: The old guard takes up the SLAC. *Current Biology, 25*(7), R271–R273.

Chen, Z. H., Chen, G., Dai, F., Wang, Y., Hills, A., Ruan, Y.-L., et al. (2017). Molecular evolution of grass stomata. *Trends in Plant Science, 22*(2), 124–139.

Chen, Z. H., Hills, A., Lim, C. K., & Blatt, M. R. (2010). Dynamic regulation of guard cell anion channels by cytosolic free Ca^{2+} concentration and protein phosphorylation. *Plant Journal, 61*(5), 816–825.

Chen, Z., Newman, I., Zhou, M., Mendham, N., Zhang, G., & Shabala, S. (2005). Screening plants for salt tolerance by measuring K^+ flux: A case study for barley. *Plant Cell and Environment, 28*(10), 1230–1246.

Chen, M., Yang, Z., Liu, J., Zhu, T., Wei, X., Fan, H., et al. (2018). Adaptation mechanism of salt excluders under saline conditions and its applications. *International Journal of Molecular Sciences, 19*.

Chen, J., Yin, H., Gu, J., Li, L., Liu, Z., Jiang, X., et al. (2015). Genomic characterization, phylogenetic comparison and differential expression of the cyclic nucleotide-gated channels gene family in pear (*Pyrus bretchneideri* Rehd.). *Genomics, 105*(1), 39–52.

Choudhury, F. K., Rivero, R. M., Blumwald, E., & Mittler, R. (2017). Reactive oxygen species, abiotic stress and stress combination. *The Plant Journal, 90*(5), 856–867.

Chung, J. S., Zhu, J. K., Bressan, R. A., Hasegawa, P. M., & Shi, H. (2008). Reactive oxygen species mediate Na^+-induced SOS1 mRNA stability in *Arabidopsis*. *The Plant Journal, 53*(3), 554–565.

Cotsaftis, O., Plett, D., Shirley, N., Tester, M., & Hrmova, M. (2012). A two-staged model of Na^+ exclusion in rice explained by 3D modeling of HKT transporters and alternative splicing. *PLoS One*, 7(7), e39865.

Cubero-Font, P., Maierhofer, T., Jaslan, J., Rosales, M. A., Espartero, J., Díaz-Rueda, P., et al. (2016). Silent S-type anion channel subunit SLAH1 gates SLAH3 open for chloride root-to-shoot translocation. *Current Biology, 26*(16), 2213–2220.

Cui, X. H., Hao, F., Chen, H., Cai, J., Chen, J., & Wang, X. (2005). solaLion and expression of an aquaporin-like gene *VfPIP1* in *Vicia faba*. *Progress in Natural Science, 15*(6), 496–501.

Cui, X. H., Hao, F. S., Chen, H., Chen, J., & Wang, X. C. (2008). Expression of the *Vicia faba VfPIP1* gene in *Arabidopsis thaliana* plants improves their drought resistance. *Journal of Plant Research, 121*(2), 207–214.

Dabbous, A., Saad, R. B., Brini, F., Farhat-Khemekhem, A., Zorrig, W., Abdely, C., et al. (2017). Over-expression of a subunit E1 of a vacuolar H^+-ATPase gene (*LmVHA-E1*) cloned from the halophyte *Lobularia maritima* improves the tolerance of *Arabidopsis thaliana* to salt and osmotic stresses. *Environmental and Experimental Botany, 137*, 128–141.

De Angeli, A., Monachello, D., Ephritikhine, G., Frachisse, J., Thomine, S., Gambale, F., et al. (2006). The nitrate/proton antiporter *AtCLCa* mediates nitrate accumulation in plant vacuoles. *Nature, 442*(7105), 939–942.

De Angeli, A., Monachello, D., Ephritikhine, G., Frachisse, J.-M., Thomine, S., Gambale, F., et al. (2009). CLC-mediated anion transport in plant cells. *Philosophical Transactions of the Royal Society B: Biological Sciences, 364*(1514), 195–201.

De Angeli, A., Zhang, J., Meyer, S., & Martinoia, E. (2013). *AtALMT9* is a malate-activated vacuolar chloride channel required for stomatal opening in *Arabidopsis*. *Nature Communications, 4*(1), 1–10.

Deinlein, U., Stephan, A. B., Horie, T., Luo, W., Xu, G., & Schroeder, J. I. (2014). Plant salt-tolerance mechanisms. *Trends in Plant Science, 19*(6), 371–379.

Demidchik, V., & Tester, M. A. (2002). Nonselective cation channels in plants. *Annual Reviews of Plant Biology, 53*, 67–107.

Demiral, T., & Türkan, I. (2005). Comparative lipid peroxidation, antioxidant defense systems and proline content in roots of two rice cultivars differing in salt tolerance. *Environmental and Experimental Botany, 53*(3), 247–257.

Detar, R. A., Barahimipour, R., Manavski, N., Schwenkert, S., Hohner, R., Bolter, B., et al. (2021). Loss of inner-envelope K^+/H^+ exchangers impairs plastid rRNA maturation and gene expression. *Plant Cell, 33*, 2479–2505.

Dreyer, I., Gomez-Porras, J. L., Riaño Pachón, D. M., Hedrich, R., & Geiger, D. (2012). Molecular evolution of slow and quick anion channels (SLACs and QUACs/ALMTs). *Frontiers in Plant Science, 3*, 263.
Durand, M., Cohen, D., Aubry, N., Buré, C., Tomášková, I., Hummel, I., et al. (2020). Element content and expression of genes of interest in guard cells are connected to spatiotemporal variations in stomatal conductance. *Plant, Cell & Environment, 43*(1), 87–102.
Falhof, J., Pedersen, J. T., Fuglsang, A. T., & Palmgren, M. (2016). Plasma membrane H^+-ATPase regulation in the center of plant physiology. *Molecular Plant, 9*(3), 323–337.
Flowers, T. J., & Colmer, T. D. (2008). Salinity tolerance in halophytes. *New Phytologist, 179*(4), 945–963.
Fraile-Escanciano, A., Kamisugi, Y., Cuming, A. C., Rodríguez-Navarro, A., & Benito, B. (2010). The SOS1 transporter of *Physcomitrella patens* mediates sodium efflux in planta. *New Phytologist, 188*(3), 750–761.
Fu, L. B., Shen, Q. F., Kuang, L. H., Wu, D. Z., & Zhang, G. P. (2019). Transcriptomic and alternative splicing analyses reveal mechanisms of the difference in salt tolerance between barley and rice. *Environmental and Experimental Botany, 166*.
Fujii, H., & Zhu, J.-K. (2009). *Arabidopsis* mutant deficient in 3 abscisic acid-activated protein kinases reveals critical roles in growth, reproduction, and stress. *Proceedings of the National Academy of Sciences of the United States of America, 106*(20), 8380–8385.
Fujita, Y., Nakashima, K., Yoshida, T., Katagiri, T., Kidokoro, S., Kanamori, N., et al. (2009). Three SnRK2 protein kinases are the main positive regulators of abscisic acid signaling in response to water stress in *Arabidopsis*. *Plant and Cell Physiology, 50*(12), 2123–2132.
Gao, Z., He, X., Zhao, B., Zhou, C., Liang, Y., Ge, R., et al. (2010). Overexpressing a putative aquaporin gene from wheat, *TaNIP*, enhances salt tolerance in transgenic *Arabidopsis*. *Plant and Cell Physiology, 51*(5), 767–775.
Garcia-Mata, C., Wang, J., Gajdanowicz, P., Gonzalez, W., Hills, A., Donald, N., et al. (2010). A minimal cysteine motif required to activate the SKOR K^+ channel of *Arabidopsis* by the reactive oxygen species H_2O_2*[S]. *Journal of Biological Chemistry, 285*(38), 29286–29294.
Geiger, D., Maierhofer, T., Al-Rasheid, K. A., Scherzer, S., Mumm, P., Liese, A., et al. (2011). Stomatal closure by fast abscisic acid signaling is mediated by the guard cell anion channel SLAH3 and the receptor RCAR1. *Science Signaling, 4*(173), ra32.
Gobert, A., Park, G., Amtmann, A., Sanders, D., & Maathuis, F. J. (2006). *Arabidopsis thaliana* cyclic nucleotide gated channel 3 forms a non-selective ion transporter involved in germination and cation transport. *Journal of Experimental Botany, 57*(4), 791–800.
Golldack, D., Li, C., Mohan, H., & Probst, N. (2014). Tolerance to drought and salt stress in plants: Unraveling the signaling networks. *Frontiers in Plant Science, 5*, 151.
Guo, H., Huang, Z., Li, M., & Hou, Z. (2020). Growth, ionic homeostasis, and physiological responses of cotton under different salt and alkali stresses. *Scientific Reports, 10*(1), 21844.
Hachez, C., Milhiet, T., Heinen, R. B., & Chaumont, F. (2017). Roles of aquaporins in stomata. *Plant Aquaporins*, 167–183.
Hanin, M., Ebel, C., Ngom, M., Laplaze, L., & Masmoudi, K. (2016). New insights on plant salt tolerance mechanisms and their potential use for breeding. *Frontiers in Plant Science, 7*, 1787.
Harada, H., Kuromori, T., Hirayama, T., Shinozaki, K., & Leigh, R. A. (2004). Quantitative trait loci analysis of nitrate storage in Arabidopsis leading to an investigation of the contribution of the anion channel gene, *AtCLC-c*, to variation in nitrate levels. *Journal of Experimental Botany, 55*(405), 2005–2014.
Hasegawa, P. M., Bressan, R. A., Zhu, J. K., & Bohnert, H. J. (2000). Plant cellular and molecular responses to high salinity. *Annual Review of Plant Biology, 51*(1), 463–499.

Hayashi, M., Inoue, S.-I., Takahashi, K., & Kinoshita, T. (2011). Immunohistochemical detection of blue light-induced phosphorylation of the plasma membrane H^+-ATPase in stomatal guard cells. *Plant and Cell Physiology*, *52*(7), 1238–1248.

Hedrich, R. (2012). Ion channels in plants. *Physiological Reviews*, *92*(4), 1777–1811.

Hedrich, R., & Geiger, D. (2017). Biology of SLAC 1-type anion channels–from nutrient uptake to stomatal closure. *New Phytologist*, *216*(1), 46–61.

Hedrich, R., & Shabala, S. (2018). Stomata in a saline world. *Current Opinion in Plant Biology*, *46*, 87–95.

Heinen, R. B., Bienert, G. P., Cohen, D., Chevalier, A. S., Uehlein, N., Hachez, C., et al. (2014). Expression and characterization of plasma membrane aquaporins in stomatal complexes of *Zea mays*. *Plant Molecular Biology*, *86*(3), 335–350.

Hentzen, A. E., Smart, L. B., Whimmers, L. E., Fang, H. H., Schroeder, J. I., & Bennett, A. B. (1996). Two plasma membrane H^+-ATPase genes expressed in guard cells of *Vicia faba* are also expressed throughout the plant. *Plant and Cell Physiology*, *37*(5), 650–659.

Hetherington, A. M., & Woodward, F. I. (2003). The role of stomata in sensing and driving environmental change. *Nature*, *424*(6951), 901–908.

Hoekenga, O. A., Maron, L. G., Piñeros, M. A., Cançado, G. M., Shaff, J., Kobayashi, Y., et al. (2006). *AtALMT1*, which encodes a malate transporter, is identified as one of several genes critical for aluminum tolerance in *Arabidopsis*. *Proceedings of the National Academy of Sciences of the United States of America*, *103*(25), 9738–9743.

Horie, T., Karahara, I., & Katsuhara, M. (2012). Salinity tolerance mechanisms in glycophytes: An overview with the central focus on rice plants. *Rice*, *5*(1), 1–18.

Horie, T., Motoda, J., Kubo, M., Yang, H., Yoda, K., Horie, R., et al. (2005). Enhanced salt tolerance mediated by AtHKT1 transporter-induced Na^+ unloading from xylem vessels to xylem parenchyma cells. *The Plant Journal*, *44*(6), 928–938.

Horie, T., Yoshida, K., Nakayama, H., Yamada, K., Oiki, S., & Shinmyo, A. (2001). Two types of HKT transporters with different properties of Na^+ and K^+ transport in *Oryza sativa*. *The Plant Journal*, *27*(2), 129–138.

Hosy, E., Vavasseur, A., Mouline, K., Dreyer, I., Gaymard, F., Porée, F., et al. (2003). The *Arabidopsis* outward K^+ channel GORK is involved in regulation of stomatal movements and plant transpiration. *Proceedings of the National Academy of Sciences of the United States of America*, *100*(9), 5549–5554.

Hou, J., Zhang, Q., Zhou, Y., Ahammed, G. J., Zhou, Y., Yu, J., et al. (2018). Glutaredoxin GRXS16 mediates brassinosteroid-induced apoplastic H_2O_2 production to promote pesticide metabolism in tomato. *Environmental Pollution*, *240*, 227–234.

Hu, R., Zhu, Y., Wei, J., Chen, J., Shi, H., Shen, G., et al. (2017). Overexpression of PP2A-C5 that encodes the catalytic subunit 5 of protein phosphatase 2A in *Arabidopsis* confers better root and shoot development under salt conditions. *Plant, Cell & Environment*, *40*(1), 150–164.

Huertas, R., Rubio, L., Cagnac, O., García-Sánchez, M. J., Alché, J. D. D., Venema, K., et al. (2013). The K^+/H^+ antiporter *LeNHX2* increases salt tolerance by improving K^+ homeostasis in transgenic tomato. *Plant, Cell & Environment*, *36*(12), 2135–2149.

Hughes, J., Hepworth, C., Dutton, C., Dunn, J. A., Hunt, L., Stephens, J., et al. (2017). Reducing stomatal density in barley improves drought tolerance without impacting on yield. *Plant Physiology*, *174*(2), 776–787.

Ighodaro, O., & Akinloye, O. (2018). First line defence antioxidants-superoxide dismutase (SOD), catalase (CAT) and glutathione peroxidase (GPX): Their fundamental role in the entire antioxidant defence grid. *Alexandria Journal of Medicine*, *54*(4), 287–293.

Imes, D., Mumm, P., Böhm, J., Al-Rasheid, K. A., Marten, I., Geiger, D., et al. (2013). Open stomata 1 (OST1) kinase controls R–type anion channel QUAC1 in *Arabidopsis* guard cells. *The Plant Journal*, *74*(3), 372–382.

Inada, M., Ueda, A., Shi, W., & Takabe, T. (2005). A stress-inducible plasma membrane protein 3 (AcPMP3) in a monocotyledonous halophyte, Aneurolepidium chinense, regulates cellular Na^+ and K^+ accumulation under salt stress. *Planta, 220*(3), 395–402.

Ito, K., & Murphy, D. (2013). Application of ggplot2 to pharmacometric graphics. CPT: Pharmacometrics & Systems. *Pharmacology, 2*, e79.

Jabnoune, M., Espeout, S., Mieulet, D., Fizames, C., Verdeil, J.-L., Conéjéro, G., et al. (2009). Diversity in expression patterns and functional properties in the rice HKT transporter family. *Plant Physiology, 150*(4), 1955–1971.

James, R. A., Blake, C., Byrt, C. S., & Munns, R. (2011). Major genes for Na^+ exclusion, *Nax1* and *Nax2* (wheat *HKT1;4* and *HKT1;5*), decrease Na^+ accumulation in bread wheat leaves under saline and waterlogged conditions. *Journal of Experimental Botany, 62*(8), 2939–2947.

Janicka-Russak, M., & Kabała, K. (2015). The role of plasma membrane H^+-ATPase in salinity stress of plants. In *Progress in botany* (pp. 77–92). Springer.

Jeanguenin, L., Lebaudy, A., Xicluna, J., Alcon, C., Hosy, E., Duby, G., et al. (2008). Heteromerization of *Arabidopsis* Kv channel α-subunits: Data and prospects. *Plant Signaling & Behavior, 3*(9), 622–625.

Jentsch, T. J., & Pusch, M. (2018). CLC chloride channels and transporters: Structure, function, physiology, and disease. *Physiological Reviews, 98*(3), 1493–1590.

Jiang, C., Belfield, E. J., Mithani, A., Visscher, A., Ragoussis, J., Mott, R., et al. (2012). ROS-mediated vascular homeostatic control of root-to-shoot soil Na delivery in *Arabidopsis*. *The EMBO Journal, 31*(22), 4359–4370.

Jiang, X., Leidi, E. O., & Pardo, J. M. (2010). How do vacuolar NHX exchangers function in plant salt tolerance? *Plant Signaling & Behavior, 5*(7), 792–795.

Jin, Y., Jing, W., Zhang, Q., & Zhang, W. (2015). Cyclic nucleotide gated channel 10 negatively regulates salt tolerance by mediating Na^+ transport in *Arabidopsis*. *Journal of Plant Research, 128*(1), 211–220.

Jossier, M., Kroniewicz, L., Dalmas, F., Le Thiec, D., Ephritikhine, G., Thomine, S., et al. (2010). The *Arabidopsis* vacuolar anion transporter, *AtCLCc*, is involved in the regulation of stomatal movements and contributes to salt tolerance. *The Plant Journal, 64*(4), 563–576.

Julkowska, M. M., & Testerink, C. (2015). Tuning plant signaling and growth to survive salt. *Trends in Plant Science, 20*(9), 586–594.

Kakar, K. U., Nawaz, Z., Kakar, K., Ali, E., Almoneafy, A. A., Ullah, R., et al. (2017). Comprehensive genomic analysis of the *CNGC* gene family in *Brassica oleracea*: Novel insights into synteny, structures, and transcript profiles. *BMC Genomics, 18*(1), 1–18.

Kaldenhoff, R. (2012). Mechanisms underlying CO_2 diffusion in leaves. *Current Opinion in Plant Biology, 15*(3), 276–281.

Kanwar, P., Samtani, H., Sanyal, S. K., Srivastava, A. K., Suprasanna, P., & Pandey, G. K. (2020). VDAC and its interacting partners in plant and animal systems: An overview. *Critical Reviews in Biotechnology, 40*(5), 715–732.

Karimi, S. M., Freund, M., Wager, B. M., Knoblauch, M., Fromm, J., Ache, M. M. H. P., et al. (2021). Under salt stress guard cells rewire ion transport and abscisic acid signaling. *New Phytologist, 231*(3), 1040–1055.

Kashtoh, H., & Baek, K. H. (2021). Structural and functional insights into the role of guard cell ion channels in abiotic stress-induced stomatal closure. *Plants, 10*(12), 2774.

Kiani-Pouya, A., Rasouli, F., Bazihizina, N., Zhang, H., Hedrich, R., & Shabala, S. (2019). A large-scale screening of quinoa accessions reveals an important role of epidermal bladder cells and stomatal patterning in salinity tolerance. *Environmental and Experimental Botany, 168*, 103885.

Kiani-Pouya, A., Rasouli, F., Rabbi, B., Falakboland, Z., Yong, M., Chen, Z. H., et al. (2020). Stomatal traits as a determinant of superior salinity tolerance in wild barley. *Journal of Plant Physiology, 245*, 153108.

Kim, T.-H., Böhmer, M., Hu, H., Nishimura, N., & Schroeder, J. I. (2010). Guard cell signal transduction network: Advances in understanding abscisic acid, CO_2, and Ca^{2+} signaling. *Annual Review of Plant Biology, 61*, 561–591.

Köhler, B., Hills, A., & Blatt, M. R. (2003). Control of guard cell ion channels by hydrogen peroxide and abscisic acid indicates their action through alternate signaling pathways. *Plant Physiology, 131*(2), 385–388.

Köhler, C., Merkle, T., & Neuhaus, G. (1999). Characterisation of a novel gene family of putative cyclic nucleotide- and calmodulin-regulated ion channels in Arabidopsis thaliana. *The Plant Journal, 18*, 97–104.

Kollist, H., Nuhkat, M., & Roelfsema, M. R. G. (2014). Closing gaps: Linking elements that control stomatal movement. *New Phytologist, 203*(1), 44–62.

Kourghi, M., Nourmohammadi, S., Pei, J. V., Qiu, J., McGaughey, S., Tyerman, S. D., et al. (2017). Divalent cations regulate the ion conductance properties of diverse classes of aquaporins. *International Journal of Molecular Sciences, 18*(11), 2323.

Kugler, A., Köhler, B., Palme, K., Wolff, P., & Dietrich, P. (2009). Salt-dependent regulation of a CNG channel subfamily in *Arabidopsis*. *BMC Plant Biology, 9*(1), 1–11.

Kurusu, T., Saito, K., Horikoshi, S., Hanamata, S., Negi, J., Yagi, C., et al. (2013). An S-type anion channel SLAC1 is involved in cryptogein-induced ion fluxes and modulates hypersensitive responses in tobacco BY-2 cells. *PLoS One, 8*(8), e70623.

Kwak, J. M., Mori, I. C., Pei, Z. M., Leonhardt, N., Torres, M. A., Dangl, J. L., et al. (2003). NADPH oxidase *AtrbohD* and *AtrbohF* genes function in ROS-dependent ABA signaling in *Arabidopsis*. *The EMBO Journal, 22*(11), 2623–2633.

Lawson, T., & Vialet-Chabrand, S. (2019). Speedy stomata, photosynthesis and plant water use efficiency. *New Phytologist, 221*(1), 93–98.

Leidi, E. O., Barragán, V., Rubio, L., El-Hamdaoui, A., Ruiz, M. T., Cubero, B., et al. (2010). The *AtNHX1* exchanger mediates potassium compartmentation in vacuoles of transgenic tomato. *The Plant Journal, 61*(3), 495–506.

Li, S., Li, X., Wei, Z., & Liu, F. (2020). ABA-mediated modulation of elevated CO_2 on stomatal response to drought. *Current Opinion in Plant Biology, 56*, 174–180.

Li, J. T., Qiu, Z. B., Zhang, X. W., & Wang, L. S. (2011). Exogenous hydrogen peroxide can enhance tolerance of wheat seedlings to salt stress. *Acta Physiologiae Plantarum, 33*(3), 835–842.

Li, Y. F., Zheng, Y., Vemireddy, L. R., Panda, S. K., Jose, S., Ranjan, A., et al. (2018). Comparative transcriptome and translatome analysis in contrasting rice genotypes reveals differential mRNA translation in salt-tolerant Pokkali under salt stress. *BMC Genomics, 19*(Suppl. 10), 935.

Ligaba, A., Maron, L., Shaff, J., Kochian, L., & Pineros, M. (2012). Maize ZmALMT2 is a root anion transporter that mediates constitutive root malate efflux. *Plant, Cell & Environment, 35*(7), 1185–1200.

Liphschitz, N., & Waisel, Y. (1982). Adaptation of plants to saline environments: Salt excretion and glandular structure. In D. N. Sen, & K. S. Rajpurohit (Eds.), *Vol. 2. Contributions to the ecology of halophytes. Tasks for vegetation science*. Dordrecht: Springer.

Liu, X., Cai, S., Wang, G., Wang, F., Dong, F., Mak, M., et al. (2017). Halophytic NHXs confer salt tolerance by altering cytosolic and vacuolar K^+ and Na^+ in *Arabidopsis* root cell. *Plant Growth Regulation, 82*(2), 333–351.

Liu, X., Fan, Y., Mak, M., Babla, M., Holford, P., Wang, F., et al. (2017). QTLs for stomatal and photosynthetic traits related to salinity tolerance in barley. *BMC Genomics, 18*(1), 1–13.

Liu, C., Fukumoto, T., Matsumoto, T., Gena, P., Frascaria, D., Kaneko, T., et al. (2013). Aquaporin OsPIP1; 1 promotes rice salt resistance and seed germination. *Plant Physiology and Biochemistry, 63*, 151–158.

Liu, X., Mak, M., Babla, M., Wang, F., Chen, G., Veljanoski, F., et al. (2014). Linking stomatal traits and expression of slow anion channel genes *HvSLAH1* and *HvSLAC1* with grain yield for increasing salinity tolerance in barley. *Frontiers in Plant Science*, *5*, 634.

Liu, C., Mao, B., Yuan, D., Chu, C., & Duan, M. (2021). Salt tolerance in rice: Physiological responses and molecular mechanisms. *The Crop Journal*.

Liu, J., Zhang, S., Dong, L., & Chu, J. (2014). Incorporation of Na^+/H^+ antiporter gene from *Aeluropus littoralis* confers salt tolerance in soybean (*Glycine max* L.). *Indian Journal of Biochemistry & Biophysics*, *51*, 58–65.

López-Pérez, L., del Carmen Martínez-Ballesta, M., Maurel, C., & Carvajal, M. (2009). Changes in plasma membrane lipids, aquaporins and proton pump of broccoli roots, as an adaptation mechanism to salinity. *Phytochemistry*, *70*(4), 492–500.

Ma, T., Liu, Q., Li, Z., & Zhang, X. (2002). Tonoplast H^+-ATPase in response to salt stress in *Populus euphratica* cell suspensions. *Plant Science*, *163*(3), 499–505.

Ma, Y., Szostkiewicz, I., Korte, A., Moes, D., Yang, Y., Christmann, A., et al. (2009). Regulators of PP2C phosphatase activity function as abscisic acid sensors. *Science*, *324*(5930), 1064–1068.

Ma, L., Zhang, H., Sun, L., Jiao, Y., Zhang, G., Miao, C., et al. (2012). NADPH oxidase *AtrbohD* and *AtrbohF* function in ROS-dependent regulation of Na^+/K^+ homeostasis in *Arabidopsis* under salt stress. *Journal of Experimental Botany*, *63*(1), 305–317.

Maathuis, F. J., Filatov, V., Herzyk, P., Krijger, G. C., Axelsen, K. B., Chen, S., et al. (2003). Transcriptome analysis of root transporters reveals participation of multiple gene families in the response to cation stress. *The Plant Journal*, *35*(6), 675–692.

Massange-Sanchez, J. A., Palmeros-Suarez, P. A., Espitia-Rangel, E., Rodriguez-Arevalo, I., Sanchez-Segura, L., Martinez-Gallardo, N. A., et al. (2016). Overexpression of grain amaranth (*Amaranthus hypochondriacus*) *AhERF* or *AhDOF* transcription factors in *Arabidopsis thaliana* increases water deficit-and salt-stress tolerance, respectively, via contrasting stress-amelioration mechanisms. *PLoS One*, *11*(10), e0164280.

Maurel, C., Boursiac, Y., Luu, D.-T., Santoni, V., Shahzad, Z., & Verdoucq, L. (2015). Aquaporins in plants. *Physiological Reviews*, *95*(4), 1321–1358.

Mekawy, A. M. M., Assaha, D. V., Yahagi, H., Tada, Y., Ueda, A., & Saneoka, H. (2015). Growth, physiological adaptation, and gene expression analysis of two Egyptian rice cultivars under salt stress. *Plant Physiology and Biochemistry*, *87*, 17–25.

Merilo, E., Jalakas, P., Kollist, H., & Brosché, M. (2015). The role of ABA recycling and transporter proteins in rapid stomatal responses to reduced air humidity, elevated CO_2, and exogenous ABA. *Molecular Plant*, *8*, 657–659.

Merlot, S., Mustilli, A. C., Genty, B., North, H., Lefebvre, V., Sotta, B., et al. (2002). Use of infrared thermal imaging to isolate *Arabidopsis* mutants defective in stomatal regulation. *The Plant Journal*, *30*(5), 601–609.

Meyer, S., Mumm, P., Imes, D., Endler, A., Weder, B., Al-Rasheid, K. A., et al. (2010). AtALMT12 represents an R-type anion channel required for stomatal movement in *Arabidopsis* guard cells. *The Plant Journal*, *63*(6), 1054–1062.

Meyer, S., Scholz-Starke, J., De Angeli, A., Kovermann, P., Burla, B., Gambale, F., et al. (2011). Malate transport by the vacuolar AtALMT6 channel in guard cells is subject to multiple regulation. *The Plant Journal*, *67*(2), 247–257.

Mian, A., Oomen, R. J., Isayenkov, S., Sentenac, H., Maathuis, F. J., & Véry, A. A. (2011). Over-expression of an Na^+-and K^+-permeable HKT transporter in barley improves salt tolerance. *The Plant Journal*, *68*(3), 468–479.

Miller, G., Suzuki, N., Ciftci-Yilmaz, S., & Mittler, R. (2010). Reactive oxygen species homeostasis and signalling during drought and salinity stresses. *Plant, Cell & Environment*, *33*(4), 453–467.

Mishra, P., Singh, A., & Roy, S. (2022). Plasma membrane H^+-ATPase in plants. In S. K. Upadhyay (Ed.), *Vol. 19. Cation transporters in plants* (pp. 357–373). Academic Press.

Mohamed, I. A., Shalby, N., Bai, C., Qin, M., Agami, R. A., Jie, K., et al. (2020). Stomatal and photosynthetic traits are associated with investigating sodium chloride tolerance of *Brassica napus* L. cultivars. *Plants*, *9*(1), 62.

Morsomme, P., & Boutry, M. (2000). The plant plasma membrane H^+-ATPase: Structure, function and regulation. *Biochimica et Biophysica Acta (BBA)-Biomembranes*, *1465*(1–2), 1–16.

Munns, R., Day, D. A., Fricke, W., Watt, M., Arsova, B., Barkla, B. J., et al. (2020). Energy costs of salt tolerance in crop plants. *New Phytologist*, *225*, 1072–1090.

Munns, R., James, R. A., Xu, B., Athman, A., Conn, S. J., Jordans, C., et al. (2012). Wheat grain yield on saline soils is improved by an ancestral Na^+ transporter gene. *Nature Biotechnology*, *30*(4), 360–364.

Muramatsu, Y., Harada, A., Ohwaki, Y., Kasahara, Y., Takagi, S., & Fukuhara, T. (2002). Salt-tolerant ATPase activity in the plasma membrane of the marine angiosperm *Zostera marina* L. *Plant and Cell Physiology*, *43*(10), 1137–1145.

Murata, Y., Mori, I. C., & Munemasa, S. (2015). Diverse stomatal signaling and the signal integration mechanism. *Annual Review of Plant Biology*, *66*, 369–392.

Nath, M., Bhatt, D., Jain, A., Saxena, S. C., Saifi, S. K., Yadav, S., et al. (2019). Salt stress triggers augmented levels of Na^+, Ca^{2+} and ROS and alter stress-responsive gene expression in roots of *CBL9* and *CIPK23* knockout mutants of *Arabidopsis thaliana*. *Environmental and Experimental Botany*, *161*, 265–276.

Nawaz, I., Iqbal, M., Bliek, M., & Schat, H. (2017). Salt and heavy metal tolerance and expression levels of candidate tolerance genes among four extremophile *Cochlearia* species with contrasting habitat preferences. *Science of the Total Environment*, *584*, 731–741.

Nawaz, Z., Kakar, K. U., Saand, M. A., & Shu, Q. Y. (2014). Cyclic nucleotide-gated ion channel gene family in rice, identification, characterization and experimental analysis of expression response to plant hormones, biotic and abiotic stresses. *BMC Genomics*, *15*(1), 1–18.

Nedelyaeva, O., Shuvalov, A., & Balnokin, Y. V. (2020). Chloride channels and transporters of the CLC family in plants. *Russian Journal of Plant Physiology*, *67*(5), 767–784.

Neeraj, K., Shubham, L., Ashwani, K., Pratima, K., Mann, A., Sarita, D., et al. (2019). Antioxidant defence in halophytes under high salinity. In *Halophytes and climate change: Adaptive mechanisms and potential uses* (pp. 196–208).

Negi, J., Matsuda, O., Nagasawa, T., Oba, Y., Takahashi, H., Kawai-Yamada, M., et al. (2008). CO_2 regulator SLAC1 and its homologues are essential for anion homeostasis in plant cells. *Nature*, *452*(7186), 483–486.

Nejat, N., & Mantri, N. (2017). Plant immune system: Crosstalk between responses to biotic and abiotic stresses the missing link in understanding plant defence. *Current Issues in Molecular Biology*, *23*(1), 1–16.

Neuhaus, C., Geilfus, C.-M., Zörb, C., & Mühling, K. H. (2013). Transcript expression of Mg-chelatase and H^+-ATPase isogenes in *Vicia faba* leaves as influenced by root and foliar magnesium supply. *Plant and Soil*, *368*(1), 41–50.

Nguyen, C. T., Agorio, A., Jossier, M., Depré, S., Thomine, S., & Filleur, S. (2016). Characterization of the chloride channel-like, *AtCLCg*, involved in chloride tolerance in *Arabidopsis thaliana*. *Plant and Cell Physiology*, *57*(4), 764–775.

Nishimura, N., Sarkeshik, A., Nito, K., Park, S. Y., Wang, A., Carvalho, P. C., et al. (2010). *PYR/PYL/RCAR* family members are major in-vivo ABI1 protein phosphatase 2C-interacting proteins in *Arabidopsis*. *The Plant Journal*, *61*(2), 290–299.

Niu, M., Huang, Y., Sun, S., Sun, J., Cao, H., Shabala, S., et al. (2018). Root respiratory burst oxidase homologue-dependent H_2O_2 production confers salt tolerance on a grafted cucumber by controlling Na^+ exclusion and stomatal closure. *Journal of Experimental Botany*, *69*(14), 3465–3476.

Olfatmiri, H., Alemzadeh, A., & Zakipour, Z. (2014). Up-regulation of plasma membrane H^+-ATPase under salt stress may enable *Aeluropus littoralis* to cope with stress. *Molecular Biology Research Communications, 3*(1), 67.

One Thousand Plant Transcriptomes Initiative. (2019). One thousand plant transcriptomes and the phylogenomics of green plants. *Nature, 574*(7780), 679–685.

Opricǎ, L., & Vochiţa, G. (2020). Enzymatic activity in halophytes. In M. N. Grigore (Ed.), *Handbook of halophytes: From molecules to ecosystems towards biosaline agriculture* (pp. 1–23). Cham: Springer International Publishing.

Oufattole, M., Arango, M., & Boutry, M. (2000). Identification and expression of three new *Nicotiana plumbaginifolia* genes which encode isoforms of a plasma-membrane H^+-ATPase, and one of which is induced by mechanical stress. *Planta, 210*(5), 715–722.

Ozgur, R., Uzilday, B., Sekmen, A. H., & Turkan, I. (2013). Reactive oxygen species regulation and antioxidant defence in halophytes. *Functional Plant Biology, 40*(9), 832–847.

Palmgren, M. G. (1998). Role of the plasma membrane H^+-ATPase. *Advances in Botanical Research, 1.*

Palmgren, M. G. (2001). Plant plasma membrane H^+-ATPases: Powerhouses for nutrient uptake. *Annual Review of Plant Biology, 52*(1), 817–845.

Pantoja, O. (2021). Recent advances in the physiology of ion channels in plants. *Annual Review of Plant Biology, 72.*

Park, S. Y., Fung, P., Nishimura, N., Jensen, D. R., Fujii, H., Zhao, Y., et al. (2009). Abscisic acid inhibits type 2C protein phosphatases via the PYR/PYL family of START proteins. *Science, 324*(5930), 1068–1071.

Pei, Z. M., Kuchitsu, K., Ward, J. M., Schwarz, M., & Schroeder, J. I. (1997). Differential abscisic acid regulation of guard cell slow anion channels in *Arabidopsis* wild-type and *abi1* and *abi2* mutants. *The Plant Cell, 9*(3), 409–423.

Pei, Z. M., Murata, Y., Benning, G., Thomine, S., Klüsener, B., Allen, G. J., et al. (2000). Calcium channels activated by hydrogen peroxide mediate abscisic acid signalling in guard cells. *Nature, 406*(6797), 731–734.

Perez, I. B., & Brown, P. J. (2014). The role of ROS signaling in cross-tolerance: From model to crop. *Frontiers in Plant Science, 5*, 754.

Perez-Romero, J. A., Mateos-Naranjo, E., Lopez-Jurado, J., Redondo-Gomez, S., & Torress-Ruiz, J. M. (2020). Importance of physiological traits vulnerability in determine halophytes tolerance to salinity excess: A comparative assessment in *Atriplex halimus. Plants (Basel), 9*(6).

Petrov, V., Hille, J., Mueller-Roeber, B., & Gechev, T. S. (2015). ROS-mediated abiotic stress-induced programmed cell death in plants. *Frontiers in Plant Science, 6*, 69.

Pilot, G., Lacombe, B. T., Gaymard, F., Chérel, I., Boucherez, J., Thibaud, J. B., et al. (2001). Guard cell inward K^+ channel activity in arabidopsis involves expression of the twin channel subunits KAT1 and KAT2. *Journal of Biological Chemistry, 276*(5), 3215–3221.

Piñeros, M. A., Cançado, G. M., & Kochian, L. V. (2008). Novel properties of the wheat aluminum tolerance organic acid transporter (TaALMT1) revealed by electrophysiological characterization in *Xenopus oocytes*: Functional and structural implications. *Plant Physiology, 147*(4), 2131–2146.

Plett, D., Safwat, G., Gilliham, M., Skrumsager Møller, I., Roy, S., Shirley, N., et al. (2010). Improved salinity tolerance of rice through cell type-specific expression of *AtHKT1;1. PLoS One, 5*(9), e12571.

Pornsiriwong, W., Estavillo, G. M., Chan, K. X., Tee, E. E., Ganguly, D., Crisp, P. A., et al. (2017). A chloroplast retrograde signal, 3′-phosphoadenosine 5′-phosphate, acts as a secondary messenger in abscisic acid signaling in stomatal closure and germination. *eLife, 6.*

Pottosin, I., Velarde-Buendía, A. M., Bose, J., Zepeda-Jazo, I., Shabala, S., & Dobrovinskaya, O. (2014). Cross-talk between reactive oxygen species and polyamines in regulation of ion transport across the plasma membrane: Implications for plant adaptive responses. *Journal of Experimental Botany, 65*(5), 1271–1283.

Qiu, J., Henderson, S. W., Tester, M., Roy, S. J., & Gilliham, M. (2016). *SLAH1*, a homologue of the slow type anion channel *SLAC1*, modulates shoot Cl^- accumulation and salt tolerance in *Arabidopsis thaliana*. *Journal of Experimental Botany, 67*(15), 4495–4505.

Rajaravindran, M., & Natarajan, S. (2012). Effects of salinity stress on growth and biochemical constituents of the halophyte *Sesuvium portulacastrum*. *International Journal of Research and in Biological Sciences, 2*, 18–25.

Rasouli, F., Kiani-Pouya, A., Shabala, L., Li, L., Tahir, A., Yu, M., et al. (2021). Salinity effects on guard cell proteome in *Chenopodium quinoa*. *International Journal of Molecular Sciences, 22*(1), 428.

Rasouli, F., Kiani-Pouya, A., Tahir, A., Shabala, L., Chen, Z. H., & Shabala, S. (2021). A comparative analysis of stomatal traits and photosynthetic responses in closely related halophytic and glycophytic species under saline conditions. *Environmental and Experimental Botany, 181*.

Rejeb, K. B., Benzarti, M., Debez, A., Bailly, C., Savouré, A., & Abdelly, C. (2015). NADPH oxidase-dependent H_2O_2 production is required for salt-induced antioxidant defense in *Arabidopsis thaliana*. *Journal of Plant Physiology, 174*, 5–15.

Rengel, Z., Bose, J., Chen, Q., & Tripathi, B. (2015). Magnesium alleviates plant toxicity of aluminium and heavy metals. *Crop and Pasture Science, 66*(12), 1298–1307.

Rodrigues, O., Reshetnyak, G., Grondin, A., Saijo, Y., Leonhardt, N., Maurel, C., et al. (2017). Aquaporins facilitate hydrogen peroxide entry into guard cells to mediate ABA-and pathogen-triggered stomatal closure. *Proceedings of the National Academy of Sciences of the United States of America, 114*(34), 9200–9205.

Rodriguez-Hernandez, M. D., Morcillo, L., & Garmendia, I. (2021). Sensitivity of quinoa cv. 'Titicaca' to low salinity conditions. *Folia Horticulturae, 33*(1), 135–145.

Roelfsema, M. R. G., & Hedrich, R. (2005). In the light of stomatal opening: New insights into 'the Watergate'. *New Phytologist, 167*(3), 665–691.

Roelfsema, M. R. G., Hedrich, R., & Geiger, D. (2012). Anion channels: Master switches of stress responses. *Trends in Plant Science, 17*(4), 221–229.

Rygol, J., Zimmermann, U., & Balling, A. (1989). Water relations of individual leaf-cells of *Mesembryanthemum crystallinum* plants grown at low and high salinity. *Journal of Membrane Biology, 107*(3), 203–212.

Saand, M. A., Xu, Y.-P., Li, W., Wang, J.-P., & Cai, X.-Z. (2015). Cyclic nucleotide gated channel gene family in tomato: Genome-wide identification and functional analyses in disease resistance. *Frontiers in Plant Science, 6*, 303.

Saito, S., & Uozumi, N. (2019). Guard cell membrane anion transport systems and their regulatory components: An elaborate mechanism controlling stress-induced stomatal closure. *Plants, 8*(1), 9.

Sanchez, D. H., Pieckenstain, F. L., Szymanski, J., Erban, A., Bromke, M., Hannah, M. A., et al. (2011). Comparative functional genomics of salt stress in related model and cultivated plants identifies and overcomes limitations to translational genomics. *PLoS One, 6*(2), e17094.

Santiago, J., Dupeux, F., Round, A., Antoni, R., Park, S.-Y., Jamin, M., et al. (2009). The abscisic acid receptor PYR1 in complex with abscisic acid. *Nature, 462*(7273), 665–668.

Sarda, X., Tousch, D., Ferrare, K., Legrand, E., Dupuis, J., Casse-Delbart, F., et al. (1997). Two TIP-like genes encoding aquaporins are expressed in sunflower guard cells. *The Plant Journal, 12*(5), 1103–1111.

Sasaki, T., Yamamoto, Y., Ezaki, B., Katsuhara, M., Ahn, S. J., Ryan, P. R., et al. (2004). A wheat gene encoding an aluminum-activated malate transporter. *The Plant Journal, 37*(5), 645–653.

Schroeder, J. I., Allen, G. J., Hugouvieux, V., Kwak, J. M., & Waner, D. (2001). Guard cell signal transduction. *Annual Review of Plant Biology*, *52*(1), 627–658.

Schroeder, J. I., Kwak, J. M., & Allen, G. J. (2001). Guard cell abscisic acid signalling and engineering drought hardiness in plants. *Nature*, *410*(6826), 327–330.

Schuurink, R. C., Shartzer, S. F., Fath, A., & Jones, R. L. (1998). Characterization of a calmodulin-binding transporter from the plasma membrane of barley aleurone. *Proceedings of the National Academy of Sciences of the United States of America*, *95*(4), 1944–1949.

Seckin, B., Turkan, I., Sekmen, A. H., & Ozfidan, C. (2010). The role of antioxidant defense systems at differential salt tolerance of *Hordeum marinum* Huds.(sea barleygrass) and *Hordeum vulgare* L.(cultivated barley). *Environmental and Experimental Botany*, *69*(1), 76–85.

Senadheera, P., Singh, R., & Maathuis, F. J. (2009). Differentially expressed membrane transporters in rice roots may contribute to cultivar dependent salt tolerance. *Journal of Experimental Botany*, *60*(9), 2553–2563.

Shabala, S. (2000). Ionic and osmotic components of salt stress specifically modulate net ion fluxes from bean leaf mesophyll. *Plant, Cell & Environment*, *23*(8), 825–837.

Shabala, S. (2013). Learning from halophytes: Physiological basis and strategies to improve abiotic stress tolerance in crops. *Annals of Botany*, *112*(7), 1209–1221.

Shabala, S., Chen, G., & Chen, Z. H. (2020). The energy cost of the tonoplast futile sodium leak. *New Phytologist*, *225*(3), 1105–1110.

Shabala, S., Hariadi, Y., & Jacobsen, S. E. (2013). Genotypic difference in salinity tolerance in quinoa is determined by differential control of xylem Na^+ loading and stomatal density. *Journal of Plant Physiology*, *170*(10), 906–914.

Shabala, S., & Newman, I. (2000). Salinity effects on the activity of plasma membrane H^+ and Ca^{2+} transporters in bean leaf mesophyll: Masking role of the cell wall. *Annals of Botany*, *85*(5), 681–686.

Shabala, S., & Pottosin, I. (2014). Regulation of potassium transport in plants under hostile conditions: Implications for abiotic and biotic stress tolerance. *Physiologia Plantarum*, *151*(3), 257–279.

Shabala, S., Wu, H., & Bose, J. (2015). Salt stress sensing and early signalling events in plant roots: Current knowledge and hypothesis. *Plant Science*, *241*, 109–119.

Sharma, T., Dreyer, I., Kochian, L., & Piñeros, M. A. (2016). The ALMT family of organic acid transporters in plants and their involvement in detoxification and nutrient security. *Frontiers in Plant Science*, *7*, 1488.

Shen, Q., Yu, J., Fu, L., Wu, L., Dai, F., Jiang, L., et al. (2018). Ionomic, metabolomic and proteomic analyses reveal molecular mechanisms of root adaption to salt stress in Tibetan wild barley. *Plant Physiology and Biochemistry*, *123*, 319–330.

Sibole, J. V., Cabot, C., Michalke, W., Poschenrieder, C., & Barceló, J. (2005). Relationship between expression of the PM H^+-ATPase, growth and ion partitioning in the leaves of salt-treated Medicago species. *Planta*, *221*(4), 557–566.

Sicilia, M. N. A., Romero, M. E. S., Rosales, M. P. R., & Venema, K. (2021). Plastidial transporters KEA1 and KEA2 at the inner envelope membrane adjust stromal pH in the dark. *New Phytologist*, *229*, 2080–2090.

Sierla, M., Rahikainen, M., Salojärvi, J., Kangasjärvi, J., & Kangasjärvi, S. (2013). Apoplastic and chloroplastic redox signaling networks in plant stress responses. *Antioxidants & Redox Signaling*, *18*(16), 2220–2239.

Silva, P., & Gerós, H. (2009). Regulation by salt of vacuolar H^+-ATPase and H+ -pyrophosphatase activities and Na^+/H^+ exchange. *Plant Signaling & Behavior*, *4*(8), 718–726.

Sondergaard, T. E., Schulz, A., & Palmgren, M. G. (2004). Energization of transport processes in plants. Roles of the plasma membrane H^+-ATPase. *Plant Physiology*, *136*(1), 2475–2482.

Song, Y., Miao, Y., & Song, C. P. (2014). Behind the scenes: The roles of reactive oxygen species in guard cells. *New Phytologist*, *201*(4), 1121–1140.

Song, J., & Wang, B. (2015). Using euhalophytes to understand salt tolerance and to develop saline agriculture: *Suaeda salsa* as a promising model. *Annals of Botany*, *115*(3), 541–553.

Sun, J., Chen, S., Dai, S., Wang, R., Li, N., Shen, X., et al. (2009). NaCl-induced alternations of cellular and tissue ion fluxes in roots of salt-resistant and salt-sensitive poplar species. *Plant Physiology*, *149*(2), 1141–1153.

Sunarpi, H. T., Motoda, J., Kubo, M., & Yang, H. (2005). Enhanced salt tolerance mediated by AtHKT1 transporter-induced Na^+ unloading from xylem vessels to xylem parenchyma cells. *The Plant Journal*, *44*, 928–938.

Suzuki, N., Koussevitzky, S., Mittler, R., & Miller, G. (2012). ROS and redox signalling in the response of plants to abiotic stress. *Plant, Cell & Environment*, *35*(2), 259–270.

Suzuki, N., Miller, G., Morales, J., Shulaev, V., Torres, M. A., & Mittler, R. (2011). Respiratory burst oxidases: The engines of ROS signaling. *Current Opinion in Plant Biology*, *14*(6), 691–699.

Suzuki, K., Yamaji, N., Costa, A., Okuma, E., Kobayashi, N. I., Kashiwagi, T., et al. (2016). *OsHKT1;4*-mediated Na^+ transport in stems contributes to Na^+ exclusion from leaf blades of rice at the reproductive growth stage upon salt stress. *BMC Plant Biology*, *16*(1), 1–15.

Szyroki, A., Ivashikina, N., Dietrich, P., Roelfsema, M. R. G., Ache, P., Reintanz, B., et al. (2001). KAT1 is not essential for stomatal opening. *Proceedings of the National Academy of Sciences of the United States of America*, *98*(5), 2917–2921.

Tisarum, R., Theerawitaya, C., Samphumphuang, T., Polispitak, K., Thongpoem, P., & Singh, H. P. (2020). Alleviation of salt stress in upland rice (*Oryza sativa* L. ssp. *indica* cv. Leum Pua) using arbuscular mycorrhizal fungi inoculation. *Frontiers in Plant Science*, 11.

Tyerman, S. D., McGaughey, S. A., Qiu, J., Yool, A. J., & Byrt, C. S. (2021). Adaptable and multifunctional ion-conducting aquaporins. *Annual Review of Plant Biology*, *72*, 703–736.

Ueno, K., Kinoshita, T., Inoue, S.-I., Emi, T., & Shimazaki, K.-I. (2005). Biochemical characterization of plasma membrane H^+-ATPase activation in guard cell protoplasts of *Arabidopsis thaliana* in response to blue light. *Plant and Cell Physiology*, *46*(6), 955–963.

Umezawa, T., Sugiyama, N., Mizoguchi, M., Hayashi, S., Myouga, F., Yamaguchi-Shinozaki, K., et al. (2009). Type 2C protein phosphatases directly regulate abscisic acid-activated protein kinases in *Arabidopsis*. *Proceedings of the National Academy of Sciences of the United States of America*, *106*(41), 17588–17593.

Vahisalu, T., Kollist, H., Wang, Y.-F., Nishimura, N., Chan, W.-Y., Valerio, G., et al. (2008). SLAC1 is required for plant guard cell S-type anion channel function in stomatal signalling. *Nature*, *452*(7186), 487–491.

van Zelm, E., Zhang, Y. X., & Testerink, C. (2020). Salt tolerance mechanisms of plants. *Annual Review of Plant Biology*, *71*, 403–433.

Vialet-Chabrand, S., Matthews, J., Brendel, O., Blatt, M., Wang, Y., Hills, A., et al. (2016). Modelling water use efficiency in a dynamic environment: An example using *Arabidopsis thaliana*. *Plant Science*, *251*, 65–74.

Vitart, V., Baxter, I., Doerner, P., & Harper, J. F. (2001). Evidence for a role in growth and salt resistance of a plasma membrane H^+-ATPase in the root endodermis. *The Plant Journal*, *27*(3), 191–201.

Wang, Y., & Chen, Z.-H. (2020). Does molecular and structural evolution shape the speedy grass stomata? *Frontiers in Plant Science*, *11*, 333.

Wang, L.-L., Chen, A.-P., Zhong, N.-Q., Liu, N., Wu, X.-M., Wang, F., et al. (2014). The *Thellungiella salsuginea* tonoplast aquaporin *TsTIP1;2* functions in protection against multiple abiotic stresses. *Plant and Cell Physiology*, *55*(1), 148–161.

Wang, M., Wang, Y., Sun, J., Ding, M., Deng, S., Hou, P., et al. (2013). Overexpression of *PeHA1* enhances hydrogen peroxide signaling in salt-stressed *Arabidopsis*. *Plant Physiology and Biochemistry*, *71*, 37–48.

Wang, L., Wu, X., Liu, Y., & Qiu, Q.-S. (2015). *AtNHX5* and *AtNHX6* control cellular K^+ and pH homeostasis in *Arabidopsis*: Three conserved acidic residues are essential for K^+ transport. *PLoS One*, *10*(12), e0144716.

Wang, B., Zhai, H., He, S., Zhang, H., Ren, Z., Zhang, D., et al. (2016). A vacuolar Na^+/H^+ antiporter gene, *IbNHX2*, enhances salt and drought tolerance in transgenic sweetpotato. *Scientia Horticulturae*, *201*, 153–166.

Wege, S., De Angeli, A., Droillard, M.-J., Kroniewicz, L., Merlot, S., Cornu, D., et al. (2014). Phosphorylation of the vacuolar anion exchanger *AtCLCa* is required for the stomatal response to abscisic acid. *Science Signaling*, 7(333), ra65.

Wei, P., Wang, L., Liu, A., Yu, B., & Lam, H. M. (2016). *GmCLC1* confers enhanced salt tolerance through regulating chloride accumulation in soybean. *Frontiers in Plant Science*, 7, 1082.

Willmer, C., & Fricker, M. (1996). The distribution of stomata. In *Stomata* (pp. 12–35). Springer.

Wrzaczek, M., Brosché, M., & Kangasjärvi, J. (2013). ROS signaling loops-production, perception, regulation. *Current Opinion in Plant Biology*, *16*(5), 575–582.

Wu, X., Ebine, K., Ueda, T., & Qiu, Q. S. (2016). *AtNHX5* and *AtNHX6* are required for the subcellular localization of the SNARE complex that mediates the trafficking of seed storage proteins in *Arabidopsis*. *PLoS One*, *11*(3), e0151658.

Wu, Q., Yang, J., Cheng, N., Hirschi, K. D., White, F. F., & Park, S. (2017). Glutaredoxins in plant development, abiotic stress response, and iron homeostasis: From model organisms to crops. *Environmental and Experimental Botany*, *139*, 91–98.

Xu, S., Dai, Z., Guo, P., Fu, X., Liu, S., Zhou, L., et al. (2021). ggtreeExtra: Compact visualization of richly annotated phylogenetic data. *Molecular Biology and Evolution*, *38*(9), 4039–4042.

Xu, M., Gruber, B. D., Delhaize, E., White, R. G., James, R. A., You, J., et al. (2015). The barley anion channel, *HvALMT1*, has multiple roles in guard cell physiology and grain metabolism. *Physiologia Plantarum*, *153*(1), 183–193.

Xu, Q., Xu, X., Zhao, Y., Jiao, K., Herbert, S. J., & Hao, L. (2008). Salicylic acid, hydrogen peroxide and calcium-induced saline tolerance associated with endogenous hydrogen peroxide homeostasis in naked oat seedlings. *Plant Growth Regulation*, *54*(3), 249–259.

Yaish, M. (2015). Proline accumulation is a general response to abiotic stress in the date palm tree (*Phoenix dactylifera* L.). *Genetics and Molecular Research*, *14*(3), 9943–9950.

Yaish, M. W., Patankar, H. V., Assaha, D. V., Zheng, Y., Al-Yahyai, R., & Sunkar, R. (2017). Genome-wide expression profiling in leaves and roots of date palm (*Phoenix dactylifera* L.) exposed to salinity. *BMC Genomics*, *18*(1), 1–17.

Yang, Y., Zhang, F., Zhao, M., An, L., Zhang, L., & Chen, N. (2007). Properties of plasma membrane H^+-ATPase in salt-treated *Populus euphratica* callus. *Plant Cell Reports*, *26*(2), 229–235.

Yong, M. T., Solis, C. A., Amatoury, S., Sellamuthu, G., Rajakani, R., Mak, M., et al. (2022). Proto kranz-like leaf traits and cellular ionic regulation are associated with salinity tolerance in a halophytic wild rice. *Stress Biology*, *2*, 8.

Yong, M. T., Solis, C. A., Rabbi, B., Huda, S., Liu, R., Zhou, M., & Chen, Z.-H. (2020). Leaf mesophyll K^+ and Cl^- fluxes and reactive oxygen species production predict rice salt tolerance at reproductive stage in greenhouse and field conditions. *Plant Growth Regulation*, *92*(1), 53–64.

Yu, W., Kan, Q., Zhang, J., Zeng, B., & Chen, Q. (2016). Role of the plasma membrane H^+-ATPase in the regulation of organic acid exudation under aluminum toxicity and phosphorus deficiency. *Plant Signaling & Behavior*, *11*(1), e1106660.

Yuan, F., Leng, B., & Wang, B. (2016). Progress in studying salt secretion from the salt glands in recretohalophytes: How do plants secrete salt? *Frontiers in Plant Science*, 7, 977.

Yuan, F., Yang, H., Xue, Y., Kong, D., Ye, R., Li, C., et al. (2014). OSCA1 mediates osmotic-stress-evoked Ca^{2+} increases vital for osmosensing in *Arabidopsis*. *Nature*, *514*(7522), 367–371.

Zhang, X., Miao, Y. C., An, G. Y., Yun, Z., Shangguan, Z. P., Gao, J. F., et al. (2001). K^+ channels inhibited by hydrogen peroxide mediate abscisic acid signaling in *Vicia* guard cells. *Cell Research*, *11*(3), 195–202.

Zeeshan, M., Lu, M., Sehar, S., Holford, P., & Wu, F. (2020). Comparison of biochemical, anatomical, morphological, and physiological responses to salinity stress in wheat and barley genotypes deferring in salinity tolerance. *Agronomy*, *10*(1), 127.

Zhang, A., Ren, H. M., Tan, Y. Q., Qi, G. N., Yao, F. Y., Wu, G. L., et al. (2016). S-type anion channels SLAC1 and SLAH3 function as essential negative regulators of inward K^+ channels and stomatal opening in *Arabidopsis*. *The Plant Cell*, *28*(4), 949–965.

Zhang, Y., Wang, L., Liu, Y., Zhang, Q., Wei, Q., & Zhang, W. (2006). Nitric oxide enhances salt tolerance in maize seedlings through increasing activities of proton-pump and Na^+/H^+ antiport in the tonoplast. *Planta*, *224*(3), 545–555.

Zhang, F., Wang, Y., Yang, Y., Wu, H., Wang, D., & Liu, J. (2007). Involvement of hydrogen peroxide and nitric oxide in salt resistance in the calluses from *Populus euphratica*. *Plant, Cell & Environment*, *30*(7), 775–785.

Zhang, Y. M., Liu, Z. H., Wen, Z. Y., Zhang, H. M., Yang, F., & Guo, X. L. (2012). The vacuolar Na^+–H^+ antiport gene *TaNHX2* confers salt tolerance on transgenic alfalfa (*Medicago sativa*). *Functional Plant Biology*, *39*(8), 708–716.

Zhao, C., Wang, Y., Chan, K. X., Marchant, D. B., Franks, P. J., Randall, D., et al. (2019). Evolution of chloroplast retrograde signaling facilitates green plant adaptation to land. *Proceedings of the National Academy of Sciences of the United States of America*, *116*(11), 5015–5020.

Zhou, L., Wang, C., Liu, R., Han, Q., Vandeleur, R. K., Du, J., et al. (2014). Constitutive overexpression of soybean plasma membrane intrinsic protein GmPIP1;6 confers salt tolerance. *BMC Plant Biology*, *14*(1), 1–13.

Zhu, J. K. (2001). Plant salt tolerance. *Trends in Plant Science*, *6*(2), 66–71.

Zhu, J. K. (2003). Regulation of ion homeostasis under salt stress. *Current Opinion in Plant Biology*, *6*(5), 441–445.

Zhu, M., Shabala, L., Cuin, T. A., Huang, X., Zhou, M., Munns, R., et al. (2016). *Nax* loci affect *SOS1*-like Na^+/H^+ exchanger expression and activity in wheat. *Journal of Experimental Botany*, *67*(3), 835–844.

Zhu, X. J., Pan, T., Zhang, X., Fan, L. G., Quintero, F. J., Zhao, H., et al. (2018). K^+ efflux antiporters 4, 5, and 6 mediate pH and K^+ homeostasis in endomembrane compartments. *Plant Physiology*, *178*(4), 1657–1678.

Zifarelli, G., & Pusch, M. (2010). CLC transport proteins in plants. *FEBS Letters*, *584*(10), 2122–2127.

CHAPTER TWO

Water use efficiency in mangroves: Conservation of water use efficiency determined by stomatal behavior across leaves, plants, and forests

Jie Liang, Graham D. Farquhar, and Marilyn C. Ball*
Plant Science Division, Research School of Biology, Australian National University, Acton, ACT, Australia
*Corresponding author: e-mail address: marilyn.ball@anu.edu.au

Contents

1. Introduction 44
2. The definitions of water use efficiency 45
3. Long-term vs. short-term water use efficiency 47
4. Instantaneous WUE vs. intrinsic WUE 49
5. Compromise between WUE and photosynthetic nitrogen use efficiency 50
6. Into the future: Response of water-use efficiency to elevated CO_2 51
7. Ecosystem-level water use efficiency of mangrove forests is comparably high as the leaf level 52
8. Unresolved complications and future directions 54
9. Conclusion 56
Acknowledgments 56
References 56

Abstract

Absorption of water from seawater with its high osmotic pressure is costly to mangroves and requires the plants to use water conservatively. Indeed, measurements of gas exchange, growth rate and carbon isotope composition have shown that mangrove species operate with higher water-use efficiency (WUE) than do most C_3 species. These characteristics of water use and water use efficiency find expression in interspecific differences in salinity tolerance, and in the structure of mangrove forests along complex gradients in salinity and aridity. Under global warming, mangroves face increasing challenges of varying salinity, humidity and temperature, all factors that affect water use and WUE. The extent to which mangroves can acclimate to changing environmental conditions by altering water-use characteristics depends on spatial (leaf, plant, and ecosystem) and temporal (short- and long-term) factors. Here, we clarify concepts of WUE at each spatiotemporal scale, compare their values across scales,

Advances in Botanical Research, Volume 103
ISSN 0065-2296
https://doi.org/10.1016/bs.abr.2022.02.017

and show consistency in maintenance of a high water use efficiency from leaves to ecosystems. We develop a framework for comparison of WUE across scales and identify areas for future research.

1. Introduction

Mangrove forests are most luxuriant and diverse along the tidal, saline coastal wetlands of the humid tropics and subtropics (Tomlinson, 2016). As halophytes, mangroves characteristically are conservative in water use (Ball & Passioura, 1995), operating with high carbon gain with respect to water expenditure. These water-use characteristics become increasingly conservative with increase in salinity and in the salt tolerance of the species (Ball, Cochrane, & Rawson, 1997; Ball, Cowan, & Farquhar, 1988). This may be a consequence of both high carbon costs of water uptake by the roots and restrictions in the rates of water usage by the leaves to maintain favorable water and salt relations (Ball, 1988; Ball et al., 1988). Maximizing carbon gain relative to water use is a whole-plant phenomenon that integrates stomatal behavior in relation to photosynthesis, physical properties and display of leaves in relation to light interception and evaporative demand, and the partitioning of carbon between structures supplying and those consuming carbon-based assimilates (Cowan, 1982; Cowan & Farquhar, 1977). Water-use characteristics thus find expression at all levels of plant form and function, ultimately contributing to variation in mangrove forest structure along natural gradients in salinity and aridity (Ball, 1988, 1996). As the improvement in leaf-level WUE may not be always translated into plant- or ecosystem-level WUE (Medrano et al., 2015), water-use efficiency for each scale should be included in comprehensive comparison of WUE among species. The present review builds on this knowledge to develop a framework enabling linkage of mangrove water use characteristics across scales from leaves to ecosystems.

The trade-off between carbon gain and water loss, water use efficiency (WUE) has been defined as the ratio of the amount of carbon assimilated and the amount of water transpired,

$$\mathrm{WUE} = \frac{\mathrm{CO_2\ fixed}}{\mathrm{H_2O\ transpired}} \tag{1}$$

Several kinds of WUE have been used to compare water use strategies of mangrove species, but there are limitations for application of each WUE.

Here, we first clarify concepts underlying the WUE terms usually used at specific spatiotemporal levels and then examine consistency in estimations of WUE between different spatiotemporal levels.

2. The definitions of water use efficiency

Plant leaves obtain CO_2 through open stomata but inevitably lose H_2O. The gas fluxes can be characterized by a one-dimensional form of Fick's first law of diffusion, i.e., conductance × concentration gradient. The assimilation rate (A) of carbon is thus $A=\frac{g}{1.6}(p_a-p_i)/P$, and transpiration rate (T) is $T=g(e_i-e_a)/P$, where P is air pressure, and $p_a - p_i$ is the drawdown of the intercellular and atmospheric CO_2 pressure, and $e_i - e_a$ is the drawdown of the intercellular and atmospheric vapor pressure. Indeed, WUE has been used interchangeably across spatiotemporal scales to refer to observations ranging from gas exchange by individual leaves for a few minutes to growth responses to precipitation through an entire season or lifecycle of the plant. However, WUE is a deceptively simple concept that requires clarification for use at different spatial and temporal scales (Table 1).

Consider, for example, that WUE calculated from leaf gas exchange does not capture leaf level adjustments in water use and water use efficiency that accompany acclimation of leaf photosynthetic capacity to short-term change in environmental conditions, nor does it capture the integrated carbon gain and water loss over longer time scales significant to growth (Medrano et al., 2015). An example is shown in Fig. 1, in which WUE in leaves of *Avicennia marina* calculated from instantaneous gas exchange data increased with gradual variation in salinity from 50 to 500 mM NaCl and then declined upon return to 50 mM NaCl (Fig. 1A). In contrast, WUEs calculated from biomass accumulation and total plant water loss during long term growth were similar under constant salinities of 50 and 250 mM NaCl and increased slightly in plants grown under 500 mM NaCl (Fig. 1B). In this example, acclimation to long term growth conditions reduced the variation in WUE between salinity treatments relative to the short-term exposure to changing salinities.

Differences in computation of WUE across scales are summarized in Table 1. These differences are particularly important when comparing values of WUE between species and environments, over short- and long-term time scales significant to daily integrated carbon gain of a leaf or to plant growth, and finally to integrated carbon/water balances relevant to ecosystem function.

Table 1 Definitions of water use efficiency (WUE).

Level	Long-term WUE	Instantaneous WUE	Intrinsic WUE
Leaf[1]	$\mathrm{WUE}_l = \frac{(1-\Phi)\int_{t1}^{t2} As\ dt}{\int_{t1}^{t2} Ts\ dt}$	$\mathrm{WUE}_l^{in} = \frac{A}{T} = \frac{p_a - p_i}{1.6(e_i - e_a)} \cong \frac{p_a - p_i}{1.6v}$	$\mathrm{IWUE}_l = \frac{A}{g} = \frac{C_a - C_i}{1.6}$
Plant[1]	$\mathrm{WUE}_p = \frac{(1-\Phi)\int_{t1}^{t2} A's'\ dt}{\int_{t1}^{t2} T's'\ dt}$	$\mathrm{WUE}_p^{in} = \frac{A'}{T'} \cong \frac{p_a' - p_i'}{1.6v}$	$\mathrm{IWUE}_p = \frac{A'}{g'}$
Ecosystem[2]	$\mathrm{WUE}_e = \frac{\int_{t1}^{t2} NPPs''\ dt}{\int_{t1}^{t2} T''s''\ dt}$	$\mathrm{WUE}_e^{in} = \frac{GPP}{T''}$	$\mathrm{IWUE}_e = \frac{GPP}{g''}$

Notes: A and T are assimilation rate and transpiration rate. The ' and '' refer to plant-level and ecosystem-level counterparts, and the subscripts l, p and e indicate the leaf-, plant-, and ecosystem-level WUE. e_i and e_a are the intercellular and atmospheric vapor pressure, and p_i and p_a are intercellular and atmospheric CO_2 pressure, $(p_a - p_i)/P = C_a - C_i$, and P is air pressure and s is leaf area. The g represents the total of boundary layer and stomatal conductance (1/resistance) to vapor, but the boundary layer resistance is much smaller than stomatal resistance especially for measurements using conventional gas exchange techniques. So g here indicates stomatal conductance for convenience, and the factor 1.6 is the ratio of diffusivities of water vapor and CO_2 in air. The Φ indicates the proportion of carbon that is fixed during the day but respired by the leaf at night and by other parts of the plant over the whole period. The net primary productivity (*NPP*) is the difference of gross primary productivity (*GPP*) and autotrophic respiration (*Ra*). The WUE_p can be estimated by the ratio of mass growth rate/water loss rate. The superscripts 1 and 2 indicate, respectively, Farquhar and Richards (1984) and Beer et al. (2009).

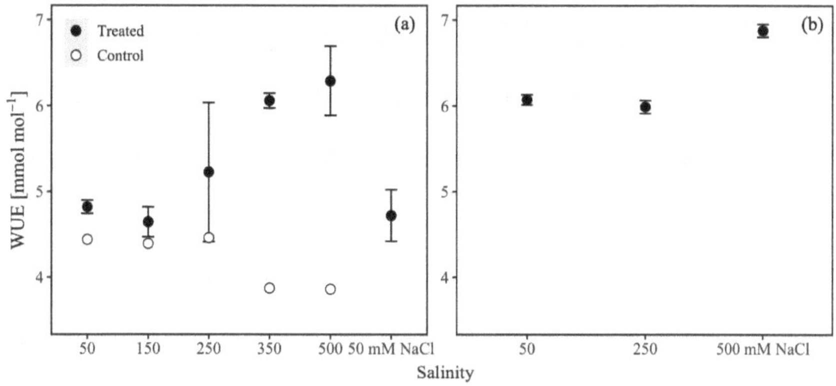

Fig. 1 Short-term (A) and long-term (B) responses of water-use efficiency (WUE) in *Avicennia marina* to variation in salinity under a constant humidity of 70% (12 mbar vpd). In (A) seedlings were grown hydroponically in nutrient solution amended with 50 mM NaCl (roughly 10% seawater salinity). Gas exchange measurements were made concurrently on fully expanded leaves under the growth conditions, and then measured repeatedly as the salinity was either gradually increased to 500 mM NaCl and then decreased gradually to the original conditions (solid circles, mean ± SE, $n=3$ seedlings) or kept at 50 mM NaCl (hollow circle, n = 1 seedling). The short-term WUE was calculated from Ball and Farquhar (1984a) as (1 − 0.35) × A/T where 0.35 is an assumed proportion of dark respiration (Φ in Table 1) to facilitate comparison with the long-term results shown in panel (B). Here, seedlings (n = 5 per salinity) were grown hydroponically for three months in nutrient solution amended with 50, 250 and 500 mM NaCl. The long-term WUE were estimated from data in Ball and Farquhar (1984b).

3. Long-term vs. short-term water use efficiency

The actual WUE is an index describing the long-term (over the course of a whole day) carbon gain per unit water loss of plants, i.e., the spatial and temporal integration of A divided by that of T (Farquhar, O'Leary, & Berry, 1982). However, the instantaneous WUE (WUE^{in}) is commonly adopted instead of the long-term WUE for two reasons: first, measurements of components of the WUE^{in} are easier to obtain than those of the long term because of the continuous monitoring required for the latter; second, the instantaneous WUEs are identical to long-term ones if the drawdown of both CO_2 ($p_a - p_i$) and water vapor ($e_i - e_a$) through a leaf stay constant during daytime (steady state) and night-time autotrophic respiration is negligible ($\Phi = 0$ in Table 1). The rare availability of steady state conditions under naturally varying field conditions (radiation, temperature and humidity) hinders direct estimations of the long-term WUE, while a short-term controlled steady state is easy to reach with a portable photosynthesis system. In this context, we can consider the WUE^{in} as an average state and then add the factor $(1-\Phi)$ to the WUEin, i.e., $(1-\Phi)$WUEin if night-term autotrophic respiration is significant (Farquhar, O'Leary, & Berry, 1982). This greatly extends applications of instantaneous WUEs as approximate estimations of long-term WUEs. In addition, eddy covariance techniques facilitate long-term continuous measurements of carbon and water fluxes at the ecosystem level, but challenges exist in calculating long-term WUEs due to difficulties in distinguishing canopy transpiration flux (T'' in Table 1) from evapotranspiration. However, T'' can be estimated by sap flow techniques (Krauss, Barr, Engel, Fuentes, & Wang, 2015) and some process-based ecological models (Liang et al., 2019; Wei, Lee, Wen, & Xiao, 2018).

Aside from the models, there are two other commonly used approaches to evaluating long-term WUEs: one is destructive measurements of plant growth and water loss (ratio of mass growth rate/water loss rate), which is usually done with potted plants growing in controlled environments (Ball, 1988); the other is obtained by measurement of carbon isotope composition (^{13}C) of plant organic matter, as clarified in experimental and theoretical studies (Farquhar et al., 1993; Farquhar, Ball, Caemmerer, & Roksandic, 1982; Farquhar, O'Leary, & Berry, 1982; Farquhar & Richards, 1984). Importantly, leaf carbon isotope composition reflects the carbon component of water use efficiency integrated over the past environment of the leaf. It therefore is a long-term concept, largely decoupled from the instantaneous WUE estimated by simultaneous gas exchange measurements (Martin et al., 2010; Medlyn et al., 2017). Farquhar, O'Leary, and Berry (1982) derived one theory relating WUE to carbon isotope composition of leaf dry matter for C_3 plants as

$$\mathrm{WUE} = \frac{(1-\Phi)C_a}{1.6\bar{D}} \frac{b-\delta_a+\delta}{b-a} \tag{2}$$

where the discrimination, a, associated with diffusion in air is 4.4‰ (Farquhar, O'Leary, & Berry, 1982). The discrimination in the carboxylation reaction, b, is 27‰ as determined from comprehensive statistics of many plants, including mangroves, with the assumption that the intercellular CO_2 mol fraction C_i is the same as that (C_c) at carboxylation sites (Farquhar, Ball, et al., 1982). More complex calculations have been developed, see Busch, Holloway-Phillips, Stuart-Williams and Farquhar, 2020. The δ_a is carbon isotope composition of atmospheric CO_2 relative to the VPVB standard (−8.7‰). The $\bar{D}$ is a weighted mean vapor mole fraction difference (v/P). In general, the greater carbon isotope ratio in mangrove leaves indicates higher water use efficiency (Farquhar, Ball, et al., 1982; Liang et al., 2018; Lin & Sternberg, 1992) than in neighboring plants growing under the same atmospheric conditions (temperature and humidity). If the plants compared grow under different environmental conditions, then Eq. (2) should be adapted with inputs of specific environment factors to estimate WUE. In addition, without **Φ** included in Eq. (2), WUE will be overestimated relative to measurements of the leaf-level instantaneous WUE (Medlyn et al., 2017).

4. Instantaneous WUE vs. intrinsic WUE

The ratio of assimilation rate over stomatal conductance, called the intrinsic WUE (IWUE), is only related to CO_2 drawdown while the instantaneous water use efficiency is related to the drawdown of both CO_2 and water vapor. This intrinsic term was first introduced to compare photosynthetic properties at a common evaporative demand (Osmond, Björkman, & Anderson, 1980). Since then, IWUE has been used in inter- and intra-species comparisons (Niinemets, Diaz-Espejo, Flexas, Galmes, & Warren, 2009), as it is thought the IWUE (Table 1) is only determined by intrinsic photosynthetic (biochemical) characteristics of vascular plants (Niinemets, Flexas, & Penuelas, 2011), such as mesophyll conductance and Rubisco properties (Flexas et al., 2016). However, IWUE and its component C_i vary with D. Consider the example shown in Fig. 2 for whole plants of *Avicennia marina* grown under different salinity x humidity conditions. The linear

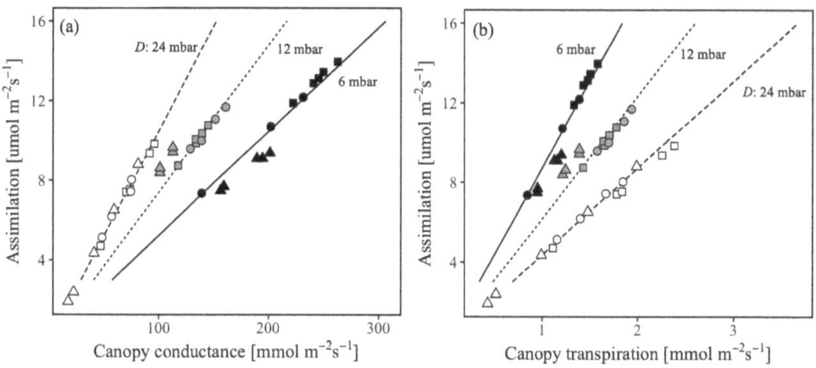

Fig. 2 Assimilation rate as a function of plant level canopy conductance (A) and canopy evaporation rate (B) in seedlings of *Avicennia marina* grown hydroponically in nutrient solution amended with 50 (diamond), 250 (circle) and 500 mM NaCl (triangle) under high, medium or low humidity where D was 6, 12 or 24 mbar, respectively. Each point within a panel represents a different plant with the same data set used for both panels. The assimilation rates were here estimated from Ball (1988) as mass growth rate × 0.45/12 assuming that the carbon content of the whole mangrove seedling is 0.45% of dry matter (Ma et al., 2018). Lines indicate best fits.

relationships between A and g for plants grown in a range of salinities under a particular D, indicate roughly constant C_i, and hence also WUE (Fig. 2A). These results showed that D had a greater effect on WUE than did salinity. Now compare the patterns in IWUE revealed by A/g (Fig. 2A) with those of actual WUE revealed by plots of A as a function of transpiration (Fig. 2B). Note that the IWUE (Fig. 2A) was greatest in plants grown under low humidity (24 mbar D), while the actual WUE (Fig. 2B) was lowest. This occurred because reduction in stomatal conductance with increasing D was not sufficient to reduce the transpiration rate. It increased with increasing D, although T under high D was not as great as it would have been if stomata had not partially closed. Nevertheless, the actual WUE declined with increasing D. This comparison underscores the importance of including estimates of water use in calculations of WUE. Specifically, comparisons of IWUE values are valid if measurements are made under common D, leaf temperature and ambient CO_2 concentration. A lack of awareness of this limitation has influenced some popular perceptions in the literature. For example, Clough and Sim (1989) measured A/g in mangroves growing naturally in forests along an aridity gradient and concluded that increase in A/g showed that WUE increased under environmental stress due to aridity. If they had plotted A/T, then they would likely have concluded the opposite.

In contrast to IWUE, which varies with D, the marginal water cost, $\lambda = \partial T/\partial A$ is constant with diurnal variations in air temperature and D over the course of a day (Cowan, 1982; Cowan & Farquhar, 1977; Farquhar, Schulze, & Kuppers, 1980). Consequently, λ is a more appropriate index for comparisons of water-use efficiency between different species or the same species growing under different atmospheric conditions. A higher λ implies greater benefit to the plant of carbon gain or lower cost to the plant of water supply to leaves. Short-term maintenance of λ can be achieved through adjustments in stomatal conductance under varying atmospheric conditions, but longer term adjustments of λ also occur with acclimation of photosynthetic or hydraulic characteristics to changing environmental conditions (Cowan, 1978; Cowan & Farquhar, 1977). Challenges remain in distinguishing short-term stomatal behavior from long-term acclimation.

5. Compromise between WUE and photosynthetic nitrogen use efficiency

In their seminal paper, Field and Mooney (1986) showed that a high WUE comes at the expense of increasing the N cost of carbon gain (i.e., PNUE), especially in C_3 species, which includes all mangroves.

Decreasing stomatal conductance reduces water loss more than assimilation rates, thereby increasing WUE for a given photosynthetic capacity, but at the expense of carbon gain. In contrast, increasing photosynthetic capacity increases both the assimilation rate and WUE for a given stomatal conductance. Enhancing photosynthetic capacity does not compromise carbon gain, but high WUE is achieved at the expense of nitrogen-use efficiency (NUE), because more nitrogen (N) must be invested in photosynthetic machinery, notably ribulose 1·5-bisphosphate carboxylase/oxygenase (Rubisco) (Evans, 1989). Hence, it follows from Field and Mooney (1986) that increasing WUE in response to increasing salinity would either compromise carbon gain or increase the N cost of carbon gain.

The trade-off between WUE and PNUE (Field & Mooney, 1986) can explain gas exchange results of field-based studies of effects of N-fertilization on individual mangrove trees growing naturally in N-limited systems. Leaf level WUE was greater in fertilized than control trees (Lovelock, Feller, McKee, Engelbrecht, & Ball, 2004; Martin et al., 2010) and increased with increasing soil water salinity (Martin et al., 2010). The N-fertilized leaves had greater N contents than controls, but there was no significant relationship between leaf N contents and photosynthetic assimilation rates (Lovelock et al., 2004; Lovelock & Feller, 2003; Martin et al., 2010). The N-fertilized leaves must have invested greater amounts of N in photosynthetic machinery to account for the maintenance of assimilation rates with reduction in stomatal conductance. However, such investment may also have been limited at high salinity by competing demands for N to meet other physiological needs. For example, osmotic adjustment of cytoplasmic compartments and other protective functions may require greater synthesis of N-containing compatible solutes with increasing salinity (Popp, 1984; Popp & Polania, 1989). Nevertheless, the greater investment of N in photosynthetic machinery by N-fertilized trees subsidized more conservative use of water without penalizing carbon gain per unit leaf area, thereby enabling limited water resources to support photosynthetic carbon gain over a larger canopy area (Martin et al., 2010).

6. Into the future: Response of water-use efficiency to elevated CO_2

Elevated atmospheric [CO_2] can enhance WUE, if a higher or similar assimilation rate can be sustained at a lower stomatal conductance. Consequently, growth responses of C_3 plants, including mangroves, could be enhanced under elevated atmospheric [CO_2] if growth in saline soils were limited by water, carbon or nitrogen. However, in a review of plant responses to elevated [CO_2] under saline conditions, Ball and Munns (1992) concluded that elevated [CO_2] would have little or no effect on growth if the soil water salinity were too high for a species to absorb water through the roots, if the carbon sink for growth was already saturated, or if saline conditions were to induce deficiencies in nutrients such as potassium. Indeed, growth of two mangrove species (*Rhizophora apiculata* and *R. stylosa*) in a multifactorial experiment, showed that WUE was enhanced by elevated [CO_2] under different combinations of salinity (125 and 350 mM NaCl) and humidity (43% and 86% relative humidity at 30°C); however, enhanced WUE did not necessarily translate into enhanced growth. Elevated [CO_2] enhanced growth when it was limited by low humidity at the leaves, but not when it was limited by high salinity at the roots (Ball et al., 1997). Similarly, Reef et al. (2015) showed that elevated [CO_2] did not alleviate limitations to growth of *Avicennia germinans* imposed by supra-optimal salinities, despite improvements in WUE. The results of these studies underscore complexities in the carbon, salt and water relations of species that are not captured in a single measure of WUE. Nevertheless, the results also offer insights into ways that future atmospheric [CO_2] could induce change in mangrove forest ecosystems along complex gradients in salinity and aridity.

7. Ecosystem-level water use efficiency of mangrove forests is comparably high as the leaf level

New observation techniques such as eddy covariance, sap flow sensors, dendrometer bands, and stable isotopes aid continuous monitoring of fluxes of energy, carbon dioxide and water vapor in mangrove ecosystems, thereby facilitating evaluation of water-carbon relations at an ecosystem level (Beer et al., 2009). According to the definition, ecosystem-level water use efficiency (WUE_e) is estimated by the ratio of gross primary production (*GPP*) to canopy transpiration (*T*) as shown in Table 1. However, availabilities of canopy transpiration data are limited for many ecosystems, particularly mangroves (Helbig et al., 2020). Instead, WUE can be estimated by alternative means using easily gaugeable variables, such as *GPP*/*ET* and *NEP*/*ET* where *NEP* is net ecosystem production. These terms were successfully applied to many upland ecosystems with two assumptions: first, evapotranspiration (*ET*) was mainly from T, and second, *NEE* was driven mainly by GPP (Beer et al., 2009; Hatfield & Dold, 2019). In contrast, those assumptions are not applicable in mangrove ecosystems where values of T/ET (Krauss et al., 2015; Liang et al., 2019) and NEP/GPP (Barr et al., 2010) are far less than one. Comparisons of different WUE_e calculations shows that *GPP/T* is significantly higher than the other estimations (*GPP/ET*, *NEP/ET*, and *NEP/T*), with the monthly-average *GPP/T* values ranging from 2.5 to 5 gC/kg H_2O in a subtropical mangrove forest (Fig. 3). These values of *GPP/T* are comparable to evergreen forests (Beer et al., 2009) but are almost double those of the commonly used term, *GPP/ET*. Use of the latter gives the misleading impression that mangrove forests are less water-use efficient than upland forest ecosystems (Xiao et al., 2013).

Calculations of *GPP/T* gave values consistent with independent measures of many parameters, such as ^{13}C (Farquhar, Ball, et al., 1982; Martin et al., 2010; Sternberg, 1992a), leaf stomatal traits (Liang et al., 2018) and leaf-level gas exchange (Naidoo & Willert, 1995; Sternberg, 1992), all of which reveal that mangroves generally have higher water use efficiency than neighboring upland trees.

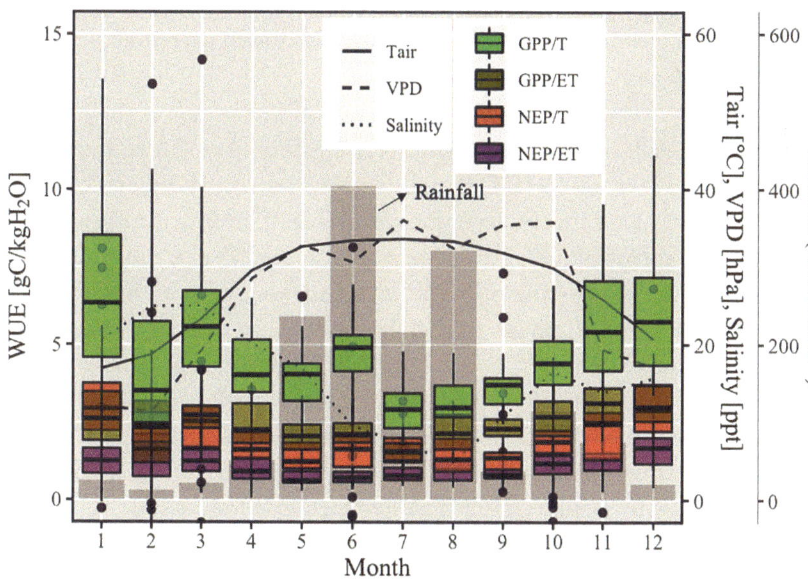

Fig. 3 Water-use efficiency of a mangrove ecosystem at Guangdong, China, 2012. The dominant mangrove species were *Aegiceras corniculatum*, *Bruguiera gymnorrhiza*, and *Avicennia marina*. The box plots were drawn from statistics of daily data in Cui et al. (2018) and Liang et al. (2019) where canopy transpiration values were modeled. The gray bars are rainfall. Salinity was measured in soil pore water, and rainfall for each month is the sum of daily values.

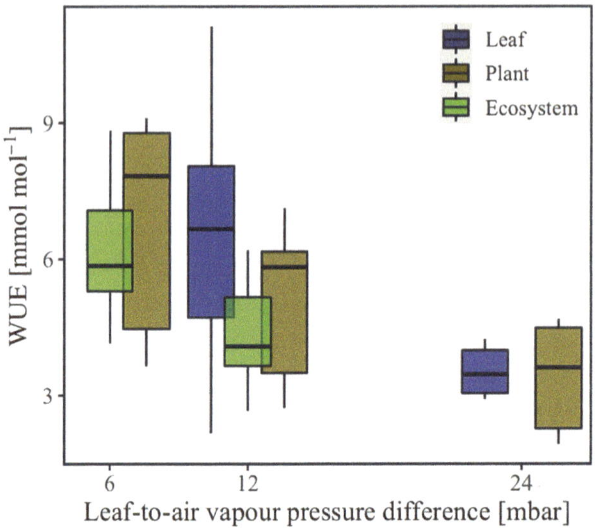

Fig. 4 Comparison of long-term (see Table 1) leaf-, plant-, and ecosystem-level water-use efficiency (WUE) under different humidity. The leaf-level WUE was calculated using data from Ball and Farquhar (1984b), and plant-level using the same data as in Fig. 2B, and ecosystem-level using data from the same source as in Fig. 3, but constrained to an irradiance of $400\,W\,m^2 <$ net radiation to match light levels across the three studies.

A striking result of the present review, is the finding of similar long-term WUE in mangroves across leaf, plant and ecosystem levels when comparisons were constrained for similar values of D (Fig. 4). A significantly higher value of leaf-level WUE (13.5 ± 1.2 mmol mol^{-1}) at D of 6 mbar (Ball and Farquhar, 1984b) was not included in Fig. 4 because of the high probability of measuring errors in leaf gas exchange measurements made under such high humidity. Further, WUE at all levels of organization declined with increasing D (Figs. 3 and 4). These patterns, admittedly compiled from a limited number of mangrove species, indicate that water use efficiencies determined through stomatal behavior are conserved across scales from individual leaves to plants, and ultimately to ecosystems.

8. Unresolved complications and future directions

There are at least three major complications that require research to further understanding of the integration of water-use efficiency across levels of individual leaves, plants and mangrove forest systems.

First, calculations of stomatal conductance require accurate estimations of vapor pressure differences between the intercellular gas spaces within a leaf and the atmosphere. While atmospheric vapor pressure can be measured, the corresponding values within leaves typically are estimated by assuming that the vapor pressure in intercellular gas spaces is equivalent to the saturation vapor pressure above pure water at leaf temperature. However, the equilibrium saturation vapor pressure declines with decreasing water potential. At the low leaf water potentials at which mangroves generally function, under-saturation of the intercellular gas spaces would lead to an underestimation of stomatal conductance for a given transpiration rate and overestimation of transpiration rates for a given stomatal conductance (Canny, Wong, Huang, & Miller, 2011; Cernusak et al., 2018; Vesala et al., 2017).

Second, foliar absorption of atmospheric water could enhance water-use efficiencies over time periods significant to carbon gain and growth. For example, Nguyen et al. (2017) showed that upper limits to leaf rehydration and turgor generation in *Avicennia marina* declined with increasing soil water salinity if roots were the sole source of water supply to shoots, suggesting increasing requirements to access atmospheric water. Indeed, a capacity for foliar water uptake has been reported in a variety of mangrove species (Bryant et al., 2021; Hayes et al., 2020) with enhancement of water status enabling turgor driven growth (Schreel, Van de Wal, Herve-Fernandez, Boeckx, & Steppe, 2019; Steppe et al., 2018), recovery of leaf hydraulic conductance (Fuenzalida et al., 2019), and maintenance of carbon gain together with protection against stem hydraulic failure (Coopman et al., 2021). Use of atmospheric water would reduce carbon costs associated with uptake of soil water by roots and its transport to leaves, potentially enhancing plant level water-use efficiencies over the long term. These effects need to be taken into account in formulation of carbon and water budgets of mangroves.

Third, integration of carbon gain and water use by forest systems requires further development of the knowledge needed to interpret complex combinations of natural isotopic signatures. The issues arise because no single isotopic signature provides all the information required to understand water use and water-use efficiencies, and because the isotopic composition of some substances, such as water, can vary because of both environmental conditions and responses of plants to them. Deciphering this complex isotopic code is fundamental for analyses of mangrove forest function under past and current conditions that are needed to develop process-based modelling tools for management of coastal mangrove resources under changing environmental conditions. For example, the oxygen isotopic composition of mangrove cellulose is related to leaf traits such as stomatal density and leaf succulence (Liang et al., 2018). Combining oxygen with carbon isotopic compositions will provide more comprehensive information of mangrove water-use efficiency and water use in response to environmental conditions.

9. Conclusion

In this review, we have clarified definitions of long-term, instantaneous, and intrinsic water-use efficiency and distinguished estimations of WUE at levels of individual leaves, plants and ecosystems. Using this framework, we showed consistency between the high WUE typically measured in mangrove leaves and those estimated over time scales significant for the growth of plants and the functioning of ecosystems. Finally, we drew attention to a few unresolved complications that affect estimations of WUE, notably effects of unsaturated intercellular vapor concentrations on estimations of conductance and transpiration, effects of foliar water uptake on long-term water-use efficiency of plants, and the need to develop methods to further exploit the information rich signatures of stable isotopes for better understanding of water use characteristics of mangrove systems under past, present and future environmental conditions.

Acknowledgments

The authors acknowledge the Ngunnawal people, the traditional custodians of the land on which this work was conducted. We pay our respects to them and their cultures and to elders both past and present. J.L. and M.C.B. thank the Australian Research Council for support through Discovery Project Grant DP180102969, and G.D.F. thanks the ARC for support through DP210103186.

References

Ball, M. C. (1988). Salinity tolerance in the mangroves Aegiceras corniculatum and Avicennia marina. I. Water use in relation to growth, carbon partitioning, and salt balance. *Functional Plant Biology, 15*(3), 447–464.

Ball, M. C. (1996). Comparative ecophysiology of mangrove forest and tropical lowland moist rainforest. In S. S. Mulkey, R. L. Chazdon, & A. P. Smith (Eds.), *Tropical forest plant ecophysiology* (pp. 461–496). New York, US: Chapman and Hall.

Ball, M. C., Cochrane, M. J., & Rawson, H. M. (1997). Growth and water use of the mangroves Rhizophora apiculata and R. stylosa in response to salinity and humidity under ambient and elevated concentrations of atmospheric CO_2. *Plant, Cell & Environment, 20*(9), 1158–1166.

Ball, M. C., Cowan, I. R., & Farquhar, G. D. (1988). Maintenance of leaf temperature and the optimisation of carbon gain in relation to water loss in a tropical mangrove forest. *Functional Plant Biology, 15*(2), 263–276.

Ball, M. C., & Farquhar, G. D. (1984a). Photosynthetic and stomatal responses of the grey Mangrove, Avicennia marina, to transient salinity conditions 1. *Plant Physiology*, *74*(1), 7–11.

Ball, M. C., & Farquhar, G. D. (1984b). Photosynthetic and stomatal responses of two Mangrove species, Aegiceras corniculatum and Avicennia marina, to long term salinity and humidity conditions. *Plant Physiology*, *74*(1), 1–6.

Ball, M. C., & Munns, R. (1992). Plant responses to salinity under elevated atmospheric concentrations of CO2. *Australian Journal of Botany*, *40*(5), 515–525.

Ball, M. C., & Passioura, J. (1995). Photosynthesis in mangroves. In E. Schulze, & M. Martyn (Eds.), *Ecophysiology of photosynthesis* (pp. 247–259). Springer.

Barr, E. V., Fuentes, J. D., Zieman, J. C., O'Halloran, T. L., Smith, T. J., & Anderson, G. H. (2010). Controls on mangrove forest-atmosphere carbon dioxide exchanges in western Everglades National Park. *Journal of Geophysical Research. Biogeosciences*, *115*(G2), 1–14.

Beer, C., Ciais, P., Reichstein, M., Baldocchi, D., Law, B. E., Papale, D., et al. (2009). Temporal and among-site variability of inherent water use efficiency at the ecosystem level. *Global Biogeochemical Cycles*, *23*(2).

Bryant, C., Fuenzalida, T. I., Zavafer, A., Nguyen, H. T., Brothers, N., Harris, R. J., et al. (2021). Foliar water uptake via cork warts in mangroves of the Sonneratia genus. *Plant, Cell & Environment*, *44*(9), 2925–2937.

Busch, F. A., Holloway-Phillips, M., Stuart-Williams, H., & Farquhar, G. D. (2020). Revisiting carbon isotope discrimination in C_3 plants shows respiration rules when photosynthesis is low. *Nature Plants*, *6*(3), 245–258.

Canny, M., Wong, S. C., Huang, C., & Miller, C. (2011). Differential shrinkage of mesophyll cells in transpiring cotton leaves: Implications for static and dynamic pools of water, and for water transport pathways. *Functional Plant Biology*, *39*(39), 91–102.

Cernusak, L. A., Ubierna, N., Jenkins, M. W., Garrity, S. R., Rahn, T., Powers, H. H., et al. (2018). Unsaturation of vapour pressure inside leaves of two conifer species. *Scientific Reports*, *8*(1), 7667.

Clough, B. F., & Sim, R. G. (1989). Changes in gas exchange characteristics and water use efficiency of mangroves in response to salinity and vapour pressure deficit. *Oecologia*, *79*, 38–44.

Coopman, R. E., Nguyen, H. T., Mencuccini, M., Oliveira, R. S., Sack, L., Lovelock, C. E., et al. (2021). Harvesting water from unsaturated atmospheres: Deliquescence of salt secreted onto leaf surfaces drives reverse sap flow in a dominant arid climate mangrove, Avicennia marina. *New Phytologist*, *231*(4), 1401–1414.

Cowan, I. R. (1978). Stomatal behaviour and environment. In *Vol. 4. Advances in botanical research* (pp. 117–228). Elsevier.

Cowan, I. R. (1982). Regulation of water use in relation to carbon gain in higher plants. In *Physiological plant ecology II* (pp. 589–613). Springer.

Cowan, I. R., & Farquhar, G. D. (1977). Stomatal function in relation to leaf metabolism and environment. *Symposia of the Society for Experimental Biology*, *31*, 471–505.

Cui, X., Liang, J., Lu, W., Chen, H., Liu, F., Lin, G., et al. (2018). Stronger ecosystem carbon sequestration potential of mangrove wetlands with respect to terrestrial forests in subtropical China. *Agricultural and Forest Meteorology*, *249*, 71–80.

Evans, J. R. (1989). Photosynthesis and nitrogen relationships in leaves of C_3 plants. *Oecologia*, *78*(1), 9–19.

Farquhar, G. D., Ball, M. C., Caemmerer, S. V., & Roksandic, Z. (1982). Effect of salinity and humidity on $\delta^{13}C$ value of halophytes—Evidence for diffusional isotope fractionation determined by the ratio of intercellular/atmospheric partial pressure of CO_2 under different environmental conditions. *Oecologia*, *52*(1), 121–124.

Farquhar, G. D., Lloyd, J., Taylor, J. A., Flanagan, L. B., Syvertsen, J. P., Hubick, K. T., et al. (1993). Vegetation effects on the isotope composition of oxygen in atmospheric CO_2. *Nature*, *363*, 439–443.

Farquhar, G. D., O'Leary, M. H., & Berry, J. A. (1982). On the Relationship Between Carbon Isotope Discrimination and the Intercellular Carbon Dioxide Concentration in Leaves. *Functional Plant Biology*, *9*(2), 121–137.

Farquhar, G. D., & Richards, R. A. (1984). Isotopic composition of plant carbon correlates with water-use efficiency of wheat genotypes. *Functional Plant Biology*, *11*(6), 539–552.

Farquhar, G., Schulze, E. D., & Kuppers, M. (1980). Responses to humidity by stomata of Nicotiana glauca L. and Corylus avellana L. are consistent with the optimization of carbon dioxide uptake with respect to water loss. *Functional Plant Biology*, 7(3), 315–327.

Field, C. H., & Mooney, H. A. (1986). The photosynthesis-nitrogen relationship in wild plants. In T. J. Givnish (Ed.), *On the economy of form and function* (pp. 25–55). Cambridge, UK: Cambridge University Press.

Flexas, J., Diaz-Espejo, A., Conesa, M. A., Coopman, R. E., Douthe, C., Gago, J., et al. (2016). Mesophyll conductance to CO_2 and Rubisco as targets for improving intrinsic water use efficiency in C_3 plants. *Plant, Cell & Environment*, *39*(5), 965–982.

Fuenzalida, T. I., Bryant, C. J., Ovington, L. I., Yoon, H. J., Oliveira, R. S., Sack, L., et al. (2019). Shoot surface water uptake enables leaf hydraulic recovery in Avicennia marina. *New Phytologist*, *224*(4), 1504–1511.

Hatfield, J. L., & Dold, C. (2019). Water-use efficiency: Advances and challenges in a changing climate. *Frontiers in Plant Science*, *10*, 103.

Hayes, M. A., Chapman, S., Jesse, A., O'Brien, E., Langley, J. A., Bardou, R., et al. (2020). Foliar water uptake by coastal wetland plants: A novel water acquisition mechanism in arid and humid subtropical mangroves. *Journal of Ecology*, *108*(6), 2625–2637.

Helbig, M., Waddington, J. M., Alekseychik, P., Amiro, B. D., Aurela, M., Barr, A. G., et al. (2020). Increasing contribution of peatlands to boreal evapotranspiration in a warming climate. *Nature Climate Change*, *10*(6), 555–560.

Krauss, K. W., Barr, J. G., Engel, V., Fuentes, J. D., & Wang, H. (2015). Approximations of stand water use versus evapotranspiration from three mangrove forests in southwest Florida, USA. *Agricultural and Forest Meteorology*, *213*, 291–303.

Liang, J., Wei, Z., Lee, X., Wright, J. S., Cui, X., Chen, H., et al. (2019). Evapotranspiration characteristics distinct to mangrove ecosystems are revealed by multiple-site observations and a modified two-source model. *Water Resources Research*, *55*(12), 11250–11273.

Liang, J., Wright, J. S., Cui, X., Sternberg, L., Gan, W., & Lin, G. (2018). Leaf anatomical traits determine the ^{18}O enrichment of leaf water in coastal halophytes. *Plant, Cell & Environment*, *41*(12), 2744–2757.

Lin, G., & Sternberg, L. (1992). Effect of growth form, salinity, nutrient and sulfide on photosynthesis, carbon isotope discrimination and growth of red mangrove (Rhizophora mangle L.). *Plant Physiology*, *19*(5), 509–517.

Lovelock, C. E., & Feller, I. C. (2003). Photosynthetic performance and resource utilization of two mangrove species coexisting in a hypersaline scrub forest. *Oecologia*, *134*(4), 455–462.

Lovelock, C. E., Feller, I. C., McKee, K. L., Engelbrecht, B. M., & Ball, M. C. (2004). The effect of nutrient enrichment on growth, photosynthesis and hydraulic conductance of dwarf mangroves in Panama. *Functional Ecology*, *18*(1), 25–33.

Ma, S., He, F., Tian, D., Zou, D., Yan, Z., Yang, Y., et al. (2018). Variations and determinants of carbon content in plants: a global synthesis. *Biogeosciences*, *15*(3), 693–702.

Martin, K. C., Bruhn, D., Lovelock, C. E., Feller, I. C., Evans, J. R., & Ball, M. C. (2010). Nitrogen fertilization enhances water-use efficiency in a saline environment. *Plant, Cell & Environment*, *33*(3), 344–357.

Medlyn, B. E., De Kauwe, M. G., Lin, Y. S., Knauer, J., Duursma, R. A., Williams, C. A., et al. (2017). How do leaf and ecosystem measures of water-use efficiency compare? *New Phytologist*, *216*(3), 758–770.

Medrano, H., Tomás, M., Martorell, S., Flexas, J., Hernández, E., Rosselló, J., et al. (2015). From leaf to whole-plant water use efficiency (WUE) in complex canopies: Limitations of leaf WUE as a selection target. *The Crop Journal, 3*(3), 220–228.
Naidoo, G., & Willert, D. J. V. (1995). Diurnal gas exchange characteristics and water use efficiency of three salt-secreting mangroves at low and high salinities. *Hydrobiologia, 295*(1-3), 13–22.
Nguyen, H. T., Meir, P., Sack, L., Evans, J. R., Oliveira, R. S., & Ball, M. C. (2017). Leaf water storage increases with salinity and aridity in the mangrove Avicennia marina: Integration of leaf structure, osmotic adjustment and access to multiple water sources. *Plant, Cell & Environment, 40*(8), 1576–1591.
Niinemets, U., Diaz-Espejo, A., Flexas, J., Galmes, J., & Warren, C. R. (2009). Role of mesophyll diffusion conductance in constraining potential photosynthetic productivity in the field. *Journal of Experimental Botany, 60*(8), 2249–2270.
Niinemets, U., Flexas, J., & Penuelas, J. (2011). Evergreens favored by higher responsiveness to increased CO_2. *Trends in Ecology & Evolution, 26*(3), 136–142.
Osmond, C. B., Björkman, O., & Anderson, D. J. (1980). Photosynthesis. In *Physiological processes in plant ecology: Toward a synthesis with atriplex* (pp. 291–377). Berlin, Heidelberg: Springer Berlin Heidelberg.
Popp, M. (1984). Chemical composition of Australian mangroves II. Low molecular weight carbohydrates. *Zeitschrift für Pflanzenphysiologie, 113*(5), 411–421.
Popp, M., & Polania, J. (1989). Compatible solutes in different organs of mangrove trees. In *Paper presented at the Annales Des Sciences Forestières.*
Reef, R., Winter, K., Morales, J., Adame, M. F., Reef, D. L., & Lovelock, C. E. (2015). The effect of atmospheric carbon dioxide concentrations on the performance of the mangrove Avicennia germinans over a range of salinities. *Physiologia Plantarum, 154*(3), 358–368.
Schreel, J. D. M., Van de Wal, B. A. E., Herve-Fernandez, P., Boeckx, P., & Steppe, K. (2019). Hydraulic redistribution of foliar absorbed water causes turgor-driven growth in mangrove seedlings. *Plant, Cell & Environment, 42*(8), 2437–2447.
Steppe, K., Vandegehuchte, M. W., Van de Wal, B. A. E., Hoste, P., Guyot, A., Lovelock, C. E., et al. (2018). Direct uptake of canopy rainwater causes turgor-driven growth spurts in the mangrove Avicennia marina. *Tree Physiology, 38*(7), 979–991.
Sternberg, G. L. A. L. D. S. L. (1992). Comparative study of water uptake and photosynthetic gas exchange between scrub and fringe red mangroves, Rhizophora mangle L. *Oecologia, 90*, 399–403.
Sternberg, G. I. L. D. S. L. (1992a). Effect of growth form, salinity, nutrient and sulfide on photosynthesis, carbon isotope discrimination and growth of red mangrove (Rhizophora mangle L.). *Plant Physiology.*
Tomlinson, P. B. (2016). *The botany of mangroves* (2nd ed.). New York: Cambridge University Press.
Vesala, T., Sevanto, S., Gronholm, T., Salmon, Y., Nikinmaa, E., Hari, P., et al. (2017). Effect of leaf water potential on internal humidity and CO_2 dissolution: Reverse transpiration and improved water use efficiency under negative pressure. *Frontiers in Plant Science, 8*, 54.
Wei, Z., Lee, X., Wen, X., & Xiao, W. (2018). Evapotranspiration partitioning for three agro-ecosystems with contrasting moisture conditions: a comparison of an isotope method and a two-source model calculation. *Agricultural and Forest Meteorology, 252*, 296–310.
Xiao, J., Sun, G., Chen, J., Chen, H., Chen, S., Dong, G., et al. (2013). Carbon fluxes, evapotranspiration, and water use efficiency of terrestrial ecosystems in China. *Agricultural and Forest Meteorology, 182-183*, 76–90.

CHAPTER THREE

Salty or sweet? Guard cell signaling and osmotic control under saline conditions

Yotam Zait and Sarah M. Assmann*
Department of Biology, Penn State University, University Park, PA, United States
*Corresponding author: e-mail address: sma3@psu.edu

Contents

1. Whole plant Na^+ responses 62
2. Overview of Na^+ transporters and their regulation 63
3. Guard cell operation in salt-stressed leaves 64
   3.1 Whole plant long-term effects (transpiration) 64
   3.2 Short-term ionic influences on stomatal conductance and stomatal apertures 65
   3.3 Guard cell ion transporters 69
   3.4 Na^+ fluxes in guard cells 72
   3.5 The role of carbohydrates in guard cell function under salt stress 74
   3.6 ABA and pH as key factors regulating salinity tolerance in guard cells 77
4. Conclusions 79
Acknowledgments 80
References 80

Abstract

Sodium ions (Na^+) negatively affect most land plant species as a consequence of both general osmotic effects and specific ion toxicity. Na^+ travels through the plant body via the transpiration stream, and this results in direct exposure of guard cells to apoplastic Na^+. This article summarizes extant literature on short- and long-term stomatal conductance responses to salinity, stomatal movements of isolated guard cells in response to NaCl, and guard cell ion content following NaCl exposure. We discuss the potential role of Na^+ as an osmoticum that itself could drive guard cell swelling and stomatal opening, possible contributions of carbohydrates in guard cell osmoregulation following NaCl exposure, and the role of ABA in stomatal responses to salinity. We highlight responses of glycophytes vs halophytes when comparative information is available. Based on our integration of published information, we propose that salt sensitivity arises in part by glycophytic guard cell uptake of Na^+ and resultant short-term stomatal opening that promotes plant desiccation and toxic accumulation of Na^+ in the leaf. However, many aspects of guard cell Na^+ sensing, transport, and response remain unknown, and new studies are required both to test this hypothesis and to address numerous knowledge gaps.

Advances in Botanical Research, Volume 103
ISSN 0065-2296
https://doi.org/10.1016/bs.abr.2022.02.009

1. Whole plant Na^+ responses

Unlike mammalian systems, where membrane potential regulation via the $3Na^+/2K^+$ ATPase is essential to cellular viability, Na^+ is a toxic ion for the majority of terrestrial plant species, including all of the major crop species. Soil salinization is increasing due to increased seawater inundation as a result of climate change-induced sea level rise and to irrigation with ground water (IPCC, 2019). Accordingly, salinity poses a global threat to agriculture and food security. In this review, we provide an overview of whole plant and cellular responses to salinity, and then focus on guard cell responses to Na^+, about which much still remains to be learned.

At the whole plant level, the response to salt stress is generally accepted to have two components: an immediate osmotic effect and an ion specific effect (Munns, 2002). Increased soil salinity decreases osmotic potential in the soil and consequently the apoplast and thus reduces water availability similarly to drought stress. This osmotic effect is independent of salt or osmolyte type and it can be short term if cellular osmotic adjustment, including synthesis of compatible solutes, occurs (reviewed in Munns and Tester, 2008). Following this rapid biophysical effect on plant water relations are ion imbalance and accumulation of toxic levels of sodium (Na^+) and chloride (Cl^-) which inhibit enzyme activity and protein synthesis and impair many metabolic processes (reviewed in Tester and Davenport (2003)). This response varies among species and depends on the plant's tolerance of high saline conditions. Halophytes are salt-tolerant plants that normally survive and complete their life cycle in saline environments such as seashores, marshes, mangrove swamps, and salt flats (Flowers, Troke, & Yeo, 1977). The other group of plants is called glycophytes and refers to all species that typically exhibit lower productivity and viability under saline conditions. In glycophytic species, the time between the first osmotic effect and second toxic phase that occurs in the shoot can be hours to days and weeks (Munns, 2002), depending on the sensitivity level of the plant species to salt and its abilities to accumulate, exclude and compartmentalize toxic ions (Binzel, Hess, Bressan, & Hasegawa, 1988; Harris, Sadras, & Tester, 2010; Munns & Passioura, 1984; Munns & Tester, 2008), as well as to detoxify reactive oxygen species (ROS) (Cuin & Shabala, 2008), which accompany exposure to Na^+ and play a role in signaling to the rest of the plant regarding root exposure to NaCl (Demidchik & Shabala, 2018).

2. Overview of Na^+ transporters and their regulation

Much remains to be learned regarding how Na^+ is sensed by plant cells, but some signaling events are known to occur quite rapidly after Na^+ exposure. For example, Na^+ induces whole-plant Ca^{2+} waves that require only a few minutes to spread throughout the entire organism (Choi, Toyota, Kim, Hilleary, & Gilroy, 2014), at least in the case of Arabidopsis seedlings. Recently the enzyme MOCA1 (monocation-induced $[Ca^{2+}]_i$ increases 1) has been implicated in this process (Jiang et al., 2019). MOCA1 is a glucuronosyltransferase needed for production of glycosyl inositol phosphorylceramide (GIPC) sphingolipids. The current model is that these sphingolipids bind Na^+ (as well as other cations) and the resultant change in cell surface potential activates plasma membrane Ca^{2+}-permeable channels that mediate Ca^{2+} influx, resulting in an initial rise in cytosolic Ca^{2+}. Subsequent transmission of the Ca^{2+} wave is dependent on Ca^{2+}-induced Ca^{2+} release from vacuoles via TPC channels in the tonoplast, a process that also requires ROS production (Evans, Choi, Gilroy, & Morris, 2016).

Given the hydrophobic nature of the lipid bilayer, movement of all ions, including Na^+, across cellular membranes is mediated by protein transport mechanisms, most of which have been initially elucidated in Arabidopsis. Na^+ first enters the root symplasm through non-selective cation channels (NSCC) that are not voltage-regulated (Volkov & Amtmann, 2006). Amelioration of resulting elevated cytosolic Na^+ concentrations is achieved by several different Na^+/H^+ (NHX) antiporters. Primary among these is NHX7/SOS1, which is localized in the plasma membrane and extrudes Na^+ into the apoplast. Although this mechanism can result in high apoplastic levels of excluded ions that could lead to cellular dehydration (i.e. a secondary osmotic effect) (Munns & Passioura, 1984), the essential role of this transporter in salinity tolerance is confirmed by the significant salt hypersensitivity of *sos1* null mutants (Wu, Ding, & Zhu, 1996). NHX1 and NHX2 are other important Na^+/H^+ antiporters that are located in the tonoplast and sequester Na^+ in the vacuole. Overexpression of tonoplast Na^+/H^+ antiporters confers salinity tolerance to Arabidopsis as well as a wide range of crop species (Bassil, Coku, & Blumwald, 2012; Blumwald, 2000). Additionally, NHX5 and NHX6 localize to vesicles in the trans-Golgi network. While their genetic knockout decreases salinity tolerance (Bassil et al., 2011), it is still unknown whether the main role of NHX5/NHX6 is to sequester Na^+ into endomembrane compartments, to allow exocytosis of Na^+ from the vesicular lumen into the vacuole upon vesicle fusion with the tonoplast, or to regulate cytosolic pH during salinity response (Bassil et al., 2012).

While these mechanisms reduce cytosolic Na^+ content, Na^+ that remains in the root symplast can transit the endodermis via plasmodesmata and enter the transpiration stream, which then carries Na^+ to the shoot. Reduction of net Na^+ content in the shoot is mediated by the Na^+ uptake transporter HKT1 via two processes. In the root, Arabidopsis HKT1 in the vascular parenchyma retrieves Na^+ from the conducting cells of the xylem, thereby minimizing Na^+ export to the shoot (Sunarpi et al., 2005). In the shoot, phloem-localized AtHKT1 loads Na^+ into sieve elements (Berthomieu et al., 2003), from whence it can be recirculated to sink tissues, including the root. One interesting question is whether these mechanisms that retain and return Na^+ to the root are adaptive because root cells are intrinsically more salt-tolerant than shoot cells, or because they minimize Na^+ buildup at the terminus of the transpiration stream, namely at the guard cells.

The importance of these transporters to Na^+ homeostasis is further confirmed by observations in halophytic relatives of Arabidopsis. For example, the NSCC channels of the salt tolerant Arabidopsis relative *Eutrema salsugineum* (previously *Thellungiella salsuginea*) have reduced permeability to Na^+ relative to Arabidopsis NSCC channels (Volkov & Amtmann, 2006), while *Eutrema salsugineum* SOS1 confers greater salt tolerance when expressed in a salinity-sensitive yeast strain than does Arabidopsis SOS1 (Jarvis, Ryu, Beilstein, & Schumaker, 2014). *HKT1*, which is a single copy gene in Arabidopsis, has been tandemly duplicated in *Eutrema salsugineum* allowing for subfunctionalization, with the paralog that favors Na^+ transport downregulated under salt stress, while the paralog that favors K^+ transport simultaneously upregulated under these conditions (reviewed in Kazachkova et al., 2018). However, the point has also been made that these halophytic transporters represent "variations on a theme" as opposed to the presence of entirely new mechanisms of dealing with Na^+ in halophytes vs glycophytes (Kazachkova et al., 2018; Shabala, 2013). One exception to this statement is the salt glands of halophytes, which extrude Na^+ onto the leaf surface (Liphschitz, Adiva-shomer-ilan, & Waisel, 1974; Liphschitz & Waisel, 1982).

3. Guard cell operation in salt-stressed leaves

3.1 Whole plant long-term effects (transpiration)

Stomata optimize photosynthetic CO_2 fixation and minimize water loss by regulation of gas exchange between the plant and the environment

(Raschke, 1975). Stomata are formed by two specialized epidermal cells, termed guard cells, which are shaped like kidney beans in most dicots and non-graminaceous monocots and like dumbbells in grasses (Meidner & Mansfield, 1968). These cells dynamically regulate their concentration of osmotically active solutes and this controls their turgor pressure and volume (Assmann, 1993). As the turgor pressure (and volume) of the guard cells increases, they bow apart to widen the pore and thereby increase gas exchange (Raschke, 1975). The movement of salt from the soil to roots and from there to the shoot depends on the rate of evaporation from plant leaves, and is mediated by transpiration, which in turn is driven by the water potential gradient along the soil-plant-atmosphere continuum (Nobel, 2020). Therefore, control of transpiration by closing stomata or reduced stomatal density can reduce salt concentration in the leaf tissues and improve salinity tolerance. However, a decrease in stomatal apertures reduces CO_2 concentrations in the leaf and, therefore, also the rates of photosynthesis and this may be a main reason for the decline in growth rates under salt stress. Therefore, only partial stomatal closure that allows sufficient photosynthesis could strike a balance between growth and the salt-carrying capacity of the shoot.

Plotting stomatal conductance data as a function of NaCl levels based on observations on multiple glycophytic and halophytic species reveals that when plants are exposed to increased levels of salt over a long-term period (weeks), glycophytes tend to reduce stomatal conductance at lower levels of NaCl while halophyte species are able to keep their stomata open under higher NaCl concentrations (Fig. 1A). However, upon compiling kinetic data on stomatal conductance over time, we have noticed that a number of glycophyte species (*Citrus sinensis L.*, *Cucumis sativus L.*, *Triticum turgidum L.*, *Solanum lycopersicum* L. and *Oryza sativa L.*) slightly increase their stomatal conductance over the first 1-3 days of exposure to salt stress, with stomatal closure occurring longer term (Fig. 1B).

3.2 Short-term ionic influences on stomatal conductance and stomatal apertures

There have been surprisingly few detailed studies on the rapid kinetics of stomatal conductance response following root exposure to NaCl. One exception is a study by Rahnama, James, Poustini, and Munns (2010) where the effect on stomatal conductance of addition of 50 mM NaCl was tested in short-term experiments of 1–2 h *in Triticum turgidum L. ssp. durum Desf.* The goal of this short-term experiment was to differentiate between the stomatal response to the osmotic stress of salinity in the root apoplast and the ion

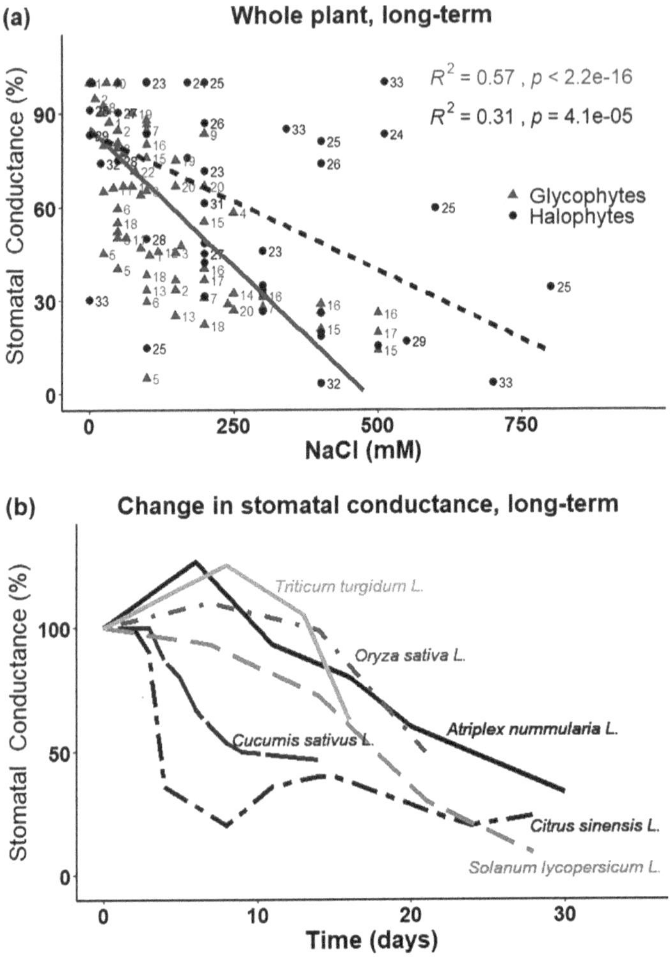

Fig. 1 (A) Responses to long term salt stress (10–30 days) of stomatal conductance of halophytes and glycophyte species. Data are derived and redrawn from the listed references. Data for glycophytes: (1) *Phoenix dactylifera L.* (Sperling, Lazarovitch, Schwartz, & Shapira, 2014), (2) *Pistacia genus L.*, (Godfrey et al., 2019) (3) *Capsicum annuum L.* (Bethke & Drew, 1992), (4) *Gossypium hirsutum L.* (Brugnoli & Lauteri, 1991), (5) *Olea europaea L.* (Tattini, Lombardini, & Gucci, 1997), (6) *Vitis vinifera* L. (Fisarakis, Chartzoulakis, & Stavrakas, 2001), (7) *Ulmus pumila L.* (Feng et al., 2014), (8) *Citrus macrophylla L.* (García-Legaz, Ortiz, Garcí-Lidón, & Cerdá, 1993), (9) *Salvadora persica L.* (Maggio, Reddy, & Joly, 2000), (10) *Ziziphus spina-christi L.* (Zait, Shtein, & Schwartz, 2019), (11) *Musa acuminata L.* (Shapira, Khadka, Israeli, Shani, & Schwartz, 2009), (12) *Medicago sativa L.* (Li, Shi, Fukuda, & Yang, 2010), (13) *Helianthus annuus L.*(Sohan, Jasoni, & Zajicek, 1999), (14) *Phaseolus vulgaris L.* (Brugnoli & Lauteri, 1991), (15) *Chenopodium album L.* (Rasouli et al., 2021), (16) *Chenopodium quinoa L.* (Rasouli et al., 2021), (17) *Beta vulgaris L.* (Rasouli et al., 2021), (18) *Oryza sativa L.* (Sultana, Ikeda, & Kashem, 2002), (19) *Solanum lycopersicum L.*

(Continued)

specific effect in the leaf apoplast. In this study, the reduction in stomatal conductance due to osmotic stress lasted for approximately 13 min. The following phase (ion specific effect) showed a moderate increase in stomatal conductance in the next 40 min. Nonetheless measurements did not last longer, and it would be interesting to repeat this type of experiment over a longer time to understand how long the increase in stomatal conductance persists. Fig. 2A demonstrates the two phases of stomatal conductance in response to salinity stress as reported in the study of Rahnama et al. (2010), differentiating between the initial osmotic phase, when stomatal conductance is decreased due to "osmotic shock," and a subsequent increase in stomatal conductance which presumably is related to the ion specific phase.

Results on epidermal peels, compiled in Fig. 2B, suggest that in glycophytes NaCl can be used as an osmoticum to support stomatal opening. Perera et al. (1994) performed a comparative study of stomatal ionic relations in isolated epidermal peels of *Commelina communis* (salt sensitive) and the halophyte *Aster tripolium* and found that Na^+ induced stomatal closure in the halophyte while promoting stomatal opening in *Commelina communis*. X-ray microanalysis after exposure to 300 mM NaCl in the incubation buffer demonstrated lower Na^+ content in guard cells of *Aster tripolium* compared to the neighboring subsidiary cells and the surrounding epidermal and mesophyll cells (Perera, De Silva, & Mansfield, 1997). On the other hand, *Commelina communis*, the salt sensitive species, accumulated higher amounts of Na^+ in the guard cells compared to neighboring cells and the other

Fig. 1—Cont'd (Amjad et al., 2014), (20) *Sorghum bicolor L.* (Netondo, Onyango, & Beck, 2004), (21) *Cucumis sativus L.* (Stępień & Kłbus, 2006), (22) *Hordeum vulgare L.* (Pérez-López, Robredo, Lacuesta, Mena-Petite, & Muñoz-Rueda, 2012). Data for halophytes: (23) *Atriplex centralasiatica L.* (Qiu & Lu, 2003), (24) *Spartina maritima L.* (Castillo et al., 2000), (25) *Sarcocornia quinqueflora L.*(Ahmed, Shabala, & Shabala, 2021), (26) *Bruguiera parviflora L.* (Parida et al., 2004), (27) *Avicennia marina L.* (Ball & Farquhar, 1984), (28) *Medicago arborea L.* (Sibole, Cabot, Poschenrieder, & Barceló, 2003), (29) *Atriplex halimus L.* (Nemat Alla, Khedr, Serag, Abu-Alnaga, & Nada, 2011), (30) *Salicornia bigelovi L.* (Ayala & O'Leary, 1995), (31) *Beta vulgaris L.* (Rasouli et al., 2021), (32) *Atriplex portulacoides L.* (Redondo-Gómez et al., 2007), (33) *Arthrocnemum macrostachyum L.* (Redondo-Gómez, Mateos-Naranjo, Figueroa, & Davy, 2010). (B) Changes in stomatal conductance over time after long-term exposure to high NaCl concentrations in the halophyte *Atriplex nummularia L.* (Alharby, Colmer, & Barrett-Lennard, 2018), and the glycophyte species *Citrus sinensis L.* (Gómes-Cadenas, Tadeo, Primo-Millo, & Talon, 1998), *Cucumis sativus L.* (Stępień & Kłbus, 2006), *Solanum lycopersicum* L. (Gharbi, Zribi, Ben, Rejeb, & Hanchi, 2018), *Triticum turgidum L.* (James, Rivelli, Munns, & Von Caemmerer, 2002), and *Oryza sativa L.* (Sultana et al., 2002).

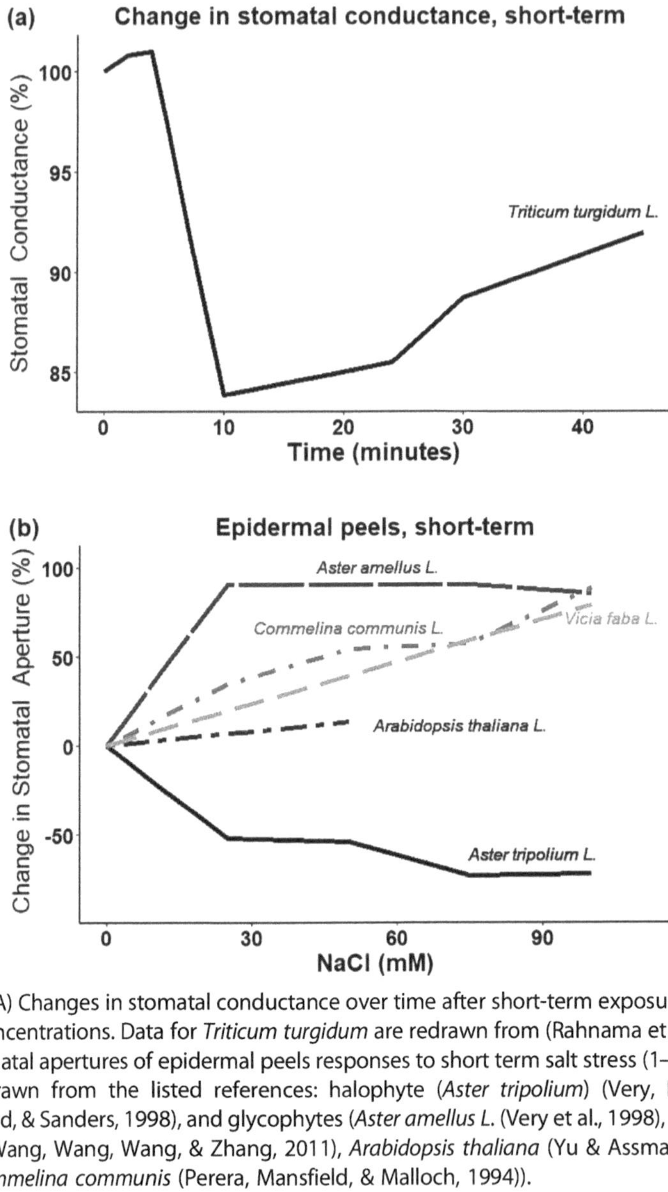

Fig. 2 (A) Changes in stomatal conductance over time after short-term exposure to high NaCl concentrations. Data for *Triticum turgidum* are redrawn from (Rahnama et al., 2010). (B) Stomatal apertures of epidermal peels responses to short term salt stress (1–3 h). Data are redrawn from the listed references: halophyte (*Aster tripolium*) (Very, Robinson, Mansfield, & Sanders, 1998), and glycophytes (*Aster amellus L.* (Very et al., 1998), *Vicia faba* (Zhao, Wang, Wang, Wang, & Zhang, 2011), *Arabidopsis thaliana* (Yu & Assmann, 2016) and *Commelina communis* (Perera, Mansfield, & Malloch, 1994)).

leaf tissues as determined by X-ray microanalysis, consistent with the Na^+-induced stomatal opening observed in this glycophyte. These findings suggest that direct guard cell sensing of Na^+ may be a salinity protection strategy in some halophytes, while Na^+ is used as an osmolyte in glycophytes.

Consistent with the above studies, NaCl was found to be equally effective as KCl in supporting light-induced stomata opening in epidermal peels of the glycophytes *Vicia faba* (Zhao et al., 2011) and *Arabidopsis thaliana* (Yu & Assmann, 2016). Also, a low concentration of Na^+ (1 mM) in the presence of KCl (25 mM) did not stimulate K^+-supported stomatal opening in Arabidopsis suggesting that Na^+ does not act as a signal to induce stomatal opening in this glycophyte (Yu & Assmann, 2016). Interestingly guard cells of *Commelina communis* were found to be hyposensitive to ABA, CO_2 and darkness in the presence of NaCl the buffer solution (Jarvis & Mansfield, 1980) suggesting that increases in guard cell Na^+ may inhibit stomatal closure signaling pathways.

The overall pattern in Fig. 2B is that glycophytic species (*Aster amellus L, Vicia faba, Arabidopsis thaliana* and *Commelina communis*) increased stomatal apertures in epidermal peels with an increase in NaCl concentration under white light illumination, while the halophyte *Aster tripolium* reduced stomatal apertures when NaCl was increased in the buffer medium (Fig. 2B). In contrast, a recent study demonstrated that Na^+ in the buffer solution supported increases in stomatal apertures in illuminated epidermal peels of salt-tolerant wild barley (*Hordeum spontaneum*) (Kiani-Pouya et al., 2020). Another epidermal peel study, conducted under green light illumination, showed that the halophyte sea beet (*Beta maritima*), and the moderately salt tolerant quinoa (*Chenopodium quinoa*) also could use Na^+ as an osmolyte supporting stomatal opening (Rasouli et al., 2021). The conflicting reports from seemingly similar protocols suggest that different halophytes may use different strategies to optimize stomatal operation under elevated salinity.

We hypothesize that, after an inevitable biophysically-induced decrease in stomatal aperture due to the lower water potential of the salinized apoplast, salt-sensitive species initially do not close their stomata in response to increase in NaCl in the apoplast (and in the short term even use Na^+ as an osmoticum and open more), and by the time stomatal closure does occur, it is "too late" and the amount of NaCl in the leaf has reached toxic levels (Fig. 3A and B).

3.3 Guard cell ion transporters

Gas exchange is dynamically regulated by guard cell control of stomatal apertures. Therefore, in order to understand how Na^+ regulates transpiration, an understanding of how Na^+ regulates guard cell physiology is essential. The basic stomatal active opening process involves extrusion of H^+ out of the

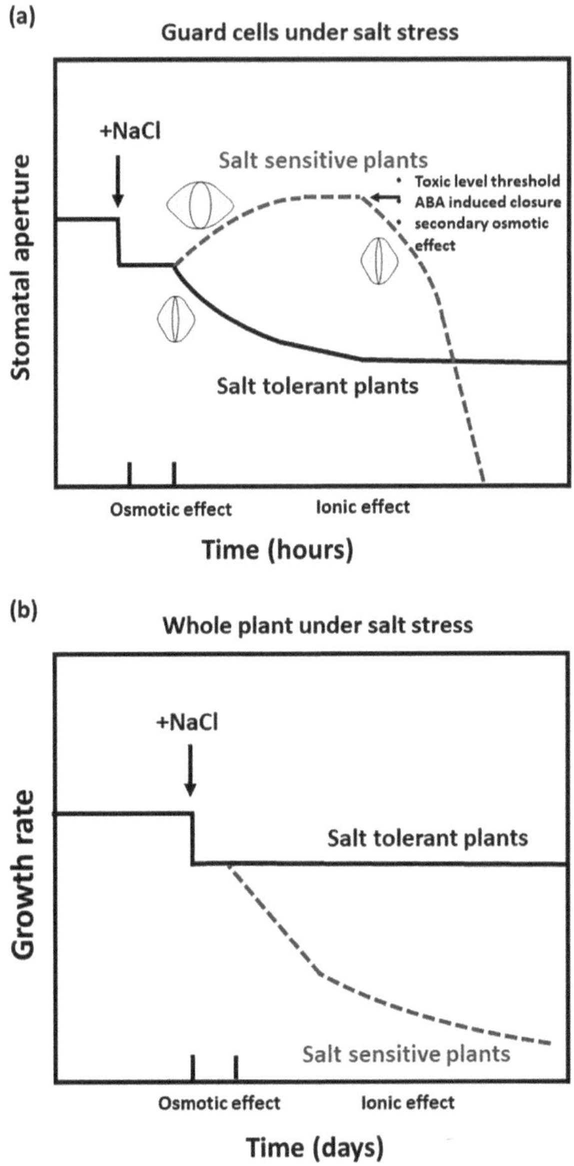

Fig. 3 (A) A proposed possible association between stomatal aperture and salt stress over short term NaCl exposure (several hours) at the guard cell level. Upon exposure to high NaCl concentration, osmotic stress induces rapid stomatal closure. Next, the Na^+ concentrations in the apoplast around the guard cells increases which leads to guard cell Na^+ influx in the glycophyte, and to guard cell Na^+ exclusion and/or extrusion in the halophyte. The decline in stomatal apertures among glycophytes is achieved

(Continued)

guard cell cytosol to the surrounding apoplast by plasma membrane H^+-ATPases, which lowers the apoplastic pH and creates an electrical gradient and driving force for K^+ uptake. The resultant membrane hyperpolarization (membrane potential becoming more negative) leads to opening of specific voltage-dependent K^+ inward channels (Schroeder, Raschke, & Neher, 1987) known as the KAT/AKT/KST type channels, and inhibition of specific K^+ outward channels, particularly GORK, which open when membrane potential becomes less negative, and so are inhibited at hyperpolarized membrane potentials (Ache et al., 2000). A main counter-anion that electrically balances K^+ is Cl^-, which enters the guard cell against its electrochemical gradient, presumably via H^+/Cl^- symport or OH^-/Cl^- antiport (Kearns & Assmann, 1993). In parallel, the hyperpolarized state of the membrane also prevents anion efflux through the S-type (slow-sustained) voltage-dependent anion channels encoded by SLAC1/SLAH3 (Vahisalu et al., 2008) and the R-type (rapid-transient) anion channels encoded by ALMT channels such as ALMT12 in Arabidopsis (Meyer et al., 2010). Under some conditions, uptake of sucrose (Talbott & Zeiger, 1996), breakdown of starch in the guard cells into glucose and fructose (Flütsch et al., 2020), and accumulation of malate—which is synthesized in the guard cell cytosol—(Allaway, 1973) can also contribute to build up the osmotic pressure needed for guard cell turgor production and stomatal opening. Stomatal closure involves increases in cytosolic Ca^{2+} (Schwartz, Ilan, & Grantz, 1988). Ca^{2+} influx (Grabov & Blatt, 1998; Thuleau, Ward, Ranjeva, & Schroeder, 1994) through ROS-activated Ca^{2+}-permeable channels (Acharya, Jeon, Zhang, & Assmann, 2013) induces guard cell membrane depolarization, which both activates outward K^+ voltage dependent channels and provides the electrochemical driving force that

Fig. 3—Cont'd either after: (1) salt has accumulated to toxic levels; (2) feed forward stomatal closure as a result of ABA accumulation in the apoplast/symplast of the guard cell; (3) secondary osmotic effect as a result of high levels of excluded ions in the apoplast causing loss of turgor and dehydration of the cells or as a result of carbohydrates accumulation in the guard cell apoplast as suggested in Section 3.6. This model suggests that at the guard cell level there are two ion specific phases in glycophytes, the first is when guard cell use Na^+ as an osmoticum for turgor production and stomatal opening, and the second is when the Na^+ eventually rises to toxic levels in the guard cell cytoplasm which leads to both inhibition of stomatal opening and promotion of stomatal closure. (B) A hypothetical model showing the initial osmotic and later ionic phases of a growth response to salinity for salt tolerant (ionic tolerant) species (solid line) and salt sensitive plants (dashed line).

allows the release of K^+ which reduces the osmotic pressure of the guard cell (Blatt, 1988; Schroeder et al., 1987).

3.4 Na^+ fluxes in guard cells

Among the ion transport mechanisms underlying guard cell movement, it is of interest to ask which transporters either mediate trans-membrane Na^+ fluxes or are regulated by Na^+. Early patch-clamp studies on *Vicia faba* guard cell protoplasts showed that voltage-activated inward (KAT1) and outward currents (GORK) were selective for K^+ over Na^+, and that K^+ permeates these channels 17 times more readily than Na^+ (Schroeder et al., 1987). In both the glycophyte *Aster amellus* and the halophyte *Aster tripolium*, inwardly-rectifying K^+ channels were shown to be even more K^+-selective than in *Vicia faba*, with a selectivity ratio of ~200 in both species (Very et al., 1998). The outward K^+ channels of the aster were less K^+ selective (selectivity ratio of 20) than the inward K^+ channels, suggesting that limited Na^+ entry through the outward channels might occur under the atypical situation of outward channel activation in the presence of a strong inwardly-directed Na^+ concentration gradient; again, there were no differences in selectivity between the glycophyte *Vicia faba* and the halophyte *Aster tripolium*.

Instead, like root cells, guard cells appear to primarily take up Na^+ via NSCC that are instantaneously activating and show little time- or voltage-dependence. Such channels have been observed in both the glycophytes *Vicia faba* (Zhao et al., 2011) and *Aster amellus* (Very et al., 1998), and in the halophyte, *Aster tripolium* (Very et al., 1998), and the two aster species show equal permeability of these channels to Na^+ vs K^+. One difference is that the magnitude of both inwardly rectifying K^+ currents and NSCC currents are larger in the halophyte aster than in the glycophyte aster; this might affect the kinetics of stomatal aperture responses but would not affect their Na^+-dependence. Na^+ uptake that is primarily mediated by non-selective cation channels (NSCC) thus appears to be a shared mechanism of both root cells and guard cells. With regard to other Na^+ transport mechanisms, there is little information concerning whether NHX and HKT transporters function in guard cells as they do in the root. One microarray study showed upregulation of the vacuolar antiporter gene *NHX6* in *Eutrema salsugineum* upon salt exposure (Karimi et al., 2021), but the functional consequences are unknown.

In guard cells of both glycophytes (*Vicia faba*, *Aster amellus*), and the halophyte *Aster tripolium*, extracellular Ca^{2+} decreases the currents through NSCC channels (Zhao et al., 2011). In both aster species, current reduction arises because Ca^{2+} decreases both the open probability of the NSCC channels and, to a lesser extent, their unitary conductance (Very et al., 1998). Similar NSCC regulation in root cells would provide a mechanistic underpinning to the observation that exposure of roots to Ca^{2+} often protects against NaCl stress (Bressan, Hasegawa, & Pardo, 1998). Further, at least in guard cells of *Vicia faba*, it has been shown that the NSCC channels are also inhibited by ROS, plausibly because ROS activate Ca^{2+}-permeable channels, and consequent elevation of cytosolic Ca^{2+} initiates a signaling cascade that has NSCC as one of its targets (Very et al., 1998; Zhao et al., 2011).

Given the similarities in glycophytes and halophytes of guard cell ion channels and their above-described regulation, is there any evidence for differential regulation of guard cell ion transport mechanisms in glycophytes vs halophytes that would account for their differential stomatal aperture responses as shown in Fig. 2B? One possible clue is provided by the transcriptome studies of Karimi et al. (2021), who observed NaCl-induced upregulation of H^+ ATPases, needed to drive stomatal opening, in microarrays of guard cells of the glycophyte *Arabidopsis thaliana* but not the halophyte *Eutrema salsugineum*. Another possible clue is provided by time courses of the magnitude of inwardly rectifying K^+ currents following Na^+ exposure of guard cell protoplasts. Very et al. (1998) observed that several hours of external Na^+ pretreatment of guard cell protoplasts shifted the activation potential of the inwardly rectifying K^+ channels to considerably more negative values in the halophyte *Aster tripolium*, while the activation potential in the glycophyte *Aster amellus* was unaltered by this treatment. A similar shift in activation potential was observed after only 20 minutes of exposure of the guard cell cytosol to Na^+ (via the patch pipette solution). Again, this sensitivity was only seen in the halophyte *Aster tripolium;* moreover, it was dependent on cytosolic Ca^{2+}, again suggesting involvement of a Ca^{2+}-dependent signaling cascade. We can speculate that consequent net reduction in guard cell K^+ only in the halophyte accounts for its reduced stomatal apertures as compared to the glycophyte (Fig. 2B).

Despite being one of the most abundant ions related to salt-stress, the direct effect of elevated Cl^- on guard cells' operation has been little studied as compared to its counterion Na^+. The majority of studies that have investigated the effect of K^+ and Na^+ on guard cell osmoregulation have used detached epidermal peels or excised leaf pieces incubated in artificial medium containing KCl vs NaCl, which does not allow identification of a direct effect of Cl^-. Karimi et al. (2021) showed that upon treatment with

NaCl the glycophytic *Arabidopsis thaliana* plants accumulate more Cl^- in the guard cell while the halophytic plant *Eutrema salsugineum* tends to exclude more Cl^- from the symplast and accumulate more K^+ in the guard cell under saline conditions. Geilfus and colleagues (Geilfus & Mühling, 2013; Geilfus, Mithöfer, Ludwig-Müller, Zörb, & Muehling, 2015) showed that increases in root apoplastic Cl^- induce transient alkalinization of the transpiration stream which appears to suppress ABA influx into the mesophyll and thus engender increased ABA concentrations in the apoplast. Alkalinization also favors the activity of the ATP-binding cassette (ABC) ABA uptake transporter (Kang et al., 2010), consistent with their observation of increased guard cell ABA accumulation and stomatal closure under these conditions (see discussion on ABA in Section 3.6). A caveat is that these experiments were deliberately performed using cystenium-Cl^-, i.e. in the absence of Na^+ ions, and so the observed phenomena may not be elicited by NaCl.

One important missing piece of the salinity puzzle is how and whether guard cell channels that are permeable to Cl^- and other anions might be regulated by Na^+ over the short- and long-term. Na^+-induced anion channel activation, if it occurred, would shift the membrane potential to less negative values, i.e. away from the voltages required to activate the inwardly-rectifying K^+ channels and toward voltages that activate outwardly-rectifying K^+ channels. It would be interesting to ascertain whether and to what extent anion channels are activated by Na^+ in glycophytes vs halophytes.

Over the longer term, guard cell gene regulation may also feed into stomatal aperture regulation. For example, in microarray analyses, Karimi et al. (2021) reported that anion transporter genes (*SLAH2,3*, *NPF7.2/NRT1.8*, *NRT3.1*, *AT2G16660*) and the K^+ efflux channel gene (*GORK1*) were upregulated, and K^+ influx channel genes *KAT1*, KAT2 and *AKT2/3* were downregulated in Arabidopsis guard cells as compared to its halophytic relative *Eutrema salsugineum*. These data were from 4h of NaCl treatment, possibly when the "toxic threshold" (Fig. 3A) is being approached.

3.5 The role of carbohydrates in guard cell function under salt stress

Early experiments showed that salinity induces carbohydrate accumulation in leaves of *Arabidopsis thaliana* (Gong, Li, Ma, Indu Rupassara, & Bohnert, 2005), wheat (*Triticum aestivum L.*; Kafi, Stewart, & Borland, 2003) tomato (*Lycopersicon esculentum*; Balibrea, Dell'Amico, Bolarín, & Pérez-Alfocea,

2000), and barley (*Hordeum vulgare*; Pérez-López et al., 2012). Salt and osmotic stress effects on carbohydrate metabolism that lead to sugar accumulation in the leaf have been proposed to result from starch hydrolysis, reduced sucrose translocation, changes in sink activity and changes in growth capacity (Balibrea et al., 2000; Krasensky & Jonak, 2012; Pérez-López et al., 2012; Villadsen, Rung, & Nielsen, 2005). For example, higher concentrations of soluble sugars and sucrose were found in rice leaves affected by NaCl, and this was attributed to enhanced osmolyte content that alleviates osmotic stress (i.e. osmotic adjustment) (Djanaguiraman, Sheeba, Shanker, Devi, & Bangarusamy, 2006). In the tomato cultivar *Lycopersicon esculentum Mill. cv. Volgogradskij*) which is considered less tolerant to salt than the cultivar *Lycopersicon esculentum Mill. cv Pera*, the sucrose content one week after 100 mM NaCl was added to the irrigation solution was elevated in both cultivars but to a lesser extent in the salt sensitive cultivar (Balibrea et al., 2000). In this study the salt-tolerant tomato cultivar accumulated more carbohydrates in the leaves and less in the roots than the salt-sensitive cultivar. Carbohydrate accumulation was also observed in the saltbush *Atriplex halimus L.* treated with either NaCl or PEG (Hassine & Lutts, 2010). Taken together, these results suggest that osmotic adjustment arising from carbohydrate accumulation is an important mechanism in both glycophytes and halophytes, with the latter perhaps utilizing it to greater effect.

However, information is missing on how accumulation of carbohydrate affects stomatal movements during salt stress. Accumulation of carbohydrates at low to moderate concentrations in the guard-cell symplast could serve as direct osmotica to participate in the buildup of guard cell turgor pressure that drives stomatal opening (Lu, Zhang, Outlaw, & Riddle, 1995; Talbott & Zeiger, 1996). It was recently reported that mesophyll-derived sugars provides carbon sources for stomatal opening via monosaccharide/proton symporters Sugar Transport Protein 1 and 4 (STP1 and STP4) in the guard cells of *Arabidopsis thaliana* (Flütsch et al., 2020). These plasma membrane hexose sugar transporters facilitated glucose and fructose uptake into guard cells during light-induced stomatal opening. It is possible that STP transporters play a similar role, promoting stomatal opening under salinity stress in the short term (Fig. 4A). Expression of the Arabidopsis transporter *STP13* was found to be upregulated in leaves within a few hours of osmotic and salt stress (Yamada et al., 2011). It was also recently reported in salt-stressed rice leaves that increased sucrose content was accompanied by elevated expression of *OsSWEET13* and *OsSWEET15* genes, encoding sugar transporters (Mathan, Singh, & Ranjan, 2021). However, potential upregulation of

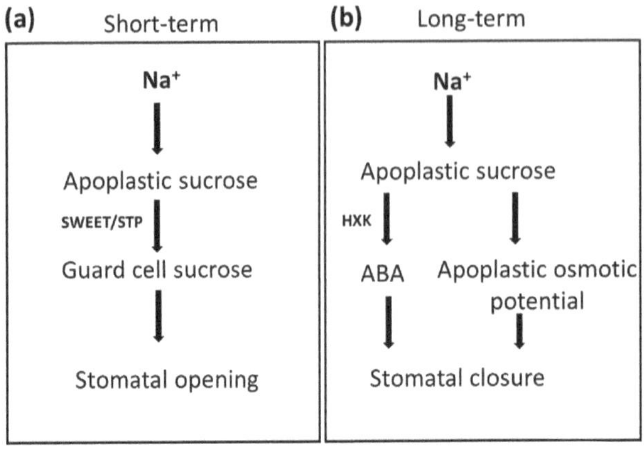

Fig. 4 Hypothetical model for signal transduction mediated by carbohydrate accumulation in the leaf apoplast under short and long-term salt-stress. (A) Upon exposure to high NaCl concentrations, there is increase in leaf carbohydrates due to a reduction in sinks for sugars. Over the short-term, low concentrations of sugars (mainly sucrose) in the leaf apoplast enter the guard cells via sugar transporters (SWEETs and/or STPs) and contribute to guard cell osmotica to induce stomatal opening. In glycophytes, this process occurs in parallel to stomatal opening resulting from guard cell Na^+ uptake. (B) After long term NaCl exposure, the concentration of sugars in the apoplast exceed a threshold level, which then promotes direct hydropassive closure of the stomata due to osmotic water loss from the guard cell to the hypertonic apoplast. Another stomatal closure pathway mediated by high sugar concentrations is attributed to guard cell ion efflux stimulated by ABA signaling via the sugar sensing enzyme HXK.

carbohydrates transporters specifically in guard cells has, to our knowledge, not been studied in the context of salinity stress.

After long-term salt stress, higher concentrations of carbohydrates in the leaf apoplast could induce secondary osmotic effects that would lead to passive stomatal closure via the water potential gradient between the guard cells and the surrounding apoplast (Fig. 4B). Another possible mechanism is that carbohydrates can cause stomatal closure via a non-osmotic mechanism via the sugar-sensing enzyme hexokinase (HXK) (Kelly et al., 2013). The study by Kelly and colleagues showed that HXK in guard cells promotes stomatal closure by triggering the abscisic acid-signaling pathway when apoplastic sugar concentrations exceed 100 mM in Arabidopsis and tomato plants. Consistent with such a mechanism, expression of Arabidopsis HXK in tobacco guard cells confers salinity tolerance, reducing leaf transpiration and Na^+ uptake as compared to the wild-type control plants (Lugassi et al., 2019).

Most recently, Franzisky et al. (2021) profiled the ionic and sucrose content of the apoplast and isolated guard cells of *Vicia faba* L. under elevated salinity, and showed that sucrose levels increased in the apoplast following NaCl treatment. Sucrose levels in guard cells under darkness were also elevated, while under illumination more minor increases in guard cell sucrose were observed as compared to the no salt control. This suggests that sucrose may not be a major osmoticum during the transient stomatal opening under illumination observed in glycophytes upon NaCl exposure (Figs. 2 and 3). The authors suggest accordingly that in addition to a potential osmotic effect of apoplastic Na^+, long-term stomatal closure under high salt conditions is mediated by sucrose feedforward signaling via HXK (Fig. 4B). Nevertheless, the role of carbohydrates in guard cell function under salt stress and its underlying mechanisms essentially remain to be elucidated.

3.6 ABA and pH as key factors regulating salinity tolerance in guard cells

The plant hormone abscisic acid (ABA) induces signaling cascades that control a range of physiological processes and environmental responses, including promotion of seed dormancy (Finkelstein, Gampala, & Rock, 2002), inhibition of plant growth and development (Kutschera & Schopfer, 1986; Negin, Yaaran, Kelly, Zait, & Moshelion, 2019), control of plant water status by reducing hydraulic conductivity (Shatil-Cohen, Attia, & Moshelion, 2011), and modulation of stomatal apertures (Jones & Mansfield, 1970; Mittelheuser & Van Steveninck, 1969). Stomata close in response to ABA, which is synthesized rapidly in response to drought and other environmental stresses (McAinsh, Brownlee, & Hetherington, 1990).

Among these environmental stresses, salt and osmotic stress are well-documented to stimulate a rapid increase in leaf ABA levels (Fujita, Fujita, Shinozaki, & Yamaguchi-Shinozaki, 2011; Geilfus et al., 2015; Sah, Reddy, & Li, 2016). Due to its important role in stomatal regulation, ABA has been considered as a possible key player in the process of stomatal closure in response to salinity (Osakabe, Yamaguchi-Shinozaki, Shinozaki, & Tran, 2014). Guard cell ABA signaling involves an ABA receptor complex, consisting of PYR/PYL/RCAR ABA receptors, phosphatases and kinases (Ma et al., 2009; Park et al., 2009). Downstream signaling is mediated through several secondary messengers, including ROS, nitric oxide (NO), and protein phosphorylation, which promote influx of Ca^{2+} from outside the cell and release from internal stores. The signaling network triggers increases in cytosolic pH and cellular depolarization, and activates outward anion (Cl^-, NO_3^-, and malate^{2-}) and K^+ efflux channels (Kim, Böhmer, Hu, Nishimura, & Schroeder, 2010; Miedema & Assmann, 1996). ABA-induced stomatal closure also includes Ca^{2+}-independent pathways which involve reduced interaction of protein kinase OPEN STOMATA 1 (OST1) with 2C-type protein phosphatases, resulting in activation of the anion channel SLAC1 and consequent anion loss (Geiger et al., 2009).

ABA acts from within guard cells (Assmann, 1994; Park et al., 2009; Schwartz, Wu, Tucker, & Assmann, 1994). ABA is both synthesized by guard cells themselves in response to environmental signals (Cornish & Zeevaart, 2008) and taken up from the apoplast (Schwartz et al., 1994). ABA is a weak acid ABA (pK_a of 4.7) and can enter the guard cell by two routes: (1) via ATP-binding cassette (ABC) transporters (Kuromori et al., 2010), or; (2) ABA in its uncharged form, ABAH, can diffuse across the plasma membrane passively from the relatively low pH environment of the apoplast to the high pH conditions of the guard cell cytosol (7.2–7.4).

The Arabidopsis ABA-deficient mutant, *aba1* (Cramer, 2002), and the tomato ABA-deficient mutant (sitiens) (Mäkelä, Munns, Colmer, & Peltonen-Sainio, 2003) are extremely sensitive to salinity compared to their corresponding wild-type plants, suggesting that ABA signaling provides key protective mechanisms to mitigate salt stress response. An interesting question is whether guard cell ABA signaling pathways are conserved between glycophytes and halophytes. Glycophytes exhibiting salt and/or osmotic induction of increased leaf ABA levels include tomato (*Solanum lycopersicum L.* Albacete et al., 2008), barley (Kefu, Munns, & King, 1991), Arabidopsis (Geilfus et al., 2015), tobacco (Boussiba & Richmond, 1976), and soybean (Roeb, Wieneke, & Fiihr, 1982). Comparative studies suggest that ABA content in the leaves of halophytes species is significantly lower than in glycophytes under both control and salt stress conditions (Clipson, Lachno, & Flowers, 1987; Hedrich & Shabala, 2018). For instance, in the saltbush (*Atriplex spongiosa*) ABA levels increased only when NaCl was increased to 150mM, while barley and cotton plants maximized ABA under half that concentration of NaCl (Kefu et al., 1991). A reduction in salt-induced ABA levels may contribute to maintenance of stomatal conductance in halophytes over long-term salinity exposure (Fig. 3B). In addition, recently Karimi et al. (2021) compared guard cell responses to short-term and long-term salinity in the glycophyte *Arabidopsis thaliana* vs the halophyte *Eutrema salsugineum*, and showed reduced ABA sensing and repressed ABA signaling in halophytic guard cells compared to the glycophytic guard cells, despite both species having similar levels of ABA in the guard cells. The reduction in ABA sensing was associated with reduced expression of the ABA receptor genes *PYL1*, *PYL4*, *PYL5* and *PYL6* in the guard cell of *Eutrema salsugineum* under saline conditions. In the long term, this could contribute to higher stomatal apertures in *Eutrema salsugineum* compared to *Arabidopsis thaliana* under elevated salinity.

Not in agreement with this concept, a study conducted on isolated epidermal peels of the halophyte *Aster tripolium*, indicated that ABA strongly inhibited stomatal opening in a dose dependent manner (Perera et al., 1994). No differences in the ABA content were detected when NaCl was added to the buffer, suggesting that the reduction of stomatal aperture caused by NaCI was not mediated by ABA. However, we know of no other reports directly assessing NaCl-ABA interactions during short term and long-term guard cell salinity response; thus, the role of ABA in both glycophytic and halophytic guard cells requires further study.

4. Conclusions

Guard cells are key regulators of salinity tolerance in plants. When thinking about the role of stomata in protection against salinity damage, it is intuitively expected that stomatal closure reduces transpiration as well accumulation of toxic ions in the plant canopy. In this review, we raise the hypothesis that guard cells do not function as anticipated to alleviate the deleterious effect of salinity. Alternatively (Figs. 2B and 3A), glycophytic stomata use Na^+ as an osmoticum and reopen after the initial osmotic shock, increasing transpiration and thus the transport of toxic ions to the above-ground organs. This will occur until either apoplastic build-up of osmotica or a yet-to-be discovered signaling mechanism causes the expected stomatal closure. Presumably, this phenomenon is not the case in halophyte species that tend to be more selective against Na^+ at the guard cell level. Given the small number of studies on halophytes that investigated isolated stomatal responses to NaCl in epidermal peels, as well as the limited electrophysiological analysis of halophytic guard cells, additional research is required to reach more decisive conclusions. Moreover, it remains unknown if guard cells utilize the transporters that allow the removal of toxic Na^+ ions from the cytosol, as occurs in roots and other plant organs. If they do, it would be informative to pursue studies of guard cell-specific knockout/overexpression of Na^+ transporters/sensors/pumps as well as transcriptome analyses, employing the glycophyte *Arabidopsis thaliana* and the halophyte *Eutrema salsugineum* as model plant species. Better understanding of guard cell responses to elevated salinity has the potential to inform strategies to improve plant productivity and growth in marginal lands and salt-affected soils.

Acknowledgments

Supported by grants NSF-MCB-17-15826 and NSF-IOS-19-20420 to S.M.A. and by USDA/NIFA post-doctoral fellowship 2021-67034-35174 to Y.Z.

References

Acharya, B. R., Jeon, B. W., Zhang, W., & Assmann, S. M. (2013). Open stomata 1 (OST1) is limiting in abscisic acid responses of Arabidopsis guard cells. *New Phytologist, 200*, 1049–1063.

Ache, P., Becker, D., Ivashikina, N., Dietrich, P., Roelfsema, M. R. G., & Hedrich, R. (2000). GORK, a delayed outward rectifier expressed in guard cells of Arabidopsis thaliana, is a K^+-selective, K^+-sensing ion channel. *FEBS Letters, 486*, 93–98.

Ahmed, H. A. I., Shabala, L., & Shabala, S. (2021). Understanding the mechanistic basis of adaptation of perennial Sarcocornia quinqueflora species to soil salinity. *Physiologia Plantarum, 172*, 1997–2010.

Albacete, A., Ghanem, M. E., Martínez-Andújar, C., Acosta, M., Sánchez-Bravo, J., Martínez, V., et al. (2008). Hormonal changes in relation to biomass partitioning and shoot growth impairment in salinized tomato (*Solanum lycopersicum* L.) plants. *Journal of Experimental Botany, 59*, 4119–4131.

Alharby, H. F., Colmer, T. D., & Barrett-Lennard, E. G. (2018). Salinization of the soil solution decreases the further accumulation of salt in the root zone of the halophyte Atriplex nummularia Lindl. growing above shallow saline groundwater. *Plant Cell and Environment, 41*, 99–110.

Allaway, W. G. (1973). Accumulation of malate in guard cells of *Vicia faba* during stomatal opening. *Planta, 110*, 63–70.

Amjad, M., Akhtar, J., Anwar-ul-Haq, M., Yang, A., Akhtar, S. S., & Jacobsen, S. E. (2014). Integrating role of ethylene and ABA in tomato plants adaptation to salt stress. *Scientia Horticulturae, 172*, 109–116.

Assmann, S. M. (1993). Signal transduction in guard cells. *Annual Review of Cell Biology, 9*, 345–375.

Assmann, S. M. (1994). Ins and outs of guard cell ABA receptors. *The Plant Cell, 6*, 1187.

Ayala, F., & O'Leary, J. W. (1995). Growth and physiology of *Salicornia bigelovii* Torr. at suboptimal salinity. *International Journal of Plant Sciences, 156*, 197–205.

Balibrea, M. E., Dell'Amico, J., Bolaŕn, M. C., & Pérez-Alfocea, F. (2000). Carbon partitioning and sucrose metabolism in tomato plants growing under salinity. *Physiologia Plantarum, 110*, 503–511.

Ball, M. C., & Farquhar, G. D. (1984). Photosynthetic and stomatal responses of the grey mangrove, Avicennia marina, to transient salinity conditions. *Plant Physiology, 74*, 7–11.

Bassil, E., Aki, O. M., Esumi, T., Tajima, H., Zhu, Z., Cagnac, O., et al. (2011). The Arabidopsis intracellular Na^+/H^+ antiporters NHX5 and NHX6 are endosome associated and necessary for plant growth and development. *Plant Cell, 23*, 224–239.

Bassil, E., Coku, A., & Blumwald, E. (2012). Cellular ion homeostasis: Emerging roles of intracellular NHX methylation and chromatin patterning Na+/H+ antiporters in plant growth and development. *Journal of Experimental Botany, 63*, 695–709.

Berthomieu, P., Conéjéro, G., Nublat, A., Brackenbury, W. J., Lambert, C., Savio, C., et al. (2003). Functional analysis of AtHKT1 in Arabidopsis shows that Na^+ recirculation by the phloem is crucial for salt tolerance. *EMBO Journal, 22*, 2004–2014.

Bethke, P. C., & Drew, M. C. (1992). Stomatal and nonstomatal components to inhibition of photosynthesis in leaves of Capsicum annuum during progressive exposure to NaCl salinity. *Plant Physiology, 99*, 219–226.

Binzel, M. L., Hess, F. D., Bressan, R. A., & Hasegawa, P. M. (1988). Intracellular compartmentation of ions in salt adapted tobacco cells. *Plant Physiology, 86*, 607–614.

Blatt, M. R. (1988). Potassium-dependent, bipolar gating of K^+ channels in guard cells. *The Journal of Membrane Biology, 102*, 235–246.

Blumwald, E. (2000). Sodium transport and salt tolerance in plants. *Current Opinion in Cell Biology, 12*, 431–434.

Boussiba, S., & Richmond, A. E. (1976). Abscisic acid and the after-effect of stress in tobacco plants. *Planta, 129*, 217–219.

Bressan, R. A., Hasegawa, P. M., & Pardo, J. M. (1998). Plants use calcium to resolve salt stress. *Trends in Plant Science, 3*, 411–412.

Brugnoli, E., & Lauteri, M. (1991). Effects of salinity on stomatal conductance, photosynthetic capacity, and carbon isotope discrimination of salt-tolerant (*Gossypium hirsutum* L.) and salt-sensitive (*Phaseolus vulgaris* L.) C3 non-halophytes. *Plant Physiology, 95*, 628–635.

Castillo, J. M., Fernández-Baco, L., Castellanos, E. M., Luque, C. J., Figueroa, M. E., & Davy, A. J. (2000). Lower limits of *Spartina densiflora* and *S. maritima* in a Mediterranean salt marsh determined by different ecophysiological tolerances. *Journal of Ecology, 88*, 801–812.

Choi, W. G., Toyota, M., Kim, S. H., Hilleary, R., & Gilroy, S. (2014). Salt stress-induced Ca^{2+} waves are associated with rapid, long-distance root-to-shoot signaling in plants. *Proceedings of the National Academy of Sciences of the United States of America, 111*, 6497–6502.

Clipson, N. J. W., Lachno, D. R., & Flowers, T. J. (1987). Salt tolerance in the halophyte *Suaeda maritima* L. Dum. growth, ion and water relations and gas exchange in response to altered salinity. *Journal of Experimental Botany, 38*, 1996–2004.

Cornish, K., & Zeevaart, J. A. D. (2008). Abscisic acid accumulation by in situ and isolated guard cells of *Pisum sativum* L. and *Vicia faba* L. in relation to water stress. *Plant Physiology, 81*, 1017–1021.

Cramer, G. R. (2002). Response of abscisic acid mutants of Arabidopsis to salinity. *Functional Plant Biology, 29*, 561–567.

Cuin, T. A., & Shabala, S. (2008). Compatible solutes mitigate damaging effects of salt stress by reducing the impact of stress-induced reactive oxygen species. *Plant Signaling & Behavior, 3*, 207–208.

Demidchik, V., & Shabala, S. (2018). Mechanisms of cytosolic calcium elevation in plants: The role of ion channels, calcium extrusion systems and NADPH oxidase-mediated 'ROS-Ca^{2+} Hub'. *Functional Plant Biology, 45*, 9–27.

Djanaguiraman, M., Sheeba, J. A., Shanker, A. K., Devi, D. D., & Bangarusamy, U. (2006). Rice can acclimate to lethal level of salinity by pretreatment with sublethal level of salinity through osmotic adjustment. *Plant and Soil, 284*, 363–373.

Evans, M. J., Choi, W. G., Gilroy, S., & Morris, R. J. (2016). A ROS-assisted calcium wave dependent on the AtRBOHD NADPH oxidase and TPC1 cation channel propagates the systemic response to salt stress. *Plant Physiology, 171*, 1771–1784.

Feng, Z. T., Deng, Y. Q., Fan, H., Sun, Q. J., Sui, N., & Wang, B. S. (2014). Effects of NaCl stress on the growth and photosynthetic characteristics of *Ulmus pumila* L. seedlings in sand culture. *Photosynthetica, 52*, 313–320.

Finkelstein, R. R., Gampala, S. S. L., & Rock, C. D. (2002). Abscisic acid signaling in seeds and seedlings. *Plant Cell, 14*, 15–45.

Fisarakis, I., Chartzoulakis, K., & Stavrakas, D. (2001). Response of Sultana vines (*V. vinifera* L.) on six rootstocks to NaCl salinity exposure and recovery. *Agricultural Water Management, 51*, 13–27.
Flowers, T. J., Troke, P. F., & Yeo, A. R. (1977). The mechanism of salt tolerance in halophytes. *Annual Review of Plant Physiology, 28*, 89–121.
Flütsch, S., Nigro, A., Conci, F., Fajkus, J., Thalmann, M., Trtílek, M., et al. (2020). Glucose uptake to guard cells via STP transporters provides carbon sources for stomatal opening and plant growth. *EMBO Reports, 21*, 1–13.
Flütsch, S., Wang, Y., Takemiya, A., Vialet-Chabrand, S. R. M., Klejchová, M., Nigro, A., et al. (2020). Guard cell starch degradation yields glucose for rapid stomatal opening in Arabidopsis. *Plant Cell, 32*, 2325–2344.
Franzisky, B. L., Geilfus, C. M., Romo-Pérez, M. L., Fehrle, I., Erban, A., Kopka, J., et al. (2021). Acclimatisation of guard cell metabolism to long-term salinity. *Plant Cell and Environment, 44*, 870–884.
Fujita, Y., Fujita, M., Shinozaki, K., & Yamaguchi-Shinozaki, K. (2011). ABA-mediated transcriptional regulation in response to osmotic stress in plants. *Journal of Plant Research, 124*, 509–525.
García-Legaz, M. F., Ortiz, J. M., Garcí-Lidón, A., & Cerdá, A. (1993). Effect of salinity on growth, ion content and CO_2 assimilation rate in lemon varieties on different rootstocks. *Physiologia Plantarum, 89*, 427–432.
Geiger, D., Scherzer, S., Mumm, P., Stange, A., Marten, I., Bauer, H., et al. (2009). Activity of guard cell anion channel SLAC1 is controlled by drought-stress signaling kinase-phosphatase pair. *Proceedings of the National Academy of Sciences of the United States of America, 106*, 21425–21430.
Geilfus, C. M., Mithöfer, A., Ludwig-Müller, J., Zörb, C., & Muehling, K. H. (2015). Chloride-inducible transient apoplastic alkalinizations induce stomata closure by controlling abscisic acid distribution between leaf apoplast and guard cells in salt-stressed *Vicia faba*. *New Phytologist, 208*, 803–816.
Geilfus, C. M., & Mühling, K. H. (2013). Ratiometric monitoring of transient apoplastic alkalinizations in the leaf apoplast of living Vicia faba plants: Chloride primes and PM-H^+-ATPase shapes NaCl-induced systemic alkalinizations. *New Phytologist, 197*, 1117–1129.
Gharbi, F., Zribi, L., Ben, D. A., Rejeb, S., & Hanchi, B. (2018). Photosynthetic responses of tomato leaves to salt and cadmium stresses: Growth and chlorophyll a fluorescence kinetic analyses. *Polish Journal of Environmental Studies, 27*, 2499–2508.
Godfrey, J. M., Ferguson, L., Sanden, B. L., Tixier, A., Sperling, O., Grattan, S. R., et al. (2019). Sodium interception by xylem parenchyma and chloride recirculation in phloem may augment exclusion in the salt tolerant *Pistacia* genus: Context for salinity studies on tree crops. *Tree Physiology, 39*, 1484–1498.
Gómes-Cadenas, A., Tadeo, F., Primo-Millo, E., & Talon, M. (1998). Involvement of abscisic acid and ethylene in the responses of citrus seedlings to salt shock. *Physiologia Plantarum, 103*, 475–484.
Gong, Q., Li, P., Ma, S., Indu Rupassara, S., & Bohnert, H. J. (2005). Salinity stress adaptation competence in the extremophile *Thellungiella halophila* in comparison with its relative Arabidopsis thaliana. *Plant Journal, 44*, 826–839.
Grabov, A., & Blatt, M. R. (1998). Membrane voltage initiates Ca^{2+} waves and potentiates Ca^{2+} increases with abscisic acid in stomatal guard cells. *Proceedings of the National Academy of Sciences of the United States of America, 95*, 4778–4783.
Harris, B. N., Sadras, V. O., & Tester, M. (2010). A water-centred framework to assess the effects of salinity on the growth and yield of wheat and barley. *Plant and Soil, 336*, 377–389.

Hassine, A. B., & Lutts, S. (2010). Differential responses of saltbush *Atriplex halimus* L. exposed to salinity and water stress in relation to senescing hormones abscisic acid and ethylene. *Journal of Plant Physiology, 167*, 1448–1456.
Hedrich, R., & Shabala, S. (2018). Stomata in a saline world. *Current Opinion in Plant Biology, 46*, 87–95.
IPCC. (2019). In J. M. P. R. Shukla, J. Skea, E. Calvo Buendia, V. Masson-Delmotte, H.-O. Pörtner, D. C. Roberts, ... K. M. Kissick (Eds.), *Climate Change and Land: An IPCC special report on climate change, desertification, land degradation, sustainable land management, food security, and greenhouse gas fluxes in terrestrial ecosystems* Cambridge University Press.
James, R. A., Rivelli, A. R., Munns, R., & Von Caemmerer, S. (2002). Factors affecting CO_2 assimilation, leaf injury and growth in salt-stressed durum wheat. *Functional Plant Biology, 29*, 1393–1403.
Jarvis, R., & Mansfield, T. (1980). Reduced stomatal responses to light, carbon dioxide and abscisic acid in the presence of sodium ions. *Plant, Cell & Environment, 3*, 279–283.
Jarvis, D. E., Ryu, C. H., Beilstein, M. A., & Schumaker, K. S. (2014). Distinct roles for SOS1 in the convergent evolution of salt tolerance in *Eutrema salsugineum* and *Schrenkiella parvula*. *Molecular Biology and Evolution, 31*, 2094–2107.
Jiang, Z., Zhou, X., Tao, M., Yuan, F., Liu, L., Wu, F., et al. (2019). Plant cell-surface GIPC sphingolipids sense salt to trigger Ca^{2+} influx. *Nature, 572*, 341–346.
Jones, R. J., & Mansfield, T. A. (1970). Suppression of stomatal opening in leaves treated with abscisic acid. *Journal of Experimental Botany, 21*, 714–719.
Kafi, M., Stewart, W. S., & Borland, A. M. (2003). Carbohydrate and proline contents in leaves, roots, and apices of salt-tolerant and salt-sensitive wheat cultivars. *Russian Journal of Plant Physiology, 50*, 155–162.
Kang, J., Hwang, J. U., Lee, M., Kim, Y. Y., Assmann, S. M., Martinoia, E., & Lee, Y. (2010). PDR-type ABC transporter mediates cellular uptake of the phytohormone abscisic acid. *Proceedings of the National Academy of Sciences of the United States of America, 107*, 2355–2360.
Karimi, S. M., Freund, M., Wager, B. M., Knoblauch, M., Fromm, J. M., Mueller, H., et al. (2021). Under salt stress guard cells rewire ion transport and abscisic acid signaling. *New Phytologist, 231*, 1040–1055.
Kazachkova, Y., Eshel, G., Pantha, P., Cheeseman, J. M., Dassanayake, M., & Barak, S. (2018). Halophytism: What have we learnt from *Arabidopsis thaliana* relative model systems? *Plant Physiology, 178*, 972–988.
Kearns, E. V., & Assmann, S. M. (1993). The guard cell-environment connection. *Plant Physiology, 102*, 711–715.
Kefu, Z., Munns, R., & King, R. (1991). Abscisic acid levels in NaCl-treated barley, cotton and saltbush. *Functional Plant Biology, 18*, 17.
Kelly, G., Moshelion, M., David-Schwartz, R., Halperin, O., Wallach, R., Attia, Z., et al. (2013). Hexokinase mediates stomatal closure. *Plant Journal, 75*, 977–988.
Kiani-Pouya, A., Rasouli, F., Rabbi, B., Falakboland, Z., Yong, M., Chen, Z. H., et al. (2020). Stomatal traits as a determinant of superior salinity tolerance in wild barley. *Journal of Plant Physiology, 245*, 153108.
Kim, T.-H., Böhmer, M., Hu, H., Nishimura, N., & Schroeder, J. I. (2010). Guard cell signal transduction network: Advances in understanding abscisic acid, CO_2, and Ca^{2+} signaling. *Annual Review of Plant Biology, 61*, 561–591.
Krasensky, J., & Jonak, C. (2012). Drought, salt, and temperature stress-induced metabolic rearrangements and regulatory networks. *Journal of Experimental Botany, 63*, 1593–1608.
Kuromori, T., Miyaji, T., Yabuuchi, H., Shimizu, H., Sugimoto, E., Kamiya, A., et al. (2010). ABC transporter AtABCG25 is involved in abscisic acid transport and responses. *Proceedings of the National Academy of Sciences of the United States of America, 107*, 2361–2366.

Kutschera, U., & Schopfer, P. (1986). In-vivo measurement of cell-wall extensibility in maize coleoptiles: Effects of auxin and abscisic acid. *Planta, 169*, 437–442.
Li, R., Shi, F., Fukuda, K., & Yang, Y. (2010). Effects of salt and alkali stresses on germination, growth, photosynthesis and ion accumulation in alfalfa (*Medicago sativa* L.). *Soil Science and Plant Nutrition, 56*, 725–733.
Liphschitz, N., Adiva-shomer-ilan, E. A., & Waisel, Y. (1974). Salt glands on leaves of rhodes grass (*Chloris gayana* Kth.). *Annals of Botany, 38*, 459–462.
Liphschitz, N., & Waisel, Y. (1982). Adaptation of plants to saline environments: Salt excretion and glandular structure. In D. N. Sen, & K. S. Rajpurohit (Eds.), *Tasks for vegetation science. The Hague, NL* (pp. 197–214). Dr. W. Junk Publishers.
Lu, P., Zhang, S. Q., Outlaw, W. H., Jr., & Riddle, K. A. (1995). Sucrose: A solute that accumulates in the guard-cell-apoplast and guard cell symplast of open stomata. *FEBS Letters, 326*, 180–184.
Lugassi, N., Yadav, B. S., Egbaria, A., Wolf, D., Kelly, G., Neuhaus, E., et al. (2019). Expression of Arabidopsis hexokinase in tobacco guard cells increases water-use efficiency and confers tolerance to drought and salt stress. *Plants, 8*.
Ma, Y., Szostkiewicz, I., Korte, A., Moes, D., Yang, Y., Christmann, A., et al. (2009). Regulators of PP2C phosphatase activity function as abscisic acid sensors. *Science, 324*, 1064–1069.
Maggio, A., Reddy, M. P., & Joly, R. J. (2000). Leaf gas exchange and solute accumulation in the halophyte *Salvadora persica* grown at moderate salinity. *Environmental and Experimental Botany, 44*, 31–38.
Mäkelä, P., Munns, R., Colmer, T. D., & Peltonen-Sainio, P. (2003). Growth of tomato and an ABA-deficient mutant (sitiens) under saline conditions. *Physiologia Plantarum, 117*, 58–63.
Mathan, J., Singh, A., & Ranjan, A. (2021). Sucrose transport in response to drought and salt stress involves ABA-mediated induction of OsSWEET13 and OsSWEET15 in rice. *Physiologia Plantarum, 171*, 620–637.
McAinsh, M. R., Brownlee, C., & Hetherington, A. M. (1990). Abscisic acid-induced elevation of guard cell cytosolic Ca^{2+} precedes stomatal closure. *Nature, 343*, 186–188.
Meidner, H., & Mansfield, T. A. (1968). *The Physiology of Stomata* (pp. 1–179). London: McGraw-Hil.
Meyer, S., Mumm, P., Imes, D., Endler, A., Weder, B., Al-Rasheid, K. A. S., et al. (2010). AtALMT12 represents an R-type anion channel required for stomatal movement in Arabidopsis guard cells. *Plant Journal, 63*, 1054–1062.
Miedema, H., & Assmann, S. M. (1996). A membrane-delimited effect of internal pH on the K^+ outward rectifier of *Vicia faba* guard cells. *Journal of Membrane Biology, 154*, 227–237.
Mittelheuser, C. J., & Van Steveninck, R. F. M. (1969). Stomatal closure and inhibition of transpiration induced by (RS)-abscisic acid. *Nature, 221*, 281–282.
Munns, R. (2002). Comparative physiology of salt and water stress. *Plant, Cell and Environment, 25*, 239–250.
Munns, R., & Passioura, J. B. (1984). Effect of prolonged exposure to NaCl on the osmotic pressure of leaf xylem sap from intact, transpiring barley plants. *Australian Journal of Plant Physiology, 11*, 497–507.
Munns, R., & Tester, M. (2008). Mechanisms of salinity tolerance. *Annual Review of Plant Biology, 59*, 651–681.
Negin, B., Yaaran, A., Kelly, G., Zait, Y., & Moshelion, M. (2019). Mesophyll abscisic acid restrains early growth and flowering but does not directly suppress photosynthesis. *Plant Physiology, 180*, 910–925.
Nemat Alla, M. M., Khedr, A. H. A., Serag, M. M., Abu-Alnaga, A. Z., & Nada, R. M. (2011). Physiological aspects of tolerance in *Atriplex halimus* L. to NaCl and drought. *Acta Physiologiae Plantarum, 33*, 547–557.

Netondo, G. W., Onyango, J. C., & Beck, E. (2004). Sorghum and Salinity: II. Gas exchange and chlorophyll fluorescence of Sorghum under salt stress. *Crop Science*, *44*, 806–811.
Nobel, P. S. (2020). *Physicochemical and environmental plant physiology*. London, UK: Academic Press.
Osakabe, Y., Yamaguchi-Shinozaki, K., Shinozaki, K., & Tran, L. S. P. (2014). ABA control of plant macroelement membrane transport systems in response to water deficit and high salinity. *New Phytologist*, *202*, 35–49.
Park, S. S.-Y., Fung, P., Nishimura, N., Jensen, D. R., Fujii, H., Zhao, Y., et al. (2009). Abscisic acid inhibits type 2C protein phosphatases via the PYR/PYL family of START proteins. *Science*, *324*, 1068–1069.
Perera, L. K. R. R., De Silva, D. L. R., & Mansfield, T. A. (1997). Avoidance of sodium accumulation by the stomatal guard cells of the halophyte *Aster tripolium*. *Journal of Experimental Botany*, *48*, 707–711.
Perera, L. K. R. R., Mansfield, T. A., & Malloch, A. J. C. (1994). Stomatal responses to sodium ions in *Aster tripolium*: A new hypothesis to explain salinity regulation in above-ground tissues. *Plant, Cell & Environment*, *17*, 335–340.
Pérez-López, U., Robredo, A., Lacuesta, M., Mena-Petite, A., & Muñoz-Rueda, A. (2012). Elevated CO_2 reduces stomatal and metabolic limitations on photosynthesis caused by salinity in Hordeum vulgare. *Photosynthesis Research*, *111*, 269–283.
Qiu, N., & Lu, C. (2003). Enhanced tolerance of photosynthesis against high temperature damage in salt-adapted halophyte *Atriplex centralasiatica* plants. *Plant, Cell and Environment*, *26*, 1137–1145.
Rahnama, A., James, R. A., Poustini, K., & Munns, R. (2010). Stomatal conductance as a screen for osmotic stress tolerance in durum wheat growing in saline soil. *Functional Plant Biology*, *37*, 255–263.
Raschke, K. (1975). Stomatal action. *Annals of Botany*, *26*, 309–340.
Rasouli, F., Kiani-Pouya, A., Tahir, A., Shabala, L., Chen, Z., & Shabala, S. (2021). A comparative analysis of stomatal traits and photosynthetic responses in closely related halophytic and glycophytic species under saline conditions. *Environmental and Experimental Botany*, *181*, 104300.
Redondo-Gómez, S., Mateos-Naranjo, E., Davy, A. J., Fernández-Muñoz, F., Castellanos, E. M., Luque, T., et al. (2007). Growth and photosynthetic responses to salinity of the salt-marsh shrub *Atriplex portulacoides*. *Annals of Botany*, *100*, 555–563.
Redondo-Gómez, S., Mateos-Naranjo, E., Figueroa, M. E., & Davy, A. J. (2010). Salt stimulation of growth and photosynthesis in an extreme halophyte, *Arthrocnemum macrostachyum*. *Plant Biology*, *12*, 79–87.
Roeb, G. W., Wieneke, J., & Fiihr, F. (1982). Effect of high NaCl concentration in the nutrient medium on transpiration, abscisic acid, cytokinin and proline content of two soybean varieties. *Journal of Plant Nutrition and Soil Science*, *145*, 103–116.
Sah, S. K., Reddy, K. R., & Li, J. (2016). Abscisic acid and abiotic stress tolerance in crop plants. *Frontiers in Plant Science*, *7*, 1–26.
Schroeder, J. I., Raschke, K., & Neher, E. (1987). Voltage dependence of K^+ channels in guard-cell protoplasts. *Proceedings of the National Academy of Sciences of the United States of America*, *84*, 4108–4112.
Schwartz, A., Ilan, N., & Grantz, D. A. (1988). Calcium effects on stomatal movement in Commelina communis L. *Plant Physiology*, *87*, 583–587.
Schwartz, A., Wu, W. H., Tucker, E. B., & Assmann, S. M. (1994). Inhibition of inward K^+ channels and stomatal response by abscisic acid: An intracellular locus of phytohormone action. *Proceedings of the National Academy of Sciences of the United States of America*, *91*, 4019–4023.
Shabala, S. (2013). Learning from halophytes: Physiological basis and strategies to improve abiotic stress tolerance in crops. *Annals of Botany*, *112*, 1209–1221.

Shapira, O., Khadka, S., Israeli, Y., Shani, U., & Schwartz, A. (2009). Functional anatomy controls ion distribution in banana leaves: Significance of Na^+ seclusion at the leaf margins. *Plant, Cell and Environment, 32*, 476–485.

Shatil-Cohen, A., Attia, Z., & Moshelion, M. (2011). Bundle-sheath cell regulation of xylem-mesophyll water transport via aquaporins under drought stress: A target of xylem-borne ABA? *Plant Journal, 67*, 72–80.

Sibole, J. V., Cabot, C., Poschenrieder, C., & Barceló, J. (2003). Efficient leaf ion partitioning, an overriding condition for abscisic acid-controlled stomatal and leaf growth responses to NaCl salinization in two legumes. *Journal of Experimental Botany, 54*, 2111–2119.

Sohan, D., Jasoni, R., & Zajicek, J. (1999). Plant-water relations of NaCl and calcium-treated sunflower plants. *Environmental and Experimental Botany, 42*, 105–111.

Sperling, O., Lazarovitch, N., Schwartz, A., & Shapira, O. (2014). Effects of high salinity irrigation on growth, gas-exchange, and photoprotection in date palms (*Phoenix dactylifera* L., cv. Medjool). *Environmental and Experimental Botany, 99*, 100–109.

Stępień, P., & Kłbus, G. (2006). Water relations and photosynthesis in *Cucumis sativus* L. leaves under salt stress. *Biologia Plantarum, 50*, 610–616.

Sultana, N., Ikeda, T., & Kashem, M. A. (2002). Effect of seawater on photosynthesis and dry matter accumulation in developing rice grains. *Photosynthetica, 40*, 115–119.

Sunarpi, H. T., Motoda, J., Kubo, M., Yang, H., Yoda, K., Horie, R., et al. (2005). Enhanced salt tolerance mediated by AtHKT1 transporter-induced Na^+ unloading from xylem vessels to xylem parenchyma cells. *Plant Journal, 44*, 928–938.

Talbott, L. D., & Zeiger, E. (1996). Central roles for potassium and sucrose in guard-cell osmoregulation. *Plant Physiology, 111*, 1051–1057.

Tattini, M., Lombardini, L., & Gucci, R. (1997). The effect of NaCl stress and relief on gas exchange properties of two olive cultivars differing in tolerance to salinity. *Plant and Soil, 197*, 87–93.

Tester, M., & Davenport, R. (2003). Na^+ tolerance and Na^+ transport in higher plants. *Annals of Botany, 91*, 503–527.

Thuleau, P., Ward, J. M., Ranjeva, R., & Schroeder, J. I. (1994). Voltage-dependent calcium-permeable channels in the plasma membrane of a higher plant cell. *EMBO Journal, 13*, 2970–2975.

Vahisalu, T., Kollist, H., Wang, Y. F., Nishimura, N., Chan, W. Y., Valerio, G., et al. (2008). SLAC1 is required for plant guard cell S-type anion channel function in stomatal signalling. *Nature, 452*, 487–491.

Very, A. A., Robinson, M. F., Mansfield, T. A., & Sanders, D. (1998). Guard cell cation channels are involved in Na^+-induced stomatal closure in a halophyte. *Plant Journal, 14*, 509–521.

Villadsen, D., Rung, J. H., & Nielsen, T. H. (2005). Osmotic stress changes carbohydrate partitioning and fructose-2,6-bisphosphate metabolism in barley leaves. *Functional Plant Biology, 32*, 1033–1043.

Volkov, V., & Amtmann, A. (2006). *Thellungiella halophila*, a salt-tolerant relative of Arabidopsis thaliana, has specific root ion-channel features supporting K^+/Na^+ homeostasis under salinity stress. *Plant Journal, 48*, 342–353.

Wu, S. J., Ding, L., & Zhu, J. K. (1996). SOS1, a genetic locus essential for salt tolerance and potassium acquisition. *Plant Cell, 8*, 617–627.

Yamada, K., Kanai, M., Osakabe, Y., Ohiraki, H., Shinozaki, K., & Yamaguchi-Shinozaki, K. (2011). Monosaccharide absorption activity of Arabidopsis roots depends on expression profiles of transporter genes under high salinity conditions. *Journal of Biological Chemistry, 286*, 43577–43586.

Yu, Y., & Assmann, S. M. (2016). The effect of NaCl on stomatal opening in Arabidopsis wild type and *agb1* heterotrimeric G-protein mutant plants. *Plant Signaling & Behavior, 11*, 14–16.
Zait, Y., Shtein, I., & Schwartz, A. (2019). Long-term acclimation to drought, salinity and temperature in the thermophilic tree *Ziziphus spina-christi*: Revealing different tradeoffs between mesophyll and stomatal conductance. *Tree Physiology, 39*, 701–716.
Zhao, X., Wang, Y. J., Wang, Y. L., Wang, X. L., & Zhang, X. (2011). Extracellular Ca^{2+} alleviates NaCl-induced stomatal opening through a pathway involving H_2O_2-blocked Na^+ influx in Vicia guard cells. *Journal of Plant Physiology, 168*, 903–910.

CHAPTER FOUR

Modulation of abscisic acid signaling for stomatal operation under salt stress conditions

Jong-Joo Cheong*
Center for Food and Bioconvergence, Seoul National University, Seoul, Republic of Korea
*Corresponding author: e-mail address: cheongjj@snu.ac.kr

Contents

1. Introduction 90
2. Modulation of ABA metabolic pathways 92
   2.1 Regulation of ABA biosynthesis 92
   2.2 Regulation of ABA catabolism 94
   2.3 ABA conjugation with glucose 95
3. Modulation of the ABA signaling network 97
   3.1 ABA transportation 97
   3.2 ABA-induced stomatal closure 98
   3.3 Regulation of PP2C gene transcription 102
   3.4 Expression of ABA-responsive genes 105
4. Modulation of ABA signaling by protein degradation 105
   4.1 Degradation of ABA signaling components 105
   4.2 Selective degradation of ABA receptors 106
   4.3 ABA-induced degradation of PP2Cs 109
   4.4 Ubiquitination of SnRK2s and transcription factors (ABI5 and AREB/ABFs) 110
5. Summary and concluding remarks 110
Acknowledgment 111
References 111

Abstract

The plant hormone abscisic acid (ABA) triggers cellular tolerance responses to osmotic stress caused by high-salinity and water-deficient conditions. ABA controls the turgor pressure of guard cells in the plant epidermis through ionic fluxes via ion transporters anchored at the plasma membrane, leading to stomatal closure to limit transpiration and prevent the loss of water. However, stomatal apertures should be reversibly opened by a light signal to uptake carbon dioxide for photosynthesis, even under osmotic stress conditions. Plants utilize various strategies to regulate ABA-mediated stomatal closure in response to stressful conditions. Basically, plants regulate the accumulation of ABA, the signaling molecule induced in response to osmotic stress, during multiple metabolic

Advances in Botanical Research, Volume 103
ISSN 0065-2296
https://doi.org/10.1016/bs.abr.2022.02.010

processes (biosynthesis, catabolism, and conjugation) that occur in different subcellular organelles. In addition, the ABA signaling pathway in guard cells is modulated by ABA-dependent physical interactions among its core components, namely ABA molecules, receptors (e.g., pyrabactin resistance [PYR]/PYR1-like/regulatory components of ABA receptors), protein phosphatases (e.g., protein phosphatase type 2C), and kinases (e.g., sucrose non-fermenting-related protein kinase 2). Moreover, ABA induces the expression of numerous genes encoding proteins exerting either positive or negative effects on its accumulation, transportation, and signaling network. In parallel, the ABA signaling network can be desensitized by degradation of its core components via the ubiquitin-26S proteasome system or endocytic/vacuolar pathway in an ABA-dependent manner. This chapter details the mechanisms by which ABA constitutes a regulatory loop to finely tune its own homeostasis and signaling network, thereby modulating the stomatal operation.

1. Introduction

Plants often encounter high-salinity conditions during their life cycle. Salts cause ionic toxicity, oxidative damage, and osmotic stress in plants, hindering their development and growth (Arif, Singh, Siddiqui, Bajguz, & Hayat, 2020; Hasegawa, Bressan, Zhu, & Bohnert, 2000). Unlike salt-tolerant halophytes that grow in regions of high-salinity (Hedrich & Shabala, 2018), most crops are glycophytes, and their growth and productivity are severely affected by salt stress. Under drought and hot weather conditions, the salt concentration in soil increases reciprocally as the soil moisture content decreases, which gives rise to a stressful environment for plants. Plants adapt to high-salinity conditions through many physiological and cellular responses.

High-salinity leads to the extensive accumulation of sodium (Na^+) and chloride (Cl^-) ions, and activates ion channels to inhibit potassium (K^+) and calcium (Ca^{2+}) uptake, thereby disturbing ion homeostasis in plant cells (Hasegawa et al., 2000; Isayenkov & Maathuis, 2019; Manishankar, Wang, Köster, Alatar, & Kudla, 2018). Such an ionic imbalance causes depolarisation of the plasma membrane and oxidative stress through the accumulation of reactive oxygen species (ROS), such as hydrogen peroxide (H_2O_2), oxygen radicals ($O_2^{\bullet}$) and hydroxyl radicals ($HO^{\bullet}$) (Arif et al., 2020). Under salt stress conditions, imported Na^+ triggers a cytosolic Ca^{2+} signal to activate the salt overly sensitive (SOS) pathway, which mediates ionic (Na^+/K^+) homeostasis by extruding Na^+ ions into the apoplast (Amin et al., 2021; Park, Kim, & Yun, 2016; Seifikalhor et al., 2019; Zhu, 2003). Mitogen-activated protein kinase (MAPK) cascades also participate in salt stress signaling in plants, mediating ionic, osmotic, and ROS homeostasis (Mishra, Tuteja, & Tuteja, 2006; Nakagami, Pitzschke, & Hirt, 2005).

In addition, high-salinity induces shrinkage of plant cells via dehydration and physical changes at the interface of the cell wall and plasma membrane, leading to osmotic stress in cells (Seifikalhor et al., 2019). Plants respond to osmotic stress caused by high-salinity and water-deficient conditions by upregulating the levels of endogenous phytohormones, including abscisic acid (ABA), brassinosteroid, ethylene, and gibberellin (Arif et al., 2020). ABA is the main plant hormone mediating cellular tolerance responses to osmotic stress (Marusig & Tombesi, 2020; Sah, Reddy, & Li, 2016), which are integrated with other ABA-independent signaling pathways (Hussain, Gomes, Yano, & Nambara, 2019; Kumar et al., 2019). ABA induces the closure of stomatal apertures on the leaf epidermis to limit transpiration, thereby preventing the loss of water (Lim, Baek, Jung, Kim, & Lee, 2015; Munemasa et al., 2015). ABA also induces the expression of numerous genes encoding enzymes that catalyze the biosynthesis of osmoprotectants, which mitigates stress and promotes plant tolerance. In addition to its role in osmotic stress tolerance, ABA plays pivotal roles in seed maturation, dormancy, germination, and embryogenesis (Chen et al., 2020; Kumar et al., 2019). However, excess ABA inhibits photosynthesis, accelerates senescence, lowers pollen fertility, and induces seed dormancy, which impairs the development and growth of plants (Chaves, Flexas, & Pinheiro, 2009; Gietler, Fidler, Labudda, & Nykiel, 2020).

Stomata are small pores on leaf surfaces formed by a pair of flanking guard cells that swell or deflate in response to turgor pressure, which is regulated through ionic fluxes via ion channels anchored at the plasma membrane (Saito & Uozumi, 2019). In guard cells, ABA molecules are perceived by soluble receptors localized at the cytosol, activating a family of protein kinases that phosphorylate the ion channels on the plasma membrane. Stomata control transpiration and carbon dioxide (CO_2) uptake by optimizing the aperture size in response to various environmental and endogenous signals (Kim, Böhmer, Hu, Nishimura, & Schroeder, 2010; Roelfsema & Hedrich, 2005). ABA and elevated CO_2 levels promote partial or complete closure of stomata, whereas light stimulates the opening of stomata to enhance CO_2 assimilation for photosynthesis (Negi, Hashimoto-Sugimoto, Kusumi, & Iba, 2014; Shimazaki, Doi, Assmann, & Kinoshita, 2007). Plants integrate both stress and light signals during the daytime, and thus frequently close and open stomatal pores to maintain their viability.

Plants utilize various strategies to finely control the ABA-mediated stomatal operation during stressful conditions. Basically, plants regulate the accumulation of ABA by several simultaneously occurring processes, namely biosynthesis, catabolism, and conjugation, which occur in different subcellular organelles (Cutler & Krochko, 1999; Nambara & Marion-Poll, 2005; Seo & Marion-Poll, 2019). In addition, the ABA signaling pathway in guard cells is modulated by ABA-dependent physical interactions among its core components: ABA molecules, receptors, protein phosphatases, and kinases (Chen et al., 2020; Jung, Nguyen, & Cheong, 2020). ABA induces the production of proteins exerting either positive or negative effects on its accumulation, transportation, and signaling network. Moreover, the ABA signaling network can be desensitized by degradation of the core components in an ABA-dependent manner (Ali, Pardo, & Yun, 2020; Coego et al., 2021). This chapter details the mechanisms by which ABA self-regulates its accumulation and signaling network to finely tune the stomatal aperture.

2. Modulation of ABA metabolic pathways

2.1 Regulation of ABA biosynthesis

ABA accumulates in higher plants under high-salinity, drought, and cold stress conditions through a Ca^{2+}-dependent phosphorylation cascade (Xiong, Lee, Ishitani, & Zhu, 2002). The ABA biosynthesis pathway is detailed in a number of review articles (Cutler & Krochko, 1999; Nambara & Marion-Poll, 2005; Seo & Marion-Poll, 2019). As summarized in Fig. 1, ABA biosynthesis is primarily initiated in plastids by dihydroxylation of β-carotene, a 40-carbon (C40) compound that originates from isopentenyl diphosphate via the 2-C-methyl-D-erythritol-4-phosphate (MEP) pathway, to produce zeaxanthin, an oxygenated carotenoid. A reverse reaction catalyzed by violaxanthin de-epoxidase (VDE) can occur in chloroplasts under high light conditions (Guan et al., 2015; Li et al., 2013; Nambara & Marion-Poll, 2005). Then, zeaxanthin is converted to all-*trans*-violaxanthin by a two-step reaction with zeaxanthin epoxidase (ZEP). All-*trans*-violaxanthin is converted to 9-*cis*-neoxanthin and 9-*cis*-violaxanthin presumably by a neoxanthin synthase (NSY) and unknown isomerases. The two C40 isomers are cleaved by a family of 9-*cis*-epoxycarotenoid dioxygenases (NCEDs) to produce the C15 *cis*-xanthoxin. The xanthoxin is transported to the cytosol and is then converted by a short-chain alcohol dehydrogenase (SCAD) to abscisic aldehyde. ABA aldehyde oxidase (AAO) catalyzes the final step, in which abscisic aldehyde

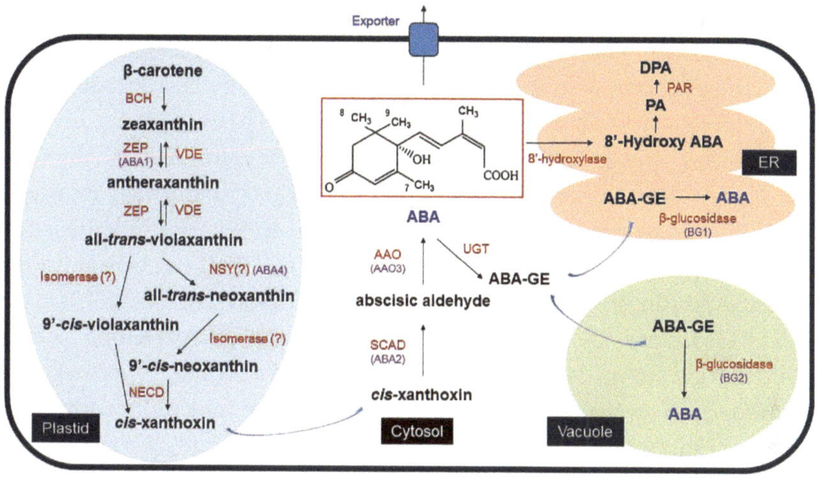

Fig. 1 ABA metabolic pathways. In the ABA biosynthesis pathway, the conversion of β-carotene (C40) to xanthoxin (C15) takes place in plastid. Xanthoxin is transported to the cytosol and converted into ABA by two enzymes, short-chain alcohol dehydrogenase (SCAD) and ABA aldehyde oxidase (AAO). In the ABA catabolic pathway, ABA is converted to 7,′ 8′, or 9′-hydroxy ABA by hydroxylases. Numerals (7–9) on the ABA molecule indicate the targets for ABA hydroxylation. The predominant catabolite 8′-hydroxy ABA is further catalyzed in the ER to phaseic acid (PA) by spontaneous isomerisation, and then to dihydrophaseic acid (DPA) by PA reductase (PAR). In the reversible conjugation pathway, ABA is glycosylated into ABA glucose ester (ABA-GE) by UDP-glucosyltransferases (UGTs). Active ABA can be generated by β-glucosidase-mediated hydrolysis of the stored ABA-GE in the ER and vacuole. Enzymes are denoted to each catalytic step: BCH, β-carotene hydroxylase; ZEP, zeaxanthin epoxidase, VDE, violaxanthin de-epoxidase; NSY, neoxanthin synthase; NCED, 9-*cis*-epoxycarotenoid dioxygenase. Enzymes identified in Arabidopsis are denoted in parentheses.

is oxidized to ABA, a C15 sesquiterpenoid. AAO requires a molybdenum cofactor (MoCo) generated by MoCo sulfurase. The Arabidopsis genes encoding ZEP (i.e., *LOS6/ABA1*), NCED (i.e., *AtNCED*), SCAD (i.e., *AtABA2*), AAO (i.e., *AtAAO3*), and MoCo sulfurase (i.e., *LOS5/ABA3*) were identified (González-Guzmán et al., 2002; Iuchi et al., 2001; North et al., 2007; Seo et al., 2000; Xiong et al., 2001; Xiong, Schumaker, & Zhu, 2002).

The expression of ABA biosynthetic genes encoding ZEP, NCED, AAO, and MoCo sulfurylase are all upregulated by high-salinity and drought conditions (Audran et al., 1998; Iuchi et al., 2001; Seo et al., 2000; Xiong, Gong, et al., 2001; Xiong, Schumaker, & Zhu, 2002). Overexpression of either one of the genes leads to increased ABA accumulation and enhances tolerance to osmotic stress (Frey, Audran, Marin, Sotta, & Marion-Poll, 1999; Iuchi et al., 2001; Thompson et al., 2000). The reaction catalyzed by NCEDs is a key regulatory step in the ABA biosynthesis pathway. Expression level of *NECD* genes is well correlated to endogenous ABA content in many plant species (Iuchi et al., 2001; Qin & Zeevaart, 2002; Schwartz, Qin, & Zeevaart, 2003; Thompson et al., 2000). In particular, transgenic Arabidopsis overexpressing *AtNCED3* exhibited an increase in endogenous ABA level and a reduction in transpiration rate from leaves (Iuchi et al., 2001).

The Arabidopsis ABA biosynthetic genes *LOS6/ABA1*, *AAO3*, and *LOS5/ABA3* are all upregulated by exogenous ABA (Xiong, Gong, et al., 2001; Xiong, Ishitani, Lee, & Zhu, 2001; Xiong, Schumaker, & Zhu, 2002). However, expression of *AtNCEDs* was not induced by ABA treatment (Xiong, Ishitani, et al., 2001). Similarly, the expression of cowpea *VuNCED1* gene was not induced by ABA application, while strongly induced under high-salinity conditions (Iuchi, Kobayashi, Yamaguchi-Shinozaki, & Shinozaki, 2000). These results are consistent with a notion that the reaction catalyzed by NCEDs is a rate-limiting step in the ABA biosynthesis pathway (Qin & Zeevaart, 1999; Thompson et al., 2000). Thus, it is inferred that a positive feedback loop through which ABA stimulates a Ca^{2+}-dependent phosphoprotein cascade that mediates an initial ABA biosynthesis may further accelerate the biosynthesis of ABA (Xiong, Lee, et al., 2002). Xiong, Ishitani, et al. (2001) identified SAD1 (supersensitive to ABA and drought 1) that could indirectly regulate the feedback loop in Arabidopsis. The *sad1* mutant plants showed enhanced ABA and osmotic stress sensitivity, reduced drought-induced ABA biosynthesis, and impaired ABA regulation of *AtAAO3* and *AtABA3* genes.

2.2 Regulation of ABA catabolism

When osmotic stress signals are diminished, ABA is catabolized into inactive metabolites (Cutler & Krochko, 1999; Nambara & Marion-Poll, 2005; Seo & Marion-Poll, 2019). Analyses of phytochrome- and ABA-deficient mutants suggest that light amplifies ABA catabolism (Kraepiel et al., 1994). Basically, ABA catabolism is accomplished by hydroxylation pathways that oxidize one of the three different methyl groups (C-7′, C-8′, and C-9′) on the ring structure of ABA molecule (Zeevaart & Creelman, 1988) (Fig. 1). In Arabidopsis, hydroxylation at the C-8′ position is predominant among the ABA catabolic pathways, and is mediated by ABA 8′-hydroxylase, a members of the cytochrome P450 (CYP) 707A family of proteins

(Krochko, Abrams, Loewen, Abrams, & Cutler, 1998; Kushiro et al., 2004; Saito et al., 2004). The 8′-hydroxylated ABA is subsequently converted into phaseic acid (PA) by spontaneous isomerisation, and then to biologically inactive dihydrophaseic acid (DPA) by PA reductase (PAR) (Cutler & Krochko, 1999). ABA degradation appears to take place in the endoplasmic reticulum (ER). ABA 8′-hydroxylase is an integral membrane protein localized to the microsomal fraction in suspension-cultured maize cells (Krochko et al., 1998). The expression of rice ABA 8′-hydroxylase-green fluorescent protein (GFP) fusion protein in onion cells is localized to the ER (Saika et al., 2007).

ABA negatively self-regulates its accumulation by inducing the expression of genes encoding catabolic enzymes (Cutler & Krochko, 1999; Krochko et al., 1998; Nambara & Marion-Poll, 2005). Arabidopsis contains four ABA 8′-hydroxylase genes, *CYP707A1* to *CYP707A4*, all of which are induced by osmotic stress (Kushiro et al., 2004; Saito et al., 2004; Umezawa et al., 2006). Okamoto et al. (2009) showed that CYP707A1 catabolizes local ABA pools in guard cells to control stomatal movement in response to high-humidity, whereas CYP707A3 functions in vascular tissues to reduce systematic ABA levels.

2.3 ABA conjugation with glucose

In addition to hydroxylation, ABA is inactivated by conjugation with glucose to form ABA glucose ester (ABA-GE) (Boyer & Zeevaart, 1982; Dietz, Sauter, Wichert, Messdaghi, & Hartung, 2000; Liu et al., 2015; Xu, Nakajima, Suzuki, & Yamaguchi, 2002) (Fig. 1). ABA-GE accumulates in the ER and vacuole (Dietz et al., 2000; Lee et al., 2006), which provides protection against its degradation and enables storage (Piotrowska & Bajguz, 2011; Xu et al., 2002). Conjugation of ABA with glucose is catalyzed by UDP-glucosyltransferases (UGTs) including Arabidopsis UGT71Bs and UGT71C5 (Dong et al., 2014; Liu et al., 2015; Priest et al., 2006). Liu et al. (2015) observed that mutation and downregulation of *UGT71C5* resulted in an increased level of ABA and reduced level of ABA-GE, causing a delay in seed germination and enhancing drought tolerance. By contrast, overexpression of *UGT71C5* resulted in a reduced level of ABA and increased level of ABA-GE, and accelerated seed germination and reduced drought tolerance.

In response to osmotic stress, stored ABA-GE undergoes hydrolysis catalyzed by β-glucosidases (BGs), which can rapidly increase pools of biologically active ABA. Lee et al. (2006) identified Arabidopsis AtBG1, which hydrolyses ABA-GE to produce active ABA in the ER. Xu et al. (2012) characterized a vacuolar BG homolog AtBG2, and observed that transgenic Arabidopsis with a T-DNA insertion in *AtBG2* was more sensitive to salt stress and dehydration, whereas *AtBG2* overexpression resulted in enhanced resistance. Burla et al. (2013) observed import of ABA-GE into vacuoles isolated from Arabidopsis mesophyll cells, which could be mediated by several, possibly multispecific, transporters involved in the general vacuolar sequestration of conjugated metabolites.

Lee et al. (2006) observed that plants with ectopic *AtBG1* accumulate higher levels of ABA and have enhanced tolerance to abiotic stress, whereas *atbg1* mutation causes a lower level of ABA accumulation, defective stomatal movement, early germination, and abiotic stress-sensitive phenotypes. Overexpression of *AtNCED3* rescued the *atbg1* mutant phenotype (Xu et al., 2012). However, loss of AtBG1 delayed dehydration-induced ABA accumulation (Han, Watanabe, Shimada, & Sakamoto, 2020), suggesting that ABA-GE hydrolysis precedes the de novo biosynthesis of ABA. Transgenic plants constitutively overexpressing *UGT71B6* exhibited massive accumulation of ABA-GE and reduced levels of the oxidative metabolites PA and DPA, but the levels of free ABA were only marginally affected (Priest et al., 2006). It appears that the reduced level of ABA in these transgenic plants is complemented by ABA-GE hydrolysis pathways. Dong et al. (2014) generated a transgenic Arabidopsis that ectopically expressed *UGT71B6* in the *atbg1* mutant background, and observed that the transgenic plant exhibited more severe ABA-deficient phenotypes. Transgenic Arabidopsis with RNA interference (RNAi)-mediated UGT knockdown that have low levels of *UGT71B6*, *UGT71B7*, and *UGT71B8* transcripts display hypersensitivity to exogenous ABA and salt stress (Dong et al., 2014). In comparison to the de novo biosynthesis of ABA, the hydrolysis of ABA-GE to ABA occurs in one step, leading to a rapid increase in ABA concentration, which confers a prompt response to osmotic stress.

AtBG1 was rapidly polymerized under dehydration conditions, which resulted in a 4-fold increase in enzymatic activity (Lee et al., 2006). It appears that the polymerization of AtBG1 provides a mechanism by which plants rapidly adjust their ABA levels in response to abiotic stress. Recently, Han et al. (2020) observed that AtBG1 (BGLU18) is localized in ER-derived bodies that respond dynamically to abiotic stress via changes in number and size. Under stress conditions, AtBG1 distribution shifted toward microsomes, which was accompanied by increasing AtBG1-mediated ABA-GE hydrolytic activity and ABA levels in leaf petioles of

Arabidopsis. Based on this result, it was proposed that dynamic changes in ER status are involved in the rapid activation of BG1-mediated ABA-GE hydrolysis, to modulate ABA homeostasis and abiotic stress responses (Han et al., 2020).

Two opposing catabolic pathways, hydroxylation and ABA-GE conjugation, contribute to the homeostasis of cellular ABA levels. Hydroxylation of ABA by CYP707As leads to irreversible degradation of ABA, whereas ABA-GE conjugation retains a metabolite that can be converted back to active ABA (Lee et al., 2006; Xu et al., 2012). Moreover, the ABA-GE (de)conjugation pathway is closely coordinated with the hydroxylation pathway at the transcriptional level. Dong et al. (2014) showed that *CYP707As* were expressed at lower levels in *UGT71B6*-overexpressing plants than in *UGT* RNAi plants with suppressed expression of *UGT71B6*, *UGT71B7*, and *UGT71B8*.

3. Modulation of the ABA signaling network

3.1 ABA transportation

ABA is mainly biosynthesised in leaf vascular tissues. The expression of ABA biosynthetic genes *AtNCEDs*, *AtABA2*, and *AtAAO3* is observed in vascular bundles in turgid tissues (Cheng et al., 2002; Koiwai, Nakaminami, Seo, Mitsuhashi, & Koshiba, 2004; Tan et al., 2003). Koiwai et al. (2004) performed immunohistochemical analyses, and found that the Arabidopsis AAO3 protein was abundantly localized in phloem companion cells and xylem parenchyma cells in vascular tissues. In addition, the AAO3 protein was also present in the guard cells of dehydrated leaves, indicating that guard cells themselves are able to synthesize ABA. In accordance with this observation, *AtNCED2* and *AtNCED3* transcripts are expressed in the guard cells of senescent tissues (Tan et al., 2003). Bauer et al. (2013) reported that ABA biosynthesis in guard cells was sufficient for stomatal closure in response to low air humidity.

The ABA molecules synthesized in vascular tissue are transported to guard cells through exporting and importing transporters located in the plasma membrane (Kuromori, Seo, & Shinozaki, 2018; Merilo et al., 2015). In Arabidopsis, three transporter families have been identified: ATP binding cassette (ABC), detoxification efflux carriers (DTX)/multidrug and toxic compound extrusion (MATE) proteins, and nitrate transporter 1/peptide transporter family (NPF). ABCG25, a member of the ABC subgroup G transporters, exports ABA to the apoplastic space in vascular tissues

(Kuromori et al., 2010). Osakabe, Yamaguchi-Shinozaki, Shinozaki, and Tran (2014) proposed that ABA is imported into guard cells via ABCG40 in response to osmotic stress. In addition, NPF4.6, also called ABA importing transporter 1 (AIT1), mediates ABA uptake into guard cells and regulates stomatal closure (Kanno et al., 2012; Shimizu, Kanno, Suzuki, Watanabe, & Seo, 2021). Zhang et al. (2014) reported that AtDTX50 is predominantly localized in the plasma membrane of guard cells, and functions as an ABA efflux carrier. In accordance with the location and function of this transporter, *dtx50* mutant plants are more tolerant to drought with lower stomatal conductance.

The treatment of plants with ABA triggers the expression of genes encoding either exporting or importing transporters. The *ABCG25* gene is expressed by ABA and abiotic stress in vascular tissues (Kuromori et al., 2010), whereas the *ABCG40* gene is expressed in guard cells (Kang et al., 2010). *AtDTX50* (Zhang et al., 2014) and *NPF4.6/AIT1* (Kanno et al., 2012; Shimizu et al., 2021) are expressed in vascular tissues and guard cells in response to exogenous ABA. Therefore, ABA regulates the expression of genes encoding all proteins that are necessary for its long-distance transportation.

3.2 ABA-induced stomatal closure

In guard cells, ABA molecules are perceived by soluble ABA receptors localized in different subcellular compartments including the cytosol and nucleus (Cotelle & Leonhardt, 2019; Dong, Park, & Hwang, 2015; Joshi-Saha, Valon, & Leung, 2011) (Fig. 2). A number of synonymous ABA receptors have been identified in Arabidopsis and termed pyrabactin resistance (PYR), PYR-like (PYL), or regulatory component of ABA receptor (RCAR) (Ma et al., 2009; Nishimura et al., 2010; Park et al., 2009; Rodriguez, Lozano-Juste, & Albert, 2019). The Arabidopsis PYR/PYL/RCAR receptor family consists of 14 members, nearly all of which (except PYL13) are capable of ABA perception (Klingler, Batelli, & Zhu, 2010). Six ABA receptor genes, *PYR1*, *PYL1*, *PYL2*, *PYL4*, *PYL5* and *PYL8/RCAR3*, are expressed in guard cells to mediate stomatal closure (Dittrich et al., 2019; González-Guzmán et al., 2012).

Nishimura et al. (2010) showed that the Arabidopsis quadruple knockout mutant *pyr1/pyl1/pyl2/pyl4* was insensitive to ABA-induced stomatal closure and ABA-mediated inhibition of stomatal opening. Moreover, González-Guzmán et al. (2012) observed that Arabidopsis mutants lacking three, four, five, or six of the receptor genes (*PYR1, PYL1, PYL2, PYL4,*

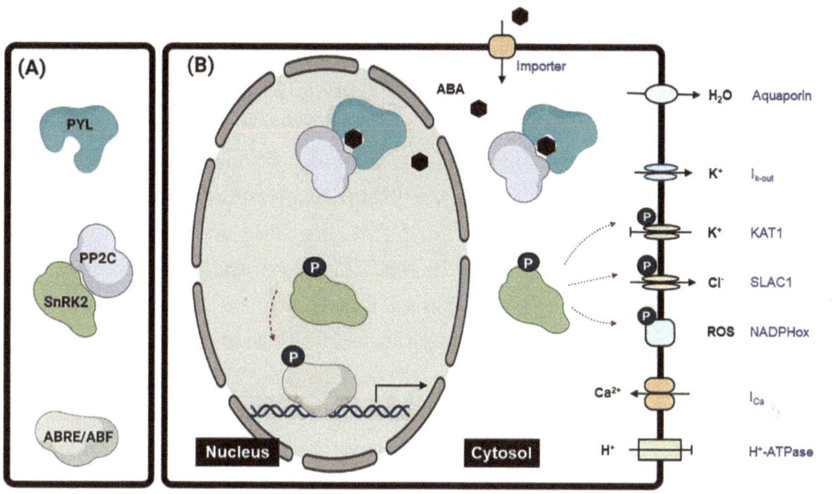

Fig. 2 ABA-signaling pathway in the guard cells. (A) Repression of ABA signaling for stomatal opening. In the absence of ABA, clade A PP2Cs physically interact with the SnRK2s to diminish kinase activity via dephosphorylation. (B) Activation of ABA signaling for stomatal closure. Under osmotic stress conditions, the PP2Cs interact with ABA receptors (PYLs) to capture ABA, which release free SnRK2s from the PP2C-SnRK2 complexes. In the cytosol, active SnRK2s subsequently phosphorylate K^+ channel 1 (KAT1), slow anion channel 1 (SLAC1) and NADPH oxidase anchored at the plasma membrane. In the nucleus, active SnRK2s phosphorylate the downstream transcription factors AREB/ABFs, which trigger the transcription of numerous ABA-responsive genes.

PYL5, and *PYL8/RCAR3*) exhibited gradually increasing stomatal conductance, suggesting that this family of receptors quantitatively regulates the stomatal aperture. However, overexpression of *PYL8/RCAR3* was enough to lead to enhanced drought resistance of Arabidopsis (Saavedra et al., 2010). Dittrich et al. (2019) reported that PYL2 is sufficient for ABA-induced response in guard cells, whereas PYL4 and PYL5 are essential in the responses to CO_2. These results suggest that response specificity is achieved by the ABA signals activating different members of the PYR/PYL/RCAR family of receptors.

In addition to receptor proteins, the clade A type 2C protein phosphatases (PP2Cs) have a direct physical interaction with ABA in the complex with ABA receptors, which is required for the high-affinity binding of ABA (Miyazono et al., 2009; Yin et al., 2009). Without ABA under non-stress conditions, PP2Cs bind to a family of protein kinases known as sucrose non-fermenting 1-related protein kinase 2 (SnRK2), thereby blocking the ABA signaling pathway (Fujii & Zhu, 2012; Umezawa et al., 2009) (Fig. 2A). Under osmotic stress conditions, PP2Cs bind to ABA receptors, functioning as co-receptors to capture ABA, which results in the release of SnRK2s in the cytosol and nucleus (Santiago et al., 2009; Umezawa et al., 2010) (Fig. 2B). Thus, PP2Cs exert dual roles, as positive and negative regulators, functioning as a switch at the center of the ABA signaling network (Fuchs, Grill, Meskiene, & Schweighofer, 2013; Jung et al., 2020). PP2C-SnRK2 and PYL-PP2C bindings are remarkably similar to each other. PP2C binds to the SnRK2 kinase domain and dephosphorylate a serine residue (Ser 175 in SnRK2.6/OST1) in the activation loop. Upon perception of ABA, receptor binds to PP2C by inserting the gate loop into the PP2C active cleft (Soon et al., 2012).

Among the nine protein phosphatases that have been classified into clade A PP2Cs in Arabidopsis, ABA insensitive 1 (ABI1), ABI2, ABA hypersensitive germination 1 (AHG1), AHG3/PP2CA, hypersensitive to ABA1 (HAB1) and HAB2 are involved in ABA signaling in response to osmotic stress (Schweighofer, Hirt, & Meskiene, 2004; Singh, Pandey, Srivastava, Tran, & Pandey, 2015; Xue et al., 2008). In particular, ABI1 and ABI2 are the main components of ABA signaling under osmotic stress conditions (Gosti et al., 1999; Rodriguez, Benning, & Grill, 1998). The dominant ABA response mutants of Arabidopsis, *abi1* and *abi2*, were originally isolated based on their ABA insensitivity, through observations of reduced seed dormancy and symptoms of withering (Koornneef, Reuling, & Karssen, 1984). However, it was later observed that knockout mutants of multiple PP2C genes exhibit significant ABA hypersensitivity (Rubio et al., 2009; Saez et al., 2006), indicating that their main functions are as negative regulators in the ABA signaling network. In addition, recessive *hab1-1* mutants exhibit enhanced ABA-mediated stomatal closure and ABA-responsive gene expression, indicating that HAB1 also acts as a negative regulator in the ABA signaling network (Hirayama & Shinozaki, 2007; Saez et al., 2004).

Each member of PP2Cs combines with a particular PYR/PYL/RCAR receptor and functions as a co-receptor of ABA in a holoreceptor complex. Szostkiewicz et al. (2010) observed that PYL8/RCAR3 conferred greater ABA sensitivity to the PP2C regulation than PYL9/RCAR1. In addition, regulation of the protein phosphatase activity by PYL9/RCAR1 and PYL8/RCAR3 was more sensitive to ABA for ABI1 than for ABI2. Thus, different combinations of PYR/PYL/RCAR receptors and PP2Cs influence ABA binding affinity, and therefore affect the ABA sensitivity of the whole plant. Tischer et al. (2017) analyzed 126 possible receptor-PP2C combinations for their capacity to regulate ABA signaling by transient expression in Arabidopsis protoplasts. They observed remarkable capability of PP2Cs to form functional holoreceptor complexes with many different receptors, overall similar PP2C regulation by individual receptors, and different potency of receptor subfamilies in regulating the ABA responses. Based on these observations, it was suggested that combinatorial assembly of the receptor-PP2C complexes act in concert to modulate and fine-tune ABA responses.

Arabidopsis contains 10 members of subgroup III SnRK2s, i.e., SnRK2.1–SnRK2.10. The triple mutation (*snrk2.2/2.3/2.6*) nearly blocks all of the main ABA responses (Fujii & Zhu, 2009; Fujita et al., 2009; Nakashima et al., 2009); thus, SnRK2.2, SnRK2.3, and SnRK2.6/OST1 are regarded as primary regulators of ABA signaling. ABA treatment or osmotic stress stimulates phosphorylation of Ser 175 in the activation loop of SnRK2.6/OST1 (Belin et al., 2006). It has been proposed that released SnRK2.6/OST1 autophosphorylates itself to recover full kinase activity (Belin et al., 2006; Fujii & Zhu, 2012). However, Lin et al. (2021) recently showed that Raf-like protein kinases (RAFs) phosphorylate SnRK2.6/OST1, and then the activated SnRK2 transphosphorylates other free SnRK2s to amplify the ABA response.

In the cytosol of guard cells, activated SnRK2s phosphorylate ion channels anchored at the plasma membrane of guard cells (Saito & Uozumi, 2019; Vahisalu et al., 2008). The SnRK2-mediated phosphorylation results in activation of slow anion channel 1 (SLAC1) to promote efflux of Cl^-, and inhibition of K^+ channel 1 (KAT1) to reduce influx of K^+. In addition to ion channels, the activated SnRK2s phosphorylate and activate NADPH oxidase, which is responsible for ABA-promoted ROS accumulation in guard cells, causing stomata to close (Chen et al., 2020; Kwak et al., 2003; Sirichandra et al., 2009). Pei et al. (2000) reported that ROS (e.g., H_2O_2) activates Ca^{2+} channels in the plasma membrane to trigger the accumulation of Ca^{2+} in the cytosol of guard cells. Under elevated Ca^{2+} concentration, efflux of Cl^- ions is enhanced by Ca^{2+}-dependent pathway to promote ABA-induced stomatal closure (Chen et al., 2020; Marusig & Tombesi, 2020). The ion displacement causes plasma membrane depolarisation and activation of the outward K^+ channel ($I_{k\text{-}out}$). For depolarization of the guard cell membrane, Hayashi, Inoue, Takahashi, and Kinoshita (2011) observed that ABA inhibits blue light-induced phosphorylation of guard cell H^+-ATPase. The overall efflux of ions induces water efflux through aquaporins by osmosis (Marusig & Tombesi, 2020), which is linked to loss of turgor in guard cells and closure of the stomatal aperture

(Fan, Zhao, & Assmann, 2004; Munemasa et al., 2015). In addition, it is notable that ABA enhances the ROS production in guard cells, whereas ROS promotes ABA biosynthesis, which constitutes a positive feedback loop to mediate stomatal closure (Mittler & Blumwald, 2015; Sah et al., 2016).

In parallel, in the nucleus of guard cells, active SnRK2s phosphorylate and thereby activate a number of basic-domain leucine zipper transcription factors termed ABA-responsive element (ABRE)-binding (AREB) proteins or ABRE-binding factors (ABFs) (Choi, Hong, Ha, Kang, & Kim, 2000; Furihata et al., 2006; Uno et al., 2000). Activated AREB/ABFs induce the transcription of ABA-responsive genes whose products are involved in a variety of osmotic stress responses (Fujita, Fujita, Shinozaki, & Yamaguchi-Shinozaki, 2011). Among the nine members of AREB/ABFs in Arabidopsis, ABF1, AREB1/ABF2, AREB2/ABF4, and ABF3 act as master transcription factors in ABA signaling for osmotic stress tolerance (Yoshida et al., 2015). Fujii et al. (2009) reconstituted ABA-triggered phosphorylation of AREB1/ABF2 in vitro by combining PYR1, ABI1, and SnRK2.6/OST1, demonstrating that these components constitute the core of the ABA signaling network.

Another bZIP transcription factor ABI5 (ABA insensitive 5) also controls expression of stress-responsive genes. Collin, Daszkowska-Golec, and Szarejko (2021) reviewed recent reports regarding the roles of ABI5 and AREB/ABFs in abiotic stress responses, and proposed that they are essential for plant adaptation to stress at different developmental stages. ABI5 is mostly active in seeds and prevents germination and post-germinative growth under stressful conditions, whereas AREB/ABFs act redundantly in regulating genes that control physiological processes in response to stress during vegetative growth.

3.3 Regulation of PP2C gene transcription

High expression of PP2Cs plays a role in a negative feedback mechanism that desensitizes plants to conditions that enhance ABA signaling (Fuchs et al., 2013; Raghavendra, Gonugunta, Christmann, & Grill, 2010). The expression of PP2C genes is suppressed or maintained at basal level under normal conditions, whereas it is enhanced in response to high ABA levels and salt stress (Fujita et al., 2011; Nguyen, Jung, & Cheong, 2019). Some MYB transcription factors, including AtMYB20 and AtMYB44, act as repressors of PP2C gene transcription in Arabidopsis. Cui et al. (2013) reported that the expression of a group of PP2C genes (*ABI1*, *ABI2* and *PP2CA*) is suppressed in *AtMYB20*-overexpressing transgenic lines, while it is enhanced in *AtMYB20*-repression lines in response to salt treatment. Jung et al. (2008) observed that transgenic Arabidopsis overexpressing *AtMYB44* exhibited increased sensitivity to ABA, more rapid ABA-induced stomatal closure, and enhanced tolerance to high-salinity and drought conditions. Microarray and northern blot analyses revealed that the salt-induced expression of a group of PP2C genes (*ABI1*, *ABI2*, *AtPP2CA*, *HAB1*, and *HAB2*) was significantly repressed in the transgenic plants.

The chromatin immunoprecipitation (ChIP) assay revealed that AtMYB44 binds to the promoters of Arabidopsis PP2C genes (*ABI1*, *ABI2*, and *HAI1*), and suppresses gene transcription in a signal-independent manner (Nguyen & Cheong, 2018). AtMYB44 contains the amino acid sequence LSLSL, a putative ethylene-responsive element binding factor-associated amphiphilic repression (EAR) motif. Yeast two-hybrid and bimolecular fluorescence complementation assays have revealed that AtMYB44 physically interacts with TOPLESS-related 1 (TPR1) and TPR3 corepressors through the EAR motif (Nguyen & Cheong, 2018). TPR repressors recruit histone deacetylases (HDAs) to compact targeted chromatin regions, leading to the repression of related genes (Kagale & Rozwadowski, 2011). Thus, these results suggest that AtMYB44 forms a complex with TPR corepressors and recruit histone deacetylases (HDAs) to suppress PP2C gene transcription (Fig. 3A). The salt-induced increase in histone H3 acetylation levels in the promoter and transcription start site (TSS) proximal regions of PP2C genes were markedly lower in *AtMYB44*-overexpressing transgenic plants than in wild-type plants. Another repressor of PP2C gene transcription AtMYB20 (Cui et al., 2013) also contains an EAR motif at the C-terminal side of the catalytic domain.

In response to salt stress, AtMYB44 binding to PP2C gene promoters is significantly reduced, whereas the gene transcription levels of the genes are increased (Nguyen et al., 2019). In parallel, Wang et al. (2019) performed ChIP assay to demonstrate that AREB/ABF transcription factors (ABF1 to ABF4) bind to the promoters of PP2C genes (*ABI1* and *ABI2*) and induce gene transcription upon exogenous ABA treatment (Fig. 3B). Supporting the positive role of AREB/ABFs in PP2C gene expression, Arabidopsis *abf3* plants displayed markedly reduced salt-induced increases in PP2C gene (*ABI1*, *ABI2*, and *HAI1*) transcription (Nguyen et al., 2019). These data indicate that AREB/ABFs contribute to negative feedback regulation of ABA signaling by activating PP2C gene expression, in addition to the

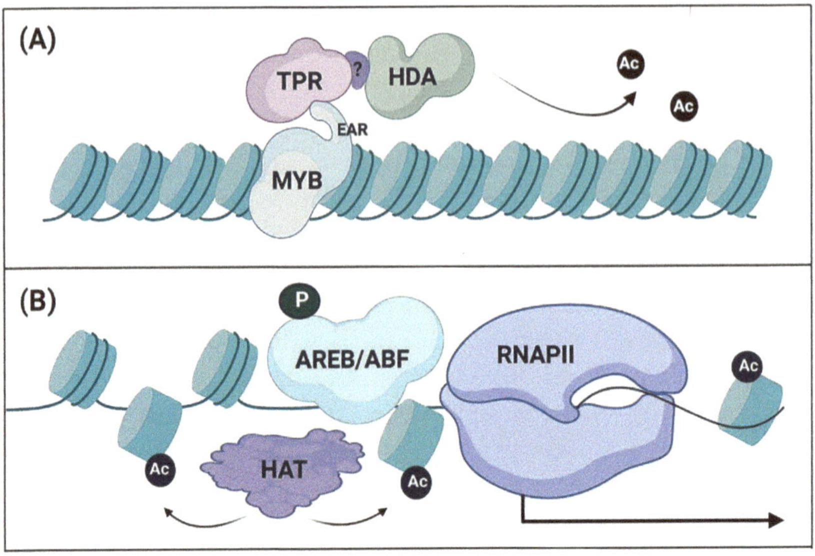

Fig. 3 A hypothetical model of chromatin remodeling for regulation of PP2C gene transcription. (A) Repression of PP2C gene transcription. Under normal conditions, EAR motif-containing MYB repressor (AtMYB44 and AtMYB20) interacts with a TOPLESS-related corepressor (TPR), which recruits histone deacetylase (HDA) to suppress PP2C gene transcription. (B) Activation of PP2C gene transcription. Under osmotic stress conditions, the repressor is released from the promoter and AREB/ABFs occupy the region. Histone acetyltransferase (HAT) complex is recruited to acetylate (Ac) histones to relax histone–DNA binding in the chromatin. Then, the RNA polymerase II (RNAPII) accesses to the open promoter region and initiates gene transcription.

ABA-induced expression of ABA-responsive genes. Therefore, AREB/ABFs play dual roles in the forward and backward regulation of ABA signaling. The AREB/ABF-mediated transcriptional upregulation of PP2Cs and PP2C-mediated inactivation of SnRK2 and AREB/ABFs constitute a tight regulatory loop in ABA signaling modulation.

Under salt stress conditions, histone H3 acetylation around the TSS regions of PP2C genes significantly increased (Nguyen et al., 2019). The level of histone acetylation, which is regulated by histone acetyltransferases (HATs) and HDAs, determines whether the chromatin is open or closed, thereby controlling access of RNA polymerase II (RNAPII) for transcriptional activation (Li et al., 2021; Luo et al., 2012). Li et al. (2019) showed that PtrAREB1-2 binds to ABRE motifs associated with *PtrNAC* genes and recruits the HAT complex, which enriches RNAPII at TSS region to initiate gene activation for drought response and tolerance in *Populus* species.

3.4 Expression of ABA-responsive genes

It appears that osmotic stress-induced expression of the responsive genes relies entirely upon ABA. Previous transcriptomic studies have shown that more than half of the genes regulated by ABA treatment are also induced under high-salinity or drought conditions, whereas ABA deficiency causes defects in the expression of those genes (Hoth et al., 2002; Seki et al., 2002; Todaka, Takahashi, Yamaguchi-Shinozaki, & Shinozaki, 2019). By analyzing the promoters of ABA-responsive genes, a conserved ABRE (PyACGTGG/TC) was identified (Busk & Pagès, 1998; Giraudat, Parcy, Bertauche, Gosti, & Leung, 1994). Subsequently, a number of AREB/ABF transcription factors were identified through yeast one-hybrid screening (Choi et al., 2000; Uno et al., 2000). Overexpression of their genes in Arabidopsis results in ABA hypersensitivity and enhances osmotic stress tolerance (Fujita et al., 2005; Kang, Choi, Im, & Kim, 2002; Kim, Kang, Cho, Park, & Kim, 2004). By contrast, the triple knockout mutant (*areb1/areb2/abf3*) displays impaired expression of ABA-responsive genes, resulting in increased sensitivity to osmotic stress (Yoshida et al., 2010).

AREB/ABF genes are markedly induced in response to exogenous ABA, high salt, or dehydration, resulting in the accumulation of endogenous AREB/ABF proteins (Fujita et al., 2005; Wang et al., 2019). Concurrently, group-A PP2C genes are highly induced in response to ABA and salt stress (Fujita et al., 2011; Nguyen et al., 2019; Xue et al., 2008). As described in the previous section, PP2C gene promoters are also targets of AREB/ABF transcription factors, which play dual roles in the forward and backward regulation of ABA signaling. In addition, *AtMYB44* transcripts accumulate to a greater degree under exogenous ABA, high-salinity, and dehydration (Jung et al., 2008, 2010), which contributes to the inhibition of PP2C gene expression and indirectly enhances ABA signaling. Therefore, ABA upregulates the genes encoding both positive and negative effectors of its own signaling network, constituting a regulatory loop that finely tunes ABA signaling homeostasis.

4. Modulation of ABA signaling by protein degradation

4.1 Degradation of ABA signaling components

Plants exposed to osmotic stress activate the ABA signaling network, triggering stomatal closure and expression of ABA-responsive genes. However, after achieving stress tolerance, plants should terminate or attenuate the ABA signaling pathway. Cellular protein homeostasis (proteostasis) can be achieved by selective degradation of the regulatory or effector molecules involved in the ABA signaling pathway (Jurkiewicz & Batoko, 2018; Sirko, Wawrzyńska, Brzywczy, & Sieńko, 2021). The ABA signaling network can be desensitized by the degradation of positive effectors, i.e., ABA receptors, SnRK2s, and ABRE/ABFs. In contrast, degradation of PP2Cs can facilitate ABA signaling by preventing excessive accumulation of the negative effectors.

Degradation of ABA signaling components mainly occurs through the ubiquitin proteasome system (UPS) (Ali et al., 2020; Coego et al., 2021; Stone, 2019). Conserved 8 kDa ubiquitin peptides are covalently attached to ubiquitin-binding proteins, and directly target proteins to the 26S proteasome for degradation (Buetow & Huang, 2016). The UPS proceeds via sequential reactions mediated by three sets of enzymes: ubiquitin-activating enzyme (E1), ubiquitin-conjugating enzyme (E2), and ubiquitin protein ligase (E3) (Hershko & Ciechanover, 1998; Komander & Rape, 2012). The Arabidopsis genome encodes numerous UPS components, representing about 6% of the total transcriptomes including 2 E1 isoforms, 37 E2 isoforms, and more than 1400 E3 isoforms (Smalle & Vierstra, 2004; Vierstra, 2009). E3 ligases recruit specific proteins, thus conferring target specificity of these enzymes.

Based on the molecular mechanism underlying the interaction and transfer of the ubiquitin moiety to their substrate, E3 ligases are classified into four main types: homologous to the E6-AP carboxyl terminus (HECT), really interesting new gene (RING), U-box, and Cullin (Cul)-RING ligases (CRLs) (Vierstra, 2009). HECT, RING, and U-box ligases act as a single subunit by directly interacting with their substrate, while CRLs are multi-subunit ligase complexes assembled using scaffold proteins (Cul1, Cul3a/3b, or Cul4) that bind the RING box-1 protein for the recruitment of E2 ubiquitin-conjugating enzymes. For multimeric CRLs, substrate adaptors are required to recognize the target proteins. Cul1-based E3 complexes are denoted as SKP1-Cullin-F box (SCF), whereas Cul3- and Cul4-based complexes are referred to as CRL3 and CRL4, respectively (Coego et al., 2021).

4.2 Selective degradation of ABA receptors

Selective degradation of ABA receptors is achieved redundantly by several different types of E3 ligases (Table 1). Zhao et al. (2017) reported that the U-box E3 ligases plant U-box 22 (PUB22) and PUB23 mediate the

Table 1 Arabidopsis E3 ligases and substrate adaptors that specifically promote the degradation of core ABA signaling components by ubiquitination.

Component	Specific target	E3-adapter (E2)	References
ABA Receptors	PYL9/RCAR1	PUB22, PUB23	Zhao et al. (2017)
	PYL9/RCAR1	CRL4-RAE1	Li et al. (2018)
	PYL8/RCAR3	SCF-RIFP1	Li et al. (2016)
	PYL4, PYL8/ RCAR3, PYL9/ RCAR1,	CRL4-DDA1	Irigoyen et al. (2014)
	PYR1, PYL4	RFA1, RFA4 (UBC26)	Fernandez et al. (2020)
	PYR1, PYL4	RSL1	Bueso et al. (2014)
PP2Cs	ABI1	PUB12, PUB13	Kong et al. (2015)
	ABI1	AIRP3 (UBC27)	Pan et al. (2020)
	PP2CA	RGLG1, RGLG5	Wu et al. (2016)
	PP2CA	RGLG1	Belda-Palazon et al. (2019)
	PP2CA	PIR1, PIR2	Baek, Lim, Luan, & Lee, 2019)
	PP2CA, ABI1, ABI2, HAB1	CRL3-BPM	Julian et al. (2019)
SnRK2s	SnRK2.3	SCFAtPP2-B11	Cheng et al. (2017)
	SnRK2.6/OST1	HOS15	Ali et al. (2019)
Transcription Factors	ABI5	AFP	Lopez-Molina, Mongrand, Kinoshita, & Chua (2003)
	ABI5	DWA1, DWA2	Lee et al. (2010)
	ABI5	ABD1	Seo et al. (2014)
	ABI5,	KEG	Liu & Stone (2013)
	ABF1, ABF3	KEG	Chen, Liu, Stone, & Callis (2013)

degradation of PYL9/RCAR1 receptors. RNA export factor 1 in Arabidopsis (AtRAE1), a WD40 repeat protein, is involved in the degradation of ABA receptor PYL9/RCAR1 and negatively regulates ABA signaling in Arabidopsis (Li et al., 2018). The Arabidopsis F-box E3 ligase RCAR3-interacting F-box protein 1 (RIFP1), an adaptor subunit of the SCF ubiquitin ligase complex, plays a negative role in ABA signaling by facilitating the degradation of PYL8/RCAR3 (Li et al., 2016). Irigoyen et al. (2014) showed that DET1-, DDB1-associated 1 (DDA1), the ubiquitin ligase substrate adaptor that provides substrate specificity for CRL4s, binds to PYL8/RCAR3, as well as PYL4 and PYL9/RCAR1, facilitating its proteasomal degradation. DDA1-mediated destabilization of PYL8/RCAR3 is counteracted by ABA, which protects PYL8/RCAR3 by limiting its ubiquitination. Fernandez et al. (2020) showed that Arabidopsis RING finger ABA-related E3 ubiquitin ligases target PYR1 and PYL4 for degradation in the cytosol and nucleus. In addition, the single-subunit RING-type E3 ubiquitin ligase RING finger of seed longevity 1 (RSL1) is localized at the plasma membrane; it targets PYR1 and PYL4 receptors and triggers clathrin-mediated endocytosis (Bueso et al., 2014).

In addition to conventional degradation of the tagged components via the UPS, a number of ABA receptors are degraded through an endocytic/vacuolar pathway that requires components of the endosomal sorting complex required for transport (ESCRT) machinery (Gao, Zhuang, Shen, & Jiang, 2017). Vacuolar protein sorting 23A (VP23A) and FYVE domain-containing protein 1 (FYVE1/FREE1) of ESCRT-I complex, and ALG-2 interacting protein-X (ALIX)-associated protein of ESCRT-III have been reported as ESCRT components that mediate degradation of ABA receptors (Coego et al., 2021) (Table 2). The ESCRT-I component VPS23A, an ubiquitin

Table 2 Arabidopsis ESCRT components that mediate the endocytic/vacuolar degradation pathway that specifically promote the degradation of ABA receptors.

Specific target	ESCRT	ESCRT component	References
PYR1, PYL4	ESCRT-I	VPS23A	Yu et al. (2016, 2020)
PYL4	ESCRT-I	FYVE1/FREE1	Belda-Palazon et al. (2016)
PYL4, PYL5, PYL8/RCAR3, PYL9/RCAR1	ESCRT-III	ALIX	García-León et al. (2019)

E2-like protein, directly binds to PYR1 and PYL4 to trigger vacuole-mediated degradation, thereby negatively regulating ABA signaling (Yu et al., 2016, 2020). Disruption of VPS23A enhances the activity of SnRK2.6/OST1 upon ABA treatment. FYVE1/FREE1 interacts with PYL4 and the E3 ubiquitin ligase RSL1, and mediates its delivery to the vacuolar degradation pathway (Belda-Palazon et al., 2016). This process involves clathrin-mediated endocytosis and trafficking of PYL4 through the ESCRT pathway, which attenuate ABA signaling. García-León et al. (2019) reported that ALIX regulates the degradation of PYL4, PYL5, PYL8/RCAR3, and PYL9/RCAR1. Partial loss-of-function *alix-1* mutants exhibit ABA hypersensitivity during stomatal closure and growth. The ALIX-1 mutant protein displays reduced interactions with VPS23A and ABA receptors, providing a negative feedback mechanism for ABA signaling.

4.3 ABA-induced degradation of PP2Cs

PP2Cs in the cytosol and nucleus are also degraded through the UPS system upon exogenous ABA application, positively modulating ABA signaling by enhancing SnRK2 kinase activity. For instance, the U-box E3 ligases PUB12 and PUB13 ubiquitinate ABI1 (Kong et al., 2015) (Table 1). Because PP2Cs interact with ABA receptors to capture ABA molecules, the ubiquitination of ABI1 is strictly receptor-dependent for the dimeric PYR1 receptor and enhanced for the monomeric PYL4 and PYL9/RCAR1 receptors. Pan et al. (2020) reported that both UBC27 (an E2) and AIRP3 (a RING E3 ligase) interact with ABI1, and the UBC27-AIRP3 ubiquitination complex modulates ABA signaling for drought tolerance by promoting the degradation of ABI1 in Arabidopsis. ABA activates the expression of *UBC27*, inhibits the proteasome degradation of UBC27, and enhances the interaction between UBC27 and ABI1 to increase PP2C activity. In addition, Wu et al. (2016) reported that the RING finger E3 ligases RGLG1 and RGLG5 regulate the turnover of PP2CA. Belda-Palazon et al. (2019) reported that ABA enhances RGLG1-PP2CA interaction and hence PP2CA degradation. Baek et al. (2019) also found that the RING finger E3 ligases PIR1 and PIR2 mediate PP2CA degradation to enhance the ABA response in Arabidopsis. In addition, Julian et al. (2019) demonstrated that BTB/POZ and MATH domain proteins (BPMs), which are substrate adaptors of the CRL3s, interact with PP2CA as well as with ABI1, ABI2, and HAB1, and modulate ABA-induced stomatal closure.

4.4 Ubiquitination of SnRK2s and transcription factors (ABI5 and AREB/ABFs)

ABA induces the expression of the *AtPP2-B11* gene encoding an F-box E3 ubiquitin ligase that specifically promotes degradation of SnRK2.3 (Table 1); this attenuates ABA signaling and the abiotic stress response in Arabidopsis (Cheng et al., 2017). Ali et al. (2019) showed that the HOS15, an E3 ligase residing in the CUL4-based E3 ubiquitin ligase complex, regulates SnRK2.6/OST1 protein stability, resulting in the desensitization of ABA signaling in Arabidopsis. Upon ABA signaling, HOS15-induced degradation of SnRK2.6/OST1 is diminished, thus promoting the ABA responses.

UPS-mediated degradation of the transcription factors (ABI5 and AREB/ABFs) responsible for ABA-responsive gene expression has also been studied in detail (Joo, Baek, Lim, & Lee, 2021). ABI five binding protein 1 (AFP1) attenuates ABA signals by targeting ABI5 for UPS-mediated proteolysis in nuclear bodies (Lopez-Molina et al., 2003). DWA1 and DWA2, two Arabidopsis DDB1 binding WD40 (DWD) protein components of CUL4-based E3 ligases, also act together as negative regulators of ABA signal transduction by directly interacting with ABI5 to mediate degradation by ubiquitination (Lee et al., 2010). Additionally, Seo et al. (2014) found that DDB1-CUL4-associated factor (DCAF)-CUL4 E3 ubiquitin ligase is assembled with ABA-hypersensitive DCAF1 (ABD1) and directly binds to and affects the stability of ABI5 in the nucleus, thereby acting as a negative regulator of ABA responses.

The RING-type E3 ligase KEEP ON GOING (KEG) mediates the cytoplasmic degradation of ABI5 in the absence of ABA (Liu & Stone, 2013; Stone, Williams, Farmer, Vierstra, & Callis, 2006). Chen et al. (2013) reported that KEG affects the proteolysis of ABF1 and ABF3 in Arabidopsis, and ABA treatment of seedlings leads to slowed proteolysis and rapid accumulation of AREB/ABFs. However, the proteolysis of AREB/ABFs and effect of ABA treatment still occurs in *keg* seedlings, which implies additional E3 ligases that redundantly mediate proteolysis of AREB/ABFs.

5. Summary and concluding remarks

High-salinity conditions caused by habitual and prolonged drought are one of the most serious threats to crop productivity. Understanding the mechanisms underlying salt stress tolerance and its application to crop breeding are now among the most pressing issues in botanical research and plant biotechnology. ABA is a crucial plant hormone that plays a key role in the salt stress tolerance response. ABA induces closure of the stomata on the leaf epidermis to limit transpiration, thereby preventing the loss of water under salt-induced osmotic stress conditions. However, the stomatal pores should be opened to uptake carbon dioxide for photosynthesis, even under stressful conditions. Thus, it is not always favorable to artificially enhance ABA biosynthesis and ABA signaling by gene modification or editing technology. Therefore, it is essential to obtain deep insight into the strategies that plants exert in nature to cope with adverse environments without affecting development and growth. Plants finely control ABA homeostasis and the ABA signaling process during and after encountering stressful conditions. ABA regulates the stomatal aperture by inducing both positive and negative mechanisms that play roles in ABA metabolism, ABA signaling, and ABA-induced gene transcription. In addition, the ABA signaling network can be desensitized by degradation of core signaling components in an ABA-dependent manner. More detailed studies on ABA-mediated salt stress tolerance are required to develop new strategies to breed higher quality crops that can withstand adverse climatic conditions. An important point that should be considered is that plants encounter not only osmotic stress, but also abnormal temperatures (heat and cold) and biotic (pathogens and insects) stresses simultaneously in the field. Thus, intercommunication between different hormone-mediated signaling pathways under multiple stresses should be more extensively investigated in the future.

Acknowledgment

This work was supported by the National Research Foundation of Korea under the Basic Science Research Program (grant number NRF-2020R1I1A1A01070089).

References

Ali, A., Kim, J. K., Jan, M., Khan, H. A., Khan, I. U., Shen, M., et al. (2019). Rheostatic control of ABA signaling through HOS15-mediated OST1 degradation. *Molecular Plant, 12*, 1447–1462.

Ali, A., Pardo, J. M., & Yun, D.-J. (2020). Desensitization of ABA-signaling: The swing from activation to degradation. *Frontiers in Plant Science, 11*, 379.

Amin, I., Rasool, S., Mir, M. A., Wani, W., Masoodi, K. Z., & Ahmad, P. (2021). Ion homeostasis for salinity tolerance in plants: A molecular approach. *Physiologia Plantarum, 171*, 578–594.

Arif, Y., Singh, P., Siddiqui, H., Bajguz, A., & Hayat, S. (2020). Salinity induced physiological and biochemical changes in plants: An omic approach towards salt stress tolerance. *Plant Physiology and Biochemistry, 156*, 64–77.

Audran, C., Borel, C., Frey, A., Sotta, B., Meyer, C., Simonneau, T., et al. (1998). Expression studies of the zeaxanthin epoxidase gene in *Nicotiana plumbaginifolia*. *Plant Physiology*, *118*, 1021–1028.

Baek, W., Lim, C. W., Luan, S., & Lee, S. C. (2019). The RING finger E3 ligases PIR1 and PIR2 mediate PP2CA degradation to enhance abscisic acid response in *Arabidopsis*. *The Plant Journal*, *100*, 473–486.

Bauer, H., Ache, P., Lautner, S., Fromm, J., Hartung, W., Al-Rasheid, K. A. S., et al. (2013). The stomatal response to reduced relative humidity requires guard cell-autonomous ABA synthesis. *Current Biology*, *23*, 53–57.

Belda-Palazon, B., Julian, J., Coego, A., Wu, Q., Zhang, X., Batistic, O., et al. (2019). ABA inhibits myristoylation and induces shuttling of the RGLG1 E3 ligase to promote nuclear degradation of PP2CA. *The Plant Journal*, *98*, 813–825.

Belda-Palazon, B., Rodriguez, L., Fernandez, M. A., Castillo, M.-C., Anderson, E. A., Gao, C., et al. (2016). FYVE1/FREE1 interacts with the PYL4 ABA receptor and mediates its delivery to the vacuolar degradation pathway. *The Plant Cell*, *28*, 2291–2311.

Belin, C., De Franco, P.-O., Bourbousse, C., Chaignepain, S., Schmitter, J. M., Vavasseur, A., et al. (2006). Identification of features regulating OST1 kinase activity and OST1 function in guard cells. *Plant Physiology*, *141*, 1316–1327.

Boyer, G. L., & Zeevaart, J. A. D. (1982). Isolation and quantitation of β-D-glucopyranosyl abscisate from leaves of *Xanthium* and spinach. *Plant Physiology*, *70*, 227–231.

Bueso, E., Rodriguez, L., Lorenzo-Orts, L., Gonzalez-Guzman, M., Sayas, E., Muñoz-Bertomeu, J., et al. (2014). The single-subunit RING-type E3 ubiquitin ligase RSL1 targets PYL4 and PYR1 ABA receptors in plasma membrane to modulate abscisic acid signaling. *The Plant Journal*, *80*, 1057–1071.

Buetow, L., & Huang, D. T. (2016). Structural insights into the catalysis and regulation of E3 ubiquitin ligases. *Nature Reviews Molecular Cell Biology*, *17*, 626–642.

Burla, B., Pfrunder, S., Nagy, R., Francisco, R. M., Lee, Y., & Martinoia, E. (2013). Vacuolar transport of abscisic acid glucosyl ester is mediated by ATP-binding cassette and proton-antiport mechanisms in Arabidopsis. *Plant Physiology*, *163*, 1446–1458.

Busk, P. K., & Pagès, M. (1998). Regulation of abscisic acid-induced transcription. *Plant Molecular Biology*, *37*, 425–435.

Chaves, M. M., Flexas, J., & Pinheiro, C. (2009). Photosynthesis under drought and salt stress: Regulation mechanisms from whole plant to cell. *Annals of Botany*, *103*, 551–560.

Chen, K., Li, G.-J., Bressan, R. A., Song, C.-P., Zhu, J.-K., & Zhao, Y. (2020). Abscisic acid dynamics, signaling, and functions in plants. *Journal of Integrative Plant Biology*, *62*, 25–54.

Chen, Y. T., Liu, H., Stone, S., & Callis, J. (2013). ABA and the ubiquitin E3 ligase KEEP ON GOING affect proteolysis of the *Arabidopsis thaliana* transcription factors ABF1 and ABF3. *The Plant Journal*, *75*, 965–976.

Cheng, W.-H., Endo, A., Zhou, L., Penney, J., Chen, H.-C., Arroyo, A., et al. (2002). A unique short-chain dehydrogenase/reductase in Arabidopsis glucose signaling and abscisic acid biosynthesis and functions. *The Plant Cell*, *14*, 2723–2743.

Cheng, C., Wang, Z., Ren, Z., Zhi, L., Yao, B., Su, C., et al. (2017). SCFAtPP2-B11 modulates ABA signaling by facilitating SnRK2.3 degradation in *Arabidopsis thaliana*. *PLoS Genetics*, *13*, e1006947.

Choi, H., Hong, J., Ha, J., Kang, J., & Kim, S. Y. (2000). ABFs, a family of ABA-responsive element binding factors. *Journal of Biological Chemistry*, *275*, 1723–1730.

Coego, A., Julian, J., Lozano-Juste, J., Pizzio, G. A., Alrefaei, A. F., & Rodriguez, P. L. (2021). Ubiquitylation of ABA receptors and protein phosphatase 2C coreceptors to modulate ABA signaling and stress response. *International Journal of Molecular Sciences*, *22*, 7103.

Collin, A., Daszkowska-Golec, A., & Szarejko, I. (2021). Updates on the role of ABSCISIC ACID INSENSITIVE 5 (ABI5) and ABSCISIC ACID-RESPONSIVE ELEMENT

BINDING FACTORs (ABFs) in ABA signaling in different developmental stages in plants. *Cell*, *10*, 1996.
Cotelle, V., & Leonhardt, N. (2019). ABA signaling in guard cells. *Advances in Botanical Research*, *92*, 115–170.
Cui, M. H., Yoo, K. S., Hyoung, S., Nguyen, H. T. K., Kim, Y. Y., Kim, H. J., et al. (2013). An Arabidopsis R2R3-MYB transcription factor, AtMYB20, negatively regulates type 2C serine/threonine protein phosphatases to enhance salt tolerance. *FEBS Letters*, *587*, 1773–1778.
Cutler, A., & Krochko, J. (1999). Formation and breakdown of ABA. *Trends in Plant Science*, *4*, 472–478.
Dietz, K.-J., Sauter, A., Wichert, K., Messdaghi, D., & Hartung, W. (2000). Extracellular β-glucosidase activity in barley involved in the hydrolysis of ABA glucose conjugate in leaves. *Journal of Experimental Botany*, *51*, 937–944.
Dittrich, M., Mueller, H. M., Bauer, H., Peirats-Llobet, M., Rodriguez, P. L., Geilfus, C. M., et al. (2019). The role of Arabidopsis ABA receptors from the PYR/PYL/RCAR family in stomatal acclimation and closure signal integration. *Nature Plants*, *5*, 1002–1011.
Dong, T., Park, Y., & Hwang, I. (2015). Abscisic acid: Biosynthesis, inactivation, homoeostasis and signaling. *Essays in Biochemistry*, *58*, 29–48.
Dong, T., Xu, Z. Y., Park, Y., Kim, D. H., Lee, Y., & Hwang, I. (2014). Abscisic acid uridine diphosphate glucosyltransferases play a crucial role in abscisic acid homeostasis in Arabidopsis. *Plant Physiology*, *165*, 277–289.
Fan, L. M., Zhao, Z., & Assmann, S. M. (2004). Guard cells: A dynamic signaling model. *Current Opinion in Plant Biology*, *7*, 537–546.
Fernandez, M. A., Belda-Palazon, B., Julian, J., Coego, A., Lozano-Juste, J., Iñigo, S., et al. (2020). RBR-type E3 ligases and the ubiquitin-conjugating enzyme UBC26 regulate abscisic acid receptor levels and signaling. *Plant Physiology*, *182*, 1723–1742.
Frey, A., Audran, C., Marin, E., Sotta, B., & Marion-Poll, A. (1999). Engineering seed dormancy by the modification of zeaxanthin epoxidase gene expression. *Plant Molecular Biology*, *39*, 1267–1274.
Fuchs, S., Grill, E., Meskiene, I., & Schweighofer, A. (2013). Type 2C protein phosphatases in plants. *FEBS Journal*, *280*, 681–693.
Fujii, H., Chinnusamy, V., Rodrigues, A., Rubio, S., Antoni, R., Park, S.-Y., et al. (2009). *In vitro* reconstitution of an ABA signaling pathway. *Nature*, *462*, 660–664.
Fujii, H., & Zhu, J.-K. (2009). Arabidopsis mutant deficient in 3 abscisic acid-activated protein kinases reveals critical roles in growth, reproduction, and stress. *Proceedings of the National Academy of Sciences of the United States of America*, *106*, 8380–8385.
Fujii, H., & Zhu, J.-K. (2012). Osmotic stress signaling via protein kinases. *Cellular and Molecular Life Sciences*, *69*, 3165–3173.
Fujita, Y., Fujita, M., Satoh, R., Maruyama, K., Parvez, M. M., Seki, M., et al. (2005). AREB1 is a transcription activator of novel ABRE-dependent ABA signaling that enhances drought stress tolerance in *Arabidopsis*. *The Plant Cell*, *17*, 3470–3488.
Fujita, Y., Fujita, M., Shinozaki, K., & Yamaguchi-Shinozaki, K. (2011). ABA-mediated transcriptional regulation in response to osmotic stress in plants. *Journal of Plant Research*, *124*, 509–525.
Fujita, Y., Nakashima, K., Yoshida, T., Katagiri, T., Kidokoro, S., Kanamori, N., et al. (2009). Three SnRK2 protein kinases are the main positive regulators of abscisic acid signaling in response to water stress in Arabidopsis. *Plant & Cell Physiology*, *50*, 2123–2132.
Furihata, T., Maruyama, K., Fujita, Y., Umezawa, T., Yoshida, R., Shinozaki, K., et al. (2006). Abscisic acid-dependent multisite phosphorylation regulates the activity of a transcription activator AREB1. *Proceedings of the National Academy of Sciences of the United States of America*, *103*, 1988–1993.

Gao, C., Zhuang, X., Shen, J., & Jiang, L. (2017). Plant ESCRT complexes: Moving beyond endosomal sorting. *Trends in Plant Science*, *22*, 986–998.
García-León, M., Cuyas, L., El-Moneim, D. A., Rodriguez, L., Belda-Palazón, B., Sanchez-Quant, E., et al. (2019). Arabidopsis ALIX regulates stomatal aperture and turnover of abscisic acid receptors. *The Plant Cell*, *31*, 2411–2429.
Gietler, M., Fidler, J., Labudda, M., & Nykiel, M. (2020). Abscisic acid—Enemy or savior in the response of cereals to abiotic and biotic stresses? *International Journal of Molecular Sciences*, *21*, 4607.
Giraudat, J., Parcy, F., Bertauche, N., Gosti, F., & Leung, J. (1994). Current advances in abscisic acid action and signaling. *Plant Molecular Biology*, *26*, 1557–1577.
González-Guzmán, M., Apostolova, N., Bellés, J. M., Barrero, J. M., Piqueras, P., Ponce, M. R., et al. (2002). The short-chain alcohol dehydrogenase ABA2 catalyzes the conversion of xanthoxin to abscisic aldehyde. *The Plant Cell*, *14*, 1833–1846.
González-Guzmán, M., Pizzio, G. A., Antoni, R., Vera-Sirera, F., Merilo, E., Bassel, G. W., et al. (2012). *Arabidopsis* PYR/PYL/RCAR receptors play a major role in quantitative regulation of stomatal aperture and transcriptional response to abscisic acid. *The Plant Cell*, *24*, 2483–2496.
Gosti, F., Beaudoin, N., Serizet, C., Webb, A. A., Vartanian, N., & Giraudat, J. (1999). ABI1 protein phosphatase 2C is a negative regulator of abscisic acid signaling. *The Plant Cell*, *11*, 1897–1909.
Guan, C., Ji, J., Zhang, X., Li, X., Jin, C., Guan, W., et al. (2015). Positive feedback regulation of a *Lycium chinense*-derived VDE gene by drought-induced endogenous ABA, and over-expression of this VDE gene improve drought-induced photo-damage in *Arabidopsis*. *Journal of Plant Physiology*, *175*, 26–36.
Han, Y., Watanabe, S., Shimada, H., & Sakamoto, A. (2020). Dynamics of the leaf endoplasmic reticulum modulate β-glucosidase-mediated stress-activated ABA production from its glucosyl ester. *Journal of Experimental Botany*, *71*, 2058–2071.
Hasegawa, P. M., Bressan, R. A., Zhu, J.-K., & Bohnert, H. J. (2000). Plant cellular and molecular responses to high salinity. *Annual Review of Plant Physiology and Plant Molecular Biology*, *51*, 463–499.
Hayashi, M., Inoue, S., Takahashi, K., & Kinoshita, T. (2011). Immunohistochemical detection of blue light-induced phosphorylation of the plasma membrane H^+-ATPase in stomatal guard cells. *Plant Cell & Physiology*, *52*, 1238–1248.
Hedrich, R., & Shabala, S. (2018). Stomata in a saline world. *Current Opinion in Plant Biology*, *46*, 87–95.
Hershko, A., & Ciechanover, A. (1998). The ubiquitin system. *Annual Review of Biochemistry*, *67*, 425–479.
Hirayama, T., & Shinozaki, K. (2007). Perception and transduction of abscisic acid signals: Keys to the function of the versatile plant hormone ABA. *Trends in Plant Science*, *12*, 343–351.
Hoth, S., Morgante, M., Sanchez, J.-P., Hanafey, M. K., Tingey, S. V., & Chua, N.-H. (2002). Genome-wide gene expression profiling in *Arabidopsis thaliana* reveals new targets of abscisic acid and largely impaired gene regulation in the *abi1-1* mutant. *Journal of Cell Science*, *115*, 4891–4900.
Hussain, S., Gomes, M. M., Yano, K., & Nambara, E. (2019). Interactions between abscisic acid and other hormones. *Advances in Botanical Research*, *92*, 255–280.
Irigoyen, M. L., Iniesto, E., Rodriguez, L., Puga, M. I., Yanagawa, Y., Pick, E., et al. (2014). Targeted degradation of abscisic acid receptors is mediated by the ubiquitin ligase substrate adaptor DDA1 in *Arabidopsis*. *The Plant Cell*, *26*, 712–728.
Isayenkov, S. V., & Maathuis, F. J. M. (2019). Plant salinity stress: Many unanswered questions remain. *Frontiers in Plant Science*, *10*, 80.

Iuchi, S., Kobayashi, M., Taji, T., Naramoto, M., Seki, M., Kato, T., et al. (2001). Regulation of drought tolerance by gene manipulation of 9-*cis*-epoxycarotenoid dioxygenase, a key enzyme in abscisic acid biosynthesis in *Arabidopsis*. *The Plant Journal, 27*, 325–333.

Iuchi, S., Kobayashi, M., Yamaguchi-Shinozaki, K., & Shinozaki, K. (2000). A stress-inducible gene for 9-cis-epoxycarotenoid dioxygenase involved in abscisic acid biosynthesis under water stress in drought-tolerant cowpea. *Plant Physiology, 123*, 553–562.

Joo, H., Baek, W., Lim, C. W., & Lee, S. C. (2021). Post-translational modifications of bZIP transcription factors in abscisic acid signaling and drought responses. *Current Genomics, 22*, 4–15.

Joshi-Saha, A., Valon, C., & Leung, J. (2011). Abscisic acid signal off the STARting block. *Molecular Plant, 4*, 562–580.

Julian, J., Coego, A., Lozano-Juste, J., Lechner, E., Wu, Q., Zhang, X., et al. (2019). The MATH-BTB BPM3 and BPM5 subunits of Cullin3-RING E3 ubiquitin ligases target PP2CA and other clade a PP2Cs for degradation. *Proceedings of the National Academy of Sciences of the United States of America, 116*, 15725–15734.

Jung, C., Nguyen, N. H., & Cheong, J.-J. (2020). Transcriptional regulation of protein phosphatase 2C genes to modulate abscisic acid signaling. *International Journal of Molecular Sciences, 21*, 9517.

Jung, C., Seo, J. S., Han, S. W., Koo, Y. J., Kim, C. H., Song, S. I., et al. (2008). Overexpression of *AtMYB44* enhances stomatal closure to confer abiotic stress tolerance in transgenic Arabidopsis. *Plant Physiology, 146*, 623–635.

Jung, C., Shim, J. S., Seo, J. S., Lee, H. Y., Kim, C. H., Choi, Y. D., et al. (2010). Non-specific phytohormonal induction of AtMYB44 and suppression of jasmonate-responsive gene activation in *Arabidopsis thaliana*. *Molecules and Cells, 29*, 71–76.

Jurkiewicz, P., & Batoko, H. (2018). Protein degradation mechanisms modulate abscisic acid signaling and responses during abiotic stress. *Plant Science, 267*, 48–54.

Kagale, S., & Rozwadowski, K. (2011). EAR motif-mediated transcriptional repression in plants: An underlying mechanism for epigenetic regulation of gene expression. *Epigenetics, 6*, 141–146.

Kang, J. Y., Choi, H. I., Im, M. Y., & Kim, S. Y. (2002). Arabidopsis basic leucine zipper proteins that mediate stress-responsive abscisic acid signaling. *The Plant Cell, 14*, 343–357.

Kang, J., Hwang, J. U., Lee, M., Kim, Y. Y., Assmann, S. M., Martinoia, E., et al. (2010). PDR-type ABC transporter mediates cellular uptake of the phytohormone abscisic acid. *Proceedings of the National Academy of Sciences of the United States of America, 107*, 2355–2360.

Kanno, Y., Hanada, A., Chiba, Y., Ichikawa, T., Nakazawa, M., Matsui, M., et al. (2012). Identification of an abscisic acid transporter by functional screening using the receptor complex as a sensor. *Proceedings of the National Academy of Sciences of the United States of America, 109*, 9653–9658.

Kim, T.-H., Böhmer, M., Hu, H., Nishimura, N., & Schroeder, J. I. (2010). Guard cell signal transduction network: Advances in understanding abscisic acid, CO_2, and Ca^{2+} signaling. *Annual Review of Plant Biology, 61*, 561–591.

Kim, S., Kang, J. Y., Cho, D. I., Park, J. H., & Kim, S. Y. (2004). ABF2, an ABRE-binding bZIP factor, is an essential component of glucose signaling and its overexpression affects multiple stress tolerance. *The Plant Journal, 40*, 75–87.

Klingler, J. P., Batelli, G., & Zhu, J. K. (2010). ABA receptors: The START of a new paradigm in phytohormone signalling. *Journal of Experimental Botany, 61*, 3199–3210.

Koiwai, H., Nakaminami, K., Seo, M., Mitsuhashi, W., Toyomasu, T., & Koshiba, T. (2004). Tissue-specific localization of an abscisic acid biosynthetic enzyme, AAO3, in Arabidopsis. *Plant Physiology, 134*, 1697–1707.

Komander, D., & Rape, M. (2012). The ubiquitin code. *Annual Review of Biochemistry, 81*, 203–229.

Kong, L., Cheng, J., Zhu, Y., Ding, Y., Meng, J., Chen, Z., et al. (2015). Degradation of the ABA co-receptor ABI1 by PUB12/13 U-box E3 ligases. *Nature Communications, 6*, 8630.

Koornneef, M., Reuling, G., & Karssen, C. M. (1984). The isolation and characterization of abscisic acid-insensitive mutants of *Arabidopsis thaliana*. *Physiologia Plantarum, 61*, 377–383.

Kraepiel, Y., Rousselin, P., Sotta, B., Kerhoas, L., Einhorn, J., Caboche, M., et al. (1994). Analysis of phytochrome- and ABA-deficient mutants suggests that ABA degradation is controlled by light in *Nicotiana plumbaginifolia*. *The Plant Journal, 6*, 665–672.

Krochko, J. E., Abrams, G. D., Loewen, M. K., Abrams, S. R., & Cutler, A. J. (1998). (+)-Abscisic acid 8′-hydroxylase is a cytochrome P450 monooxygenase. *Plant Physiology, 118*, 849–860.

Kumar, M., Kesawat, M. S., Ali, A., Lee, S.-C., Gill, S. S., & Kim, H. U. (2019). Integration of abscisic acid signaling with other signaling pathways in plant stress responses and development. *Plants, 8*, 592.

Kuromori, T., Miyaji, T., Yabuuchi, H., Shimizu, H., Sugimoto, E., Kamiya, A., et al. (2010). ABC transporter AtABCG25 is involved in abscisic acid transport and responses. *Proceedings of the National Academy of Sciences of the United States of America, 107*, 2361–2366.

Kuromori, T., Seo, M., & Shinozaki, K. (2018). ABA transport and plant water stress responses. *Trends in Plant Science, 23*, 513–522.

Kushiro, T., Okamoto, M., Nakabayashi, K., Yamagishi, K., Kitamura, S., Asami, T., et al. (2004). The *Arabidopsis* cytochrome P450 CYP707A encodes ABA 8′-hydroxylases: Key enzymes in ABA catabolism. *EMBO Journal, 23*, 1647–1656.

Kwak, J. M., Mori, I., Pei, Z. M., Leonhardt, N., Torres, M. A., Dangl, J. L., et al. (2003). NADPH oxidase *AtrbohD* and *AtrbohF* genes function in ROS-dependent ABA signaling in *Arabidopsis*. *EMBO Journal, 22*, 2623–2633.

Lee, K. H., Piao, H. L., Kim, H.-Y., Choi, S. M., Jiang, F., Hartung, W., et al. (2006). Activation of glucosidase via stress-induced polymerization rapidly increases active pools of abscisic acid. *Cell, 126*, 1109–1120.

Lee, J. H., Yoon, H. J., Terzaghi, W., Martinez, C., Dai, M., Li, J., et al. (2010). DWA1 and DWA2, two Arabidopsis DWD protein components of CUL4-based E3 ligases, act together as negative regulators in ABA signal transduction. *The Plant Cell, 22*, 1716–1732.

Li, S., He, X., Gao, Y., Zhou, C., Chiang, V. L., & Li, W. (2021). Histone acetylation changes in plant response to drought stress. *Genes, 12*, 1409.

Li, S., Lin, Y.-C. J., Wang, P., Zhang, B., Li, M., Chen, S., et al. (2019). The AREB1 transcription factor influences histone acetylation to regulate drought responses and tolerance in *Populus trichocarpa*. *The Plant Cell, 31*, 663–686.

Li, D., Zhang, L., Li, X., Kong, X., Wang, X., Li, Y., et al. (2018). AtRAE1 is involved in degradation of ABA receptor RCAR1 and negatively regulates ABA signalling in *Arabidopsis*. *Plant, Cell & Environment, 41*, 231–244.

Li, Y., Zhang, L., Li, D., Liu, Z., Wang, J., Li, X., et al. (2016). The Arabidopsis F-box E3 ligase RIFP1 plays a negative role in abscisic acid signalling by facilitating ABA receptor RCAR3 degradation. *Plant, Cell & Environment, 39*, 571–582.

Li, X., Zhao, W., Sun, X., Huang, H., Kong, L., Niu, D., et al. (2013). Molecular cloning and characterization of violaxanthin de-epoxidase (*CsVDE*) in cucumber. *PLoS One, 8*, e64383.

Lim, C. W., Baek, W., Jung, J., Kim, J.-H., & Lee, S. C. (2015). Function of ABA in stomatal defense against biotic and drought stresses. *International Journal of Molecular Sciences.*, *16*, 15251–15270.
Lin, Z., Li, Y., Wang, Y., Liu, X., Ma, L., Zhang, Z., et al. (2021). Initiation and amplification of SnRK2 activation in abscisic acid signaling. *Nature Communications*, *12*, 2456.
Liu, H., & Stone, S. L. (2013). Cytoplasmic degradation of the *Arabidopsis* transcription factor ABSCISIC ACID INSENSITIVE 5 is mediated by the RING-type E3 ligase KEEP ON GOING. *Journal of Biological Chemistry*, *288*, 20267–20279.
Liu, Z., Yan, J.-P., Li, D.-K., Luo, Q., Yan, Q., Liu, Z.-B., et al. (2015). UDP-glucosyltransferase71C5, a major glucosyltransferase, mediates abscisic acid homeostasis in Arabidopsis. *Plant Physiology*, *167*, 1659–1670.
Lopez-Molina, L., Mongrand, S., Kinoshita, N., & Chua, N.-H. (2003). AFP is a novel negative regulator of ABA signaling that promotes ABI5 protein degradation. *Genes & Development*, *17*, 410–418.
Luo, M., Liu, X., Singh, P., Cui, Y., Zimmerli, L., & Wu, K. (2012). Chromatin modifications and remodeling in plant abiotic stress responses. *Biochimica et Biophysica Acta (BBA)—Gene Regulatory Mechanisms*, *1819*, 129–136.
Ma, Y., Szostkiewicz, I., Korte, A., Moes, D., Yang, Y., Christmann, A., et al. (2009). Regulators of PP2C phosphatase activity function as abscisic acid sensors. *Science*, *324*, 1064–1068.
Manishankar, P., Wang, N., Köster, P., Alatar, A. A., & Kudla, J. (2018). Calcium signaling during salt stress and in the regulation of ion homeostasis. *Journal of Experimental Botany*, *69*, 4215–4226.
Marusig, D., & Tombesi, S. (2020). Abscisic acid mediates drought and salt stress responses in *Vitis vinifera*—A review. *International Journal of Molecular Sciences*, *21*, 8648.
Merilo, E., Jalakas, P., Laanemets, K., Mohammadi, O., Hõrak, H., Kollist, H., et al. (2015). Abscisic acid transport and homeostasis in the context of stomatal regulation. *Molecular Plant*, *8*, 1321–1333.
Mishra, N. S., Tuteja, R., & Tuteja, N. (2006). Signaling through MAP kinase networks in plants. *Archives in Biochemistry and Biophysics*, *452*, 55–68.
Mittler, R., & Blumwald, E. (2015). The roles of ROS and ABA in systemic acquired acclimation. *The Plant Cell*, *27*, 64–70.
Miyazono, K., Miyakawa, T., Sawano, Y., Kubota, K., Kang, H. J., Asano, A., et al. (2009). Structural basis of abscisic acid signalling. *Nature*, *462*, 609–614.
Munemasa, S., Hauser, F., Park, J., Waadt, R., Brandt, B., & Schroeder, J. I. (2015). Mechanisms of abscisic acid-mediated control of stomatal aperture. *Current Opinion in Plant Biology*, *28*, 154–162.
Nakagami, H., Pitzschke, A., & Hirt, H. (2005). Emerging MAP kinase pathways in plant stress signaling. *Trends in Plant Science*, *10*, 339–346.
Nakashima, K., Fujita, Y., Kanamori, N., Katagiri, T., Umezawa, T., Kidokoro, S., et al. (2009). Three Arabidopsis SnRK2 protein kinases, SRK2D/SnRK2.2, SRK2E/SnRK2.6/OST1 and SRK2I/SnRK2.3, involved in ABA signaling are essential for the control of seed development and dormancy. *Plant & Cell Physiology*, *50*, 1345–1363.
Nambara, E., & Marion-Poll, A. (2005). Abscisic acid biosynthesis and catabolism. *Annual Review of Plant Biology*, *56*, 165–185.
Negi, J., Hashimoto-Sugimoto, M., Kusumi, K., & Iba, K. (2014). New approaches to the biology of stomatal guard cells. *Plant & Cell Physiology*, *55*, 241–250.
Nguyen, N. H., & Cheong, J.-J. (2018). AtMYB44 interacts with TOPLESS-RELATED corepressors to suppress *protein phosphatase 2C* gene transcription. *Biochemical and Biophysical Research Communications.*, *499*, 1039–1043.
Nguyen, N. H., Jung, C., & Cheong, J.-J. (2019). Chromatin remodeling for the transcription of type 2C protein phosphatase genes in response to salt stress. *Plant Physiology and Biochemistry*, *141*, 325–331.

Nishimura, N., Sarkeshik, A., Nito, K., Park, S.-Y., Wang, A., Carvalho, P. C., et al. (2010). PYR/PYL/RCAR family members are major *in-vivo* ABI1 protein phosphatase 2C-interacting proteins in Arabidopsis. *The Plant Journal*, *61*, 290–299.

North, H. M., De Almeida, A., Boutin, J.-P., Frey, A., To, A., Botran, L., et al. (2007). The Arabidopsis ABA-deficient mutant *aba4* demonstrates that the major route for stress-induced ABA accumulation is via neoxanthin isomers. *The Plant Journal*, *50*, 810–824.

Okamoto, M., Tanaka, Y., Abrams, S. R., Kamiya, Y., Seki, M., & Nambara, E. (2009). High humidity induces abscisic acid 8′-hydroxylase in stomata and vasculature to regulate local and systemic abscisic acid responses in Arabidopsis. *Plant Physiology*, *149*, 825–834.

Osakabe, Y., Yamaguchi-Shinozaki, K., Shinozaki, K., & Tran, L. P. (2014). ABA control of plant macroelement membrane transport systems in response to water deficit and high salinity. *New Phytologist*, *202*, 35–49.

Pan, W., Lin, B., Yang, X., Liu, L., Xia, R., Li, J., et al. (2020). The UBC27-AIRP3 ubiquitination complex modulates ABA signaling by promoting the degradation of ABI1 in Arabidopsis. *Proceedings of the National Academy of Sciences of the United States of America*, *117*, 27694–27702.

Park, S. Y., Fung, P., Nishimura, N., Jensen, D. R., Fujii, H., Zhao, Y., et al. (2009). Abscisic acid inhibits type 2C protein phosphatases via the PYR/PYL family of START proteins. *Science*, *324*, 1068–1071.

Park, H. J., Kim, W.-Y., & Yun, D.-J. (2016). A new insight of salt stress signaling in plant. *Molecules and Cells*, *39*, 447–459.

Pei, Z. M., Murata, Y., Benning, G., Thomine, S., Klüsener, B., Allen, G. J., et al. (2000). Calcium channels activated by hydrogen peroxide mediate abscisic acid signalling in guard cells. *Nature*, *406*, 731–734.

Piotrowska, A., & Bajguz, A. (2011). Conjugates of abscisic acid, brassinosteroids, ethylene, gibberellins, and jasmonates. *Phytochemistry*, *72*, 2097–2112.

Priest, D. M., Ambrose, S. J., Vaistij, F. E., Elias, L., Higgins, G. S., Ross, A. R., et al. (2006). Use of the glucosyltransferase UGT71B6 to disturb abscisic acid homeostasis in *Arabidopsis thaliana*. *The Plant Journal*, *46*, 492–502.

Qin, X., & Zeevaart, J. A. D. (1999). The 9-*cis*-epoxycarotenoid cleavage reaction is the key regulatory step of abscisic acid biosynthesis in water-stressed bean. *Proceedings of the National Academy of Sciences of the United States of America*, *96*, 15354–15361.

Qin, X. Q., & Zeevaart, J. A. D. (2002). Overexpression of a 9-cis-epoxycarotenoid dioxygenase gene in *Nicotiana plumbaginifolia* increases abscisic acid and phaseic acid levels and enhances drought tolerance. *Plant Physiology*, *128*, 544–551.

Raghavendra, A. S., Gonugunta, V. K., Christmann, A., & Grill, E. (2010). ABA perception and signalling. *Trends in Plant Science*, *15*, 395–401.

Rodriguez, P. L., Benning, G., & Grill, E. (1998). ABI2, a second protein phosphatase 2C involved in abscisic acid signal transduction in *Arabidopsis*. *FEBS Letters*, *421*, 185–190.

Rodriguez, P. L., Lozano-Juste, J., & Albert, A. (2019). PYR/PYL/RCAR ABA receptors. *Advances in Botanical Research*, *92*, 51–82.

Roelfsema, M. R. G., & Hedrich, R. (2005). In the light of stomatal opening: New insights into 'the Watergate'. *New Phytologist*, *167*, 665–691.

Rubio, S., Rodrigues, A., Saez, A., Dizon, M. B., Galle, A., Kim, T., et al. (2009). Triple loss of function of protein phosphatases type 2C leads to partial constitutive response to endogenous abscisic acid. *Plant Physiology*, *150*, 1345–1355.

Saavedra, X., Modrego, A., Rodríguez, D., González-García, M. P., Sanz, L., Nicolás, G., et al. (2010). The nuclear interactor PYL8/RCAR3 of *Fagus sylvatica* FsPP2C1 is a positive regulator of abscisic acid signaling in seeds and stress. *Plant Physiology*, *152*, 133–150.

Saez, A., Apostolova, N., Gonzalez-Guzman, M., Gonzalez-Garcia, M. P., Nicolas, C., Lorenzo, O., et al. (2004). Gain-of-function and loss-of-function phenotypes of the protein phosphatase 2C HAB1 reveal its role as a negative regulator of abscisic acid signalling. *The Plant Journal*, *37*, 354–369.

Saez, A., Robert, N., Maktabi, M. H., Schroeder, J. I., Serrano, R., & Rodriguez, P. L. (2006). Enhancement of abscisic acid sensitivity and reduction of water consumption in Arabidopsis by combined inactivation of the protein phosphatases type 2C ABI1 and HAB1. *Plant Physiology*, *141*, 1389–1399.

Sah, S. K., Reddy, K. R., & Li, J. (2016). Abscisic acid and abiotic stress tolerance in crop plants. *Frontiers in Plant Science*, *7*, 571.

Saika, H., Okamoto, M., Miyoshi, K., Kushiro, T., Shinoda, S., Jikumaru, Y., et al. (2007). Ethylene promotes submergence-induced expression of *OsABA8ox1*, a gene that encodes ABA 8′-hydroxylase in rice. *Plant & Cell Physiology*, *48*, 287–298.

Saito, S., Hirai, N., Matsumoto, C., Ohigashi, H., Ohta, D., et al. (2004). Arabidopsis CYP707As encode (+)-abscisic acid 8 -hydroxylase, a key enzyme in the oxidative catabolism of abscisic acid. *Plant Physiology*, *134*, 1439–1449.

Saito, S., & Uozumi, N. (2019). Guard cell membrane anion transport systems and their regulatory components: An elaborate mechanism controlling stress-induced stomatal closure. *Plants*, *8*, 9.

Santiago, J., Rodrigues, A., Saez, A., Rubio, S., Antoni, R., Dupeux, F., et al. (2009). Modulation of drought resistance by the abscisic acid receptor PYL5 through inhibition of clade A PP2Cs. *The Plant Journal*, *60*, 575–588.

Schwartz, S. H., Qin, X., & Zeevaart, J. A. (2003). Elucidation of the indirect pathway of abscisic acid biosynthesis by mutants, genes, and enzymes. *Plant Physiology*, *131*, 1591–1601.

Schweighofer, A., Hirt, H., & Meskiene, I. (2004). Plant PP2C phosphatases: Emerging functions in stress signaling. *Trends in Plant Science*, *9*, 236–243.

Seifikalhor, M., Aliniaeifard, S., Shomali, A., Azad, N., Hassani, B., Lastochkina, O., et al. (2019). Calcium signaling and salt tolerance are diversely entwined in plants. *Plant Signaling & Behavior*, *14*, 1665455.

Seki, M., Ishida, J., Narusaka, M., Fujita, M., Nanjo, T., Umezawa, T., et al. (2002). Monitoring the expression pattern of around 7,000 Arabidopsis genes under ABA treatments using a full-length cDNA microarray. *Functional & Integrative Genomics*, *2*, 282–291.

Seo, K.-I., Lee, J.-H., Nezames, C. D., Zhong, S., Song, E., Byun, M.-O., et al. (2014). ABD1 is an Arabidopsis DCAF substrate receptor for CUL4-DDB1-based E3 ligases that acts as a negative regulator of abscisic acid signaling. *The Plant Cell*, *26*, 695–711.

Seo, M., & Marion-Poll, A. (2019). Abscisic acid metabolism and transport. *Advances in Botanical Research*, *92*, 1–49.

Seo, M., Peeters, A. J., Koiwai, H., Oritani, T., Marion-Poll, A., Zeevaart, J. A., et al. (2000). The Arabidopsis aldehyde oxidase 3 (AAO3) gene product catalyzes the final step in abscisic acid biosynthesis in leaves. *Proceedings of the National Academy of Sciences of the United States of America*, *97*, 12908–12913.

Shimazaki, K.-I., Doi, M., Assmann, S. M., & Kinoshita, T. (2007). Light regulation of stomatal movement. *Annual Review of Plant Biology*, *58*, 219–247.

Shimizu, T., Kanno, Y., Suzuki, H., Watanabe, S., & Seo, M. (2021). Arabidopsis NPF4.6 and NPF5.1 control leaf stomatal aperture by regulating abscisic acid transport. *Genes*, *12*, 885.

Singh, A., Pandey, A., Srivastava, A. K., Tran, L.-S., & Pandey, G. K. (2015). Plant protein phosphatases 2C: From genomic diversity to functional multiplicity and importance in stress management. *Critical Review of Biotechnology*, *36*, 1023–1035.

Sirichandra, C., Gu, D., Hu, H.-C., Davanture, M., Lee, S., Djaoui, M., et al. (2009). Phosphorylation of the Arabidopsis AtrbohF NADPH oxidase by OST1 protein kinase. *FEBS Letters, 583*, 2982–2986.

Sirko, A., Wawrzyńska, A., Brzywczy, J., & Sieńko, M. (2021). Control of ABA signaling and crosstalk with other hormones by the selective degradation of pathway components. *International Journal of Molecular Sciences, 22*, 4638.

Smalle, J., & Vierstra, R. D. (2004). The ubiquitin 26S proteasome proteolytic pathway. *Annual Review of Plant Biology, 55*, 555–590.

Soon, F.-F., Ng, L.-M., Zhou, X. E., West, G. M., Kovach, A., Tan, M. H. E., et al. (2012). Molecular mimicry regulates ABA signaling by SnRK2 kinases and PP2C phosphatases. *Science, 335*, 85–88.

Stone, S. L. (2019). Role of the ubiquitin proteasome system in plant response to abiotic stress. *International Review of Cell and Molecular Biology, 343*, 65–110.

Stone, S. L., Williams, L. A., Farmer, L. M., Vierstra, R. D., & Callis, J. (2006). KEEP ON GOING, a RING E3 ligase essential for Arabidopsis growth and development, is involved in abscisic acid signaling. *The Plant Cell, 18*, 3415–3428.

Szostkiewicz, I., Richter, K., Kepka, M., Demmel, S., Ma, Y., Korte, A., et al. (2010). Closely related receptor complexes differ in their ABA selectivity and sensitivity. *The Plant Journal, 61*, 25–35.

Tan, B.-C., Joseph, L. M., Deng, W.-T., Liu, L. J., Li, Q.-B., Cline, K., et al. (2003). Molecular characterization of the *Arabidopsis* 9-*cis* epoxycarotenoid dioxygenase gene family. *The Plant Journal, 35*, 44–56.

Thompson, A. J., Jackson, A. C., Symonds, R. C., Mulholland, B. J., Dadswell, A. R., Blake, P. S., et al. (2000). Ectopic expression of a tomato 9-*cis*-epoxycarotenoid dioxygenase gene causes over-production of abscisic acid. *The Plant Journal, 23*, 363–374.

Tischer, S. V., Wunschel, C., Papacek, M., Kleigrewe, K., Hofmann, T., Christmann, A., et al. (2017). Combinatorial interaction network of abscisic acid receptors and coreceptors from *Arabidopsis thaliana. Proceedings of the National Academy of Sciences of the United States of America, 114*, 10280–10285.

Todaka, D., Takahashi, F., Yamaguchi-Shinozaki, K., & Shinozaki, K. (2019). ABA-responsive gene expression in response to drought stress: Cellular regulation and long-distance signaling. *Advances in Botanical Research, 92*, 83–113.

Umezawa, T., Nakashima, K., Miyakawa, T., Kuromori, T., Tanokura, M., Shinozaki, K., et al. (2010). Molecular basis of the core regulatory network in ABA responses: Sensing, signaling and transport. *Plant & Cell Physiology, 51*, 1821–1839.

Umezawa, T., Okamoto, M., Kushiro, T., Nambara, E., Oono, Y., Seki, M., et al. (2006). CYP707A3, a major ABA 8′-hydroxylase involved in dehydration and rehydration response in *Arabidopsis thaliana. The Plant Journal, 46*, 171–182.

Umezawa, T., Sugiyama, N., Mizoguchi, M., Hayashi, S., Myouga, F., Yamaguchi-Shinozaki, K., et al. (2009). Type 2C protein phosphatases directly regulate abscisic acid-activated protein kinases in Arabidopsis. *Proceedings of the National Academy of Sciences of the United States of America, 106*, 17588–17593.

Uno, Y., Furihata, T., Abe, H., Yoshida, R., Shinozaki, K., & Yamaguchi-Shinozaki, K. (2000). Arabidopsis basic leucine zipper transcription factors involved in an abscisic acid-dependent signal transduction pathway under drought and high-salinity conditions. *Proceedings of the National Academy of Sciences of the United States of America, 97*, 11632–11637.

Vahisalu, T., Kollist, H., Wang, Y. F., Nishimura, N., Chan, W. Y., Valerio, G., et al. (2008). SLAC1 is required for plant guard cell S-type anion channel function in stomatal signalling. *Nature, 452*, 487–491.

Vierstra, R. D. (2009). The ubiquitin-26S proteasome system at the nexus of plant biology. *Nature Reviews Molecular Cell Biology, 10*, 385–397.

Wang, X., Guo, C., Peng, J., Li, C., Wan, F., Zhang, S., et al. (2019). ABRE-BINDING FACTORS play a role in the feedback regulation of ABA signaling by mediating rapid ABA induction of ABA co-receptor genes. *New Phytologist, 221*, 341–355.

Wu, Q., Zhang, X., Peirats-Llobet, M., Belda-Palazon, B., Wang, X., Cui, S., et al. (2016). Ubiquitin ligases RGLG1 and RGLG5 regulate abscisic acid signaling by controlling the turnover of phosphatase PP2CA. *The Plant Cell, 28*, 2178–2196.
Xiong, L., Gong, Z., Rock, C. D., Subramanian, S., Guo, Y., Xu, W., et al. (2001). Modulation of abscisic acid signal transduction and biosynthesis by an Sm-like protein in *Arabidopsis*. *Developmental Cell, 1*, 771–781.
Xiong, L., Ishitani, M., Lee, H., & Zhu, J.-K. (2001). The Arabidopsis *LOS5/ABA3* locus encodes a molybdenum cofactor sulfurase and modulates cold stress- and osmotic stress-responsive gene expression. *The Plant Cell, 13*, 2063–2083.
Xiong, L., Lee, H., Ishitani, M., & Zhu, J.-K. (2002). Regulation of osmotic stress-responsive gene expression by the *LOS6/ABA1* locus in *Arabidopsis*. *Journal of Biological Chemistry, 277*, 8588–8596.
Xiong, L., Schumaker, K. S., & Zhu, J.-K. (2002). Cell signaling during cold, drought, and salt stress. *The Plant Cell, 14*, S165–S183.
Xu, Z. Y., Lee, K. H., Dong, T., Jeong, J. C., Jin, J. B., Kanno, Y., et al. (2012). A vacuolar β-glucosidase homolog that possesses glucose-conjugated abscisic acid hydrolyzing activity plays an important role in osmotic stress responses in *Arabidopsis*. *The Plant Cell, 24*, 2184–2199.
Xu, Z. J., Nakajima, M., Suzuki, Y., & Yamaguchi, I. (2002). Cloning and characterization of the abscisic acid-specific glucosyltransferase gene from adzuki bean seedlings. *Plant Physiology, 129*, 1285–1295.
Xue, T., Wang, D., Zhang, S., Ehlting, J., Ni, F., Jakab, S., et al. (2008). Genome-wide and expression analysis of protein phosphatase 2C in rice and Arabidopsis. *BMC Genomics, 9*, 550.
Yin, P., Fan, H., Hao, Q., Yuan, X., Wu, D., Pang, Y., et al. (2009). Structural insights into the mechanism of abscisic acid signaling by PYL proteins. *Nature Structural & Molecular Biology, 16*, 1230–1236.
Yoshida, T., Fujita, Y., Maruyama, K., Mogami, J., Todaka, D., Shinozaki, K., et al. (2015). Four *Arabidopsis* AREB/ABF transcription factors function predominantly in gene expression downstream of SnRK2 kinases in abscisic acid signalling in response to osmotic stress. *Plant, Cell & Environment, 38*, 35–49.
Yoshida, T., Fujita, Y., Sayama, H., Kidokoro, S., Maruyama, K., Mizoi, J., et al. (2010). AREB1, AREB2, and ABF3 are master transcription factors that cooperatively regulate ABRE-dependent ABA signaling involved in drought stress tolerance and require ABA for full activation. *The Plant Journal, 61*, 672–685.
Yu, F., Cao, X., Liu, G., Wang, Q., Xia, R., Zhang, X., et al. (2020). ESCRT-I component VPS23A is targeted by E3 ubiquitin ligase XBAT35 for proteasome-mediated degradation in modulating ABA signaling. *Molecular Plant, 13*, 1556–1569.
Yu, F., Lou, L., Tian, M., Li, Q., Ding, Y., Cao, X., et al. (2016). ESCRT-I component VPS23A affects ABA signaling by recognizing ABA receptors for endosomal degradation. *Molecular Plant, 9*, 1570–1582.
Zeevaart, J. A. D., & Creelman, R. A. (1988). Metabolism and physiology of abscisic acid. *Annual Review of Plant Physiology and Plant Molecular Biology, 39*, 439–473.
Zhang, H., Zhu, H., Pan, Y., Yu, Y., Luan, S., & Li, L. (2014). A DTX/MATE-type transporter facilitates abscisic acid efflux and modulates ABA sensitivity and drought tolerance in *Arabidopsis*. *Molecular Plant, 7*, 1522–1532.
Zhao, J., Zhao, L., Zhang, M., Zafar, S. A., Fang, J., Li, M., et al. (2017). *Arabidopsis* E3 ubiquitin ligases PUB22 and PUB23 negatively regulate drought tolerance by targeting ABA receptor PYL9 for degradation. *International Journal of Molecular Sciences, 18*, E1841.
Zhu, J.-K. (2003). Regulation of ion homeostasis under salt stress. *Current Opinion in Plant Biology, 6*, 441–445.

CHAPTER FIVE

Impact of nutritional imbalance on guard cell metabolism and stomata regulation under saline hyperosmotic conditions

Christian Zörb[a,*], Bastian L. Franzisky[b], Patrick P. Lehr[a], Robin Kosch[c], Michael Altenbuchinger[c], and Christoph-Martin Geilfus[b]
[a]Quality of Plant Products, Institute of Crop Sciences, University of Hohenheim, Stuttgart, Germany
[b]Department of Soil Science and Plant Nutrition, Hochschule Geisenheim University, Geisenheim, Germany
[c]Department of Medical Bioinformatics, University Medical Center Göttingen, Göttingen, Germany
*Corresponding author: e-mail address: christian.zoerb@uni-hohenheim.de

Contents

1. Introduction 124
2. The possible role of magnesium in stomatal regulation under sodium-salinity 125
3. The possible role of salt stress-induced GABA production as metabolic signal involved in reduction of stomatal aperture 129
4. The possible role of sulfur in guard cell metabolism under salt stress 132
5. Conclusion 134
References 135

Abstract

Stomata closure under adverse conditions leads to reduced CO_2 assimilation and yield penalty. This review focuses on advances in stomata research especially for crops growing at hyperosmotic saline environments. Possible new aspects of nutritional imbalances under saline hyperosmotic conditions on guard cell metabolism which were deduced from the recent literature were focused. In particular possible effects of high Na^+ concentration on GABA shunt which may regulate tonoplast anion channel activity and thus stomatal aperture were discussed. The extent to which stress-induced GABA production influences stomatal behavior in interaction with other signaling pathways remains an interesting topic that requires further research. Another example is magnesium and its role in stomata regulation. Up to date understanding about the relevance of a salinity-induced reduction of magnesium for stomatal opening is rudimentary. Few reports have linked the magnesium-nutritional status to transpiration, doing so by contemplating a role of mesophyll-derived CO_2- or sugar-signals. Further research is needed to test whether stomatal opening is delayed when photosynthetic electron transport or the ability to extrude protons are impaired due to salinity-induced deficiency of magnesium. Furthermore, we discuss an opinion that sulfur might be involved in guard cell

Advances in Botanical Research, Volume 103
ISSN 0065-2296
https://doi.org/10.1016/bs.abr.2022.02.011

regulation in crop plants. Our article may provoke further research to broaden the knowledge of the underlying guard cell physiology.

Abbreviations

CO_2 carbon dioxide
GABA gamma amino butyric acid
GC guard cells

1. Introduction

Stomata are tiny pores formed by opposingly arranged guard cells, which operate as gatekeepers at the boundary of atmosphere and plant interior. Guard cells (GC) mediate the regulation of stomatal aperture in response intrinsic and environmental cues *via* transient solute accumulation (Jezek & Blatt, 2017) and thus balance the trade-off between transpirational water loss and CO_2 intake as precursor for photosynthesis (Lawson & Blatt, 2014; McAusland et al., 2016). Inorganic ions such as K^+, Cl^-, and NO_3^- as well as organic solutes contribute to the GC osmotic adjustment (Lawson & Matthews, 2020). The provision of energy equivalents by the GC metabolism is pivotal to energize the active solute transport and biosynthesis of organic solutes such as malate. In response to stress events such as drought or salinity, plants undergo fast physiological changes to mitigate or prevent deleterious stress effects. Within minutes, plants close their stomatal pores to retain hydration (Geilfus, Mithofer, Ludwig-Muller, Zörb, & Mühling, 2015; Lawson & Blatt, 2014), an adaptation that is crucial for the survival of salt-sensitive plants such as most of our crops in saline environments (Hedrich & Shabala, 2018).

Climate change is expected to lead to increased soil salinization and more frequent drought periods. With high evapotranspiration rates, soluble salts will further concentrate in the soil. This will be exacerbated by the need for irrigation, leading to further input of salts into agricultural soils (Kotula, Caparros, Zörb, Colmer, & Flowers, 2020). With salinization, the fertility of the soil decreases and the nutrient uptake of plants is hindered by the prevailing excess of salt ions (Zörb & Geilfus, 2019). High Na^+ in soil solution disrupts uptake of other cations such as potassium (Zörb, Senbayram, & Peiter, 2014), calcium (Gardner, 2016), and magnesium (Jezek, Geilfus, Bayer, & Mühling, 2015). Similarly, high soil chloride concentrations can compete with the uptake of anions such as nitrate and phosphate (Geilfus, 2018). Also, there is evidence of potential antagonism between chloride and sulfate uptake. The resulting nutritional imbalance, in combination with the adverse effects of the accumulation of unwanted salt ions (*e.g.*, Na^+ and to high Cl^-) in plant tissues comprise plant metabolism and thus growth. The response of plant metabolism at the level of leaves is extensively studied in many species (Sanchez, Siahpoosh, Roessner, Udvardi, & Kopka, 2008). However, the importance of stress-induced nutritional imbalances for guard cell metabolism received less attention so far, although a dysfunction of the stomata would inevitably have serious negative effects on the plant survival due to their importance for the maintenance of the water balance. We provide evidence and arguments for future research and discussion on the underlying mechanisms of possible regulation processes under hyperosmotic saline conditions in guard cells of crop plants.

In this review article, we aim to shed the light on three aspects of stress-related nutritional imbalances and their potential consequences for guard cell metabolism and stomata regulation: (*i*) Mg deficiency, (*ii*) Na^+ accumulation, and (*iii*) S deficiency.

2. The possible role of magnesium in stomatal regulation under sodium-salinity

There are numerous reports witnessing the antagonist effect between Na^+ and magnesium (Mg^{2+}) uptake in crops under NaCl salinity. NaCl salinity results in a decrease of the amount of Mg^{2+} in Greek olive cultivars (Loupassaki, Chartzoulakis, Digalaki, & Androulakis, 2002). Intracellular accumulation of sodium correlates with a reduction of Mg^{2+} in maize. A more detailed analysis by Shahzad, Zörb, Geilfus, and Mühling (2013) on *Vicia faba* confirmed that the symplastic concentration of Mg^{2+} decreased while the apoplastic concentration of Mg^{2+} increased with an increasing dose of NaCl. Since the symplastic fraction dominates over the apoplastic fraction with regard to quantity, total concentration of Mg^{2+} decreased in the leaves (Fig. 1A). With increasing NaCl concentration in irrigation water, the amount of Mg^{2+} in the tissues of pomegranate (*Punica granatum* L.) decreased (Naeini, Khoshgoftarmanesh, Lessani, & Fallahi, 2005). In soybean (*Glycine max* L.), NaCl stress reduced the accumulation Mg^{2+} in the leaves (Essa, 2002). This is in line with a report that salt stress in strawberry can be mitigated by foliar application of Mg^{2+} (Yildirim, Karlidag, & Turan, 2009). Reproductive development of salt-stressed *Plantago crassifolia*

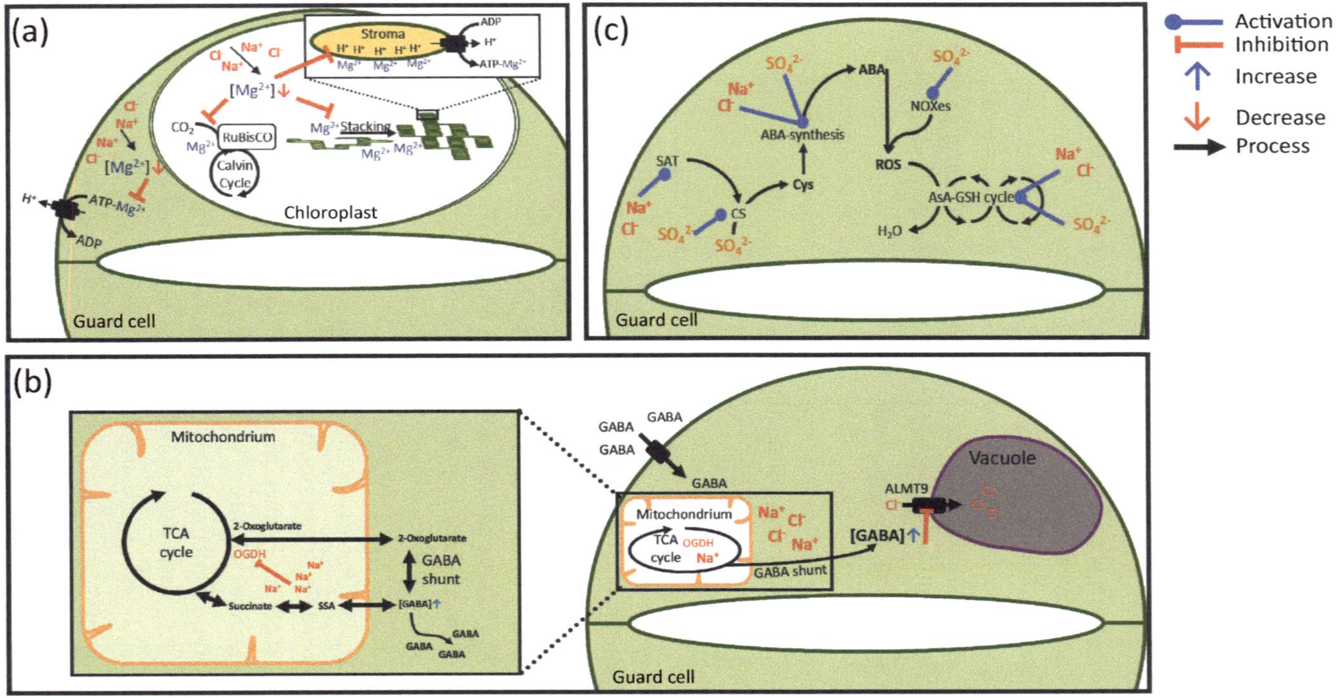

Fig. 1 Possible roles of (A) magnesium, (B) GABA, and (C) sulfur on guard cell metabolism under salt stress. Processes in guard cell chloroplasts, mitochondria, and vacuoles are depicted. GABA, gamma-aminobutyric acid; TCA cycle, tricarboxylic acids cycle; OGDH, 2-oxoglutarate dehydrogenase; SSA, succinic semi-aldehyde; ALMT9, malate-activated vacuolar chloride channel; NOXes, NADPH oxidases; ROS, reactive oxygen species; AsA-GSH cycle, ascorbate glutathione cycle; SAT, serine acetyltransferases; CS, cysteine synthase.

(thick-leaved plantain) can also be mitigated by Mg^{2+} (Grigore, Boscaiu, Llinares, & Vicente, 2012). What literature does not reveal is whether or not such an NaCl-induced decrease in the leaf content of Mg^{2+} is relevant for the effects that are exerted by Mg^{2+} for stomatal regulation.

Few studies indicate that lack of Mg^{2+} reduces transpiration (Jezek et al., 2015), stomatal conductance (Sun & Payn, 1999), and water-use efficiency (Tränkner et al., 2016). A reduction in transpiration of Mg^{2+}-deficient leaves might be related to the apoplastic concentration of CO_2. Consumption of CO_2 through chloroplastic CO_2 fixation in red light-illuminated mesophyll cells reduces the leaf apoplastic CO_2 concentration, to which stomata respond by widening the aperture for enhancing uptake of CO_2 (Roelfsema, Hanstein, Felle, & Hedrich, 2002). Since mesophyll CO_2 fixation drops under Mg^{2+} deficiency (Mehne-Jakobs, 1996), apoplastic CO_2 might not drop under Mg^{2+} deficiency because less CO_2 is fixed by mesophyll photosynthesis. This might influence guard cell regulation, leading to a reduced stomatal conductance. However, it was also reported that stomatal conductance decreased under Mg^{2+} deficiency, even though apoplastic CO_2 concentration remained constant (Tang, Li, & Chen, 2012). Findings from Messinger, Buckley, and Mott (2006) support the idea of stomata-regulating signals that act independently from a mesophyll-driven consumption of apoplastic CO_2. Their data indicate that red light promotes stomatal opening although internal CO_2 remains unchanged. Studies on experimental systems such as epidermal stripes or GC protoplasts that operate independently from mesophyll-derived signals substantiated the idea of the existence of GC-internal red light-signals that govern stomatal control, acting fully independent from the mesophyll photosynthesis (Lawson & Matthews, 2020). For instance, a study on *V. faba* epidermal stripes indicated that a red light-induced accumulation of sucrose in GC is functional for stomatal opening. The sucrose might account as additional osmotica in GC, being relevant for stomatal opening (Lawson, 2009). In addition to the osmotic role, sucrose can be catabolized in the GC for producing ATP required to energize stomatal opening. In GC, sucrose can be derived photosynthetically from the Mg^{2+}-dependent Calvin cycle. A reason why Calvin cycle is inhibited in the absence of Mg^{2+} is that Mg^{2+} promotes ribulose-1,5-bisphosphate carboxylase/oxygenase activity by mediating the binding of CO_2 to this enzyme (Fig. 1A).

The plasma membrane (PM) proton (H^+)-translocating ATPase (P-type PM-H^+-ATPase) is another enzyme that depends on Mg^{2+}. Binding of Mg^{2+} is required for the H^+ pumping activity, because Mg^{2+} is part of the enzyme's substrate Mg^{2+}-ATP (note: not free ATP). The optimal concentration of

Mg^{2+}-ATP serves as the substrate, whereas free ATP inhibits activity. Deficiency of Mg^{2+} reduces the activity of ATPase (Hanstein et al., 2011). The activity of the PM-H^+-ATPase is important for photosynthesis and GC functioning. For instance, at dawn, extrusion of H^+ out of the GCs is relevant for GC swelling, *viz.* for light-induced stomatal opening to facilitate intake of CO_2 for fixation in the Calvin cycle (Inoue & Kinoshita, 2017). As for most ATP-dependent enzymes, Mg^{2+} is a cofactor required for the hydrolysis of the high-energy phosphoanhydride bond. The ratio of Mg^{2+}-ATP complex to ATP determines PM-H^+-ATPase activity. At high concentrations of ATP without bound Mg^{2+} and as a result of Mg^{2+} deficiency (Hanstein et al., 2011), a decrease in activity is observed that illustrates the relevance of Mg^{2+} availability for ATP-dependent processes. At predawn, the GC-intrinsic ATP pool that fuels initiation of stomatal opening is at first fed from breakdown of GC-located triacylglycerols (McLachlan et al., 2016). At postdawn, GC photosynthesis is likely to take over production of ATP (Geilfus, Lan, & Carpentier, 2018). The red light-induced stomatal opening is partially fueled with ATP from GC-internal chloroplastic photosynthetic electron transport. This was early suggested by Serrano, Zeiger, and Hagiwara (1988) and is supported by studies on both GC protoplasts and epidermal stripes from *V. faba* (Olsen, Pratt, Gump, Kemper, & Tallman, 2002). Salinity might affect the production of ATP from GC-internal chloroplastic photosynthetic electron transport because this process largely depends on Mg^{2+}. These considerations assume that sodium-salinity lowers the amount of available Mg^{2+}. As central atom of the chlorophyll molecule, it is relevant for the excitation of the electrons during conversion of light energy into chemical energy (Milne, Toker, Rubio, & Nielsen, 2015). This is one reason why photosynthetic light reaction is impaired under Mg^{2+} deficiency (Hermans, Johnson, Strasser, & Verbruggen, 2004). Other reason is the relevance of Mg^{2+} for the establishment of the H^+-gradient across the thylakoid membrane during light-driven production of photosynthetic ATP in the chloroplasts. This is because the light-induced transfer of H^+ into the thylakoid lumen is counterbalanced by Mg^{2+} efflux into the chloroplast stroma (Höhner, Aboukila, Kunz, & Venema, 2016). In addition, Mg^{2+} ensures correct stacking of thylakoid membranes, which is important for light harvesting. In consequence, Mg^{2+} deficiency leads to disorganized thylakoid membranes, *i.e.* unstacking, which decreases nonphotochemical quenching of chlorophyll fluorescence (Goss, Oroszi, & Wilhelm, 2007). Thus, activities of PSI and PSII are affected by Mg^{2+} deficiency through a loss of antenna and reaction centers, respectively (Hermans et al., 2004).

Despite the multifaceted function of Mg^{2+} in cellular metabolism, the significance of salt-induced Mg^{2+} deficiency on guard cells and stomatal regulation, *i.e.* potential delay of osmotic adjustment and thus stomatal movements, is sparsely studied.

3. The possible role of salt stress-induced GABA production as metabolic signal involved in reduction of stomatal aperture

Salt and drought stress lead to modulation of certain metabolites such as the nonproteinogenic amino acid gamma-aminobutyric acid (GABA). The fast closing of stomatal pores to reduce transpirational water loss is an important response to drought or salt stress that contributes to the maintenance of plant water status and thus survival. GABA negatively regulates the major anion transporter, the "aluminum-activated malate transporter 9" (ALMT-9) which is involved in the light-induced Cl^- uptake into GC vacuoles, when Cl^- accumulates in the cytosol (Xu, Long, et al., 2021). A GABA-mediated attenuation of ALMT9 activity results in a suppressed stomatal opening, leading to reduced maximal aperture, and thus leading to less transpirational water loss (Fig. 1B). The GABA induced effect on anion transport regulation was observed throughout a selection of mono- and dicot species, indicating a conserved physiological signaling mechanism coordinating metabolism and stomatal aperture (Xu, Long, et al., 2021) to enhance water-use-efficiency and thus drought resilience of plants (Xu, Sai, & Gilliham, 2021). However, the accumulation of GABA is triggered by a variety of stressors and thus might also play a role in regulating stomata adjustments in response to other stresses than drought stress that require regulation of water flow, such as salt stress. GABA is thought to be involved in stress mitigation, for example, by regulating pH, balancing the carbon-to-nitrogen ratio and signaling that modifies protein activity (Ramesh, Tyerman, Gilliham, & Xu, 2017). Salt stress-induced accumulation of GABA occurs in many species, whereas the extent and speed depend on various factors such as plant age, stress dose, or relative salt tolerance (Kinnersley & Turano, 2000; Shelp, Bown, & Zarei, 2017).

GABA can be synthesized from putrescine *via* polyamine catabolism (reviewed by Shelp et al., 2017), or potentially from proline during oxidative stress conditions (Signorelli, Dans, Coitiño, Borsani, & Monza, 2015). The majority of the GABA produced is attributed to the activity of glutamate decarboxylase (GAD) that catalyzes the irreversible decarboxylation of glutamic acid to GABA (Che-Othman, Millar, & Taylor, 2017). Therefore, GAD is a proposed determinant of the activity of the GABA shunt pathway (Bown & Shelp, 2020). Enzymatic activity of GAD increases with both cytosolic H^+ and Ca^{2+} concentrations. Hence, stimuli affecting H^+ and Ca^{2+} levels such as cold, heat, drought, salinity or mechanical and osmotic stress lead to increased GABA biosynthesis in the cytosol (Kinnersley & Turano, 2000). Metabolization of cytosolic GABA occurs after import into the mitochondrion *via* conversion to succinic semi-aldehyde (SSA) and succinate, which can be fed back into the mitochondrial tricarboxylic acid cycle (TCA cycle) (Fig. 1B). However, GABA metabolization is suppressed under stress due to inhibition of SSA dehydrogenase by stress-related increases in redox potential (Shelp et al., 2017). In addition to the H^+ and Ca^{2+} -dependent activation, GABA shunt was found to be induced by ionic imbalance resulting from severe salt stress in wheat leaves (Che-Othman, Jacoby, Millar, & Taylor, 2020). The accumulation of Na^+, especially in the mitochondria, is thought to inhibit 2-oxoglutarate dehydrogenase, which is involved in the TCA cycle (Fig. 1B). By activating the GABA shunt, carbon skeletons other than glycolysis-derived pyruvate might be fed into mitochondrial respiration, bypassing the metabolic reactions hindered by excess salt ions in mitochondria. Thus, GABA shunt activity increases as a result of salt ion concentration-dependent enzyme inhibition in salt loaded plant tissues. Under high salt conditions, there might be a constant salt ion-dependent activation of the GABA shunt leading to production of GABA in the cytosol.

In a view of a potential stress-induced GABA production within GC, Xu, Long, et al. (2021) demonstrated that a 1.25-fold GABA increase (analyzed in a whole leaf), was sufficient to trigger the stomata closure effect. Could such increases in GABA content occur in response to increased salt ion concentrations resulting from salt stress? Under salinity, GC are also affected by high Na^+ and Cl^- concentrations. Elemental analysis of GC by means of energy dispersive X-ray revealed an about 15-fold increase in Na^+ concentration in both *Arabidopsis thaliana* and the more salt-tolerant relative *Thellungiella salsuginea*, while chloride concentrations increased threefold only in *A. thaliana* (Karimi et al., 2021). Mechanically isolated GC from *V. faba* exhibited fivefold increases in Na^+ concentrations (Franzisky et al., 2021). In comparison with the 80-fold increase in Na^+ that triggered significant reductions in wheat TCA cycle enzyme activities (Che-Othman et al., 2020), the increases observed in GC are significantly lower. Accordingly, GABA did not increase in *V. faba* GC under high salt, while the expected GABA accumulation was present in leaves (Franzisky et al., 2021). Analysis of the GC metabolic profile by using a mixed graphical model (Lee & Hastie, 2015), which estimates associations between continuous and discrete variables, confirmed the negative association between guard cells GABA content and salt treatment under these experimental conditions (analyzed according to Altenbuchinger et al., 2019, based on data of Franzisky et al., 2021). Unfortunately, there are no other data on GC metabolome to support this model. Moreover, changes in GC protein abundance such as available for the salt-tolerant species quinoa (Rasouli et al., 2021) and sugar beet (Rasouli et al., 2020) provide limited information about GABA shunt activity because the inhibitory effect of high Na^+ on TCA cycle enzymes is posttranslationally regulated (Che-Othman et al., 2020). A potential role of GC exogenous GABA triggering the suppression of stomatal opening is evidenced by experiments in which GABA is fed into leaves *via* the petiole and overproduced in leaves by complementation of *gad2* mutants with *GAD2* under control of the 35S promoter (Xu, Long, et al., 2021). Both treatments were able to recover the drought sensitivity phenotype of the *gad2 A. thaliana* mutant, demonstrating that exogenous GABA affects stomatal regulation (Xu, Sai, et al., 2021). Thus, salt stress-induced GABA production from other cells than GC, *e.g.* mesophyll, might be involved in the tuning of stomatal aperture under salinity. The overproduction of GABA achieved by the 35S:*GAD2* complementation resulted in sixfold increased total leaf GABA content. Similar or higher salt stress-related increases in leaf GABA content were reported for various species and NaCl doses, for example, 11-fold GABA increase in soybean treated with 50 mM for 6 days (Xing, Jun, Hau, & Liang, 2007) or 10-fold in *A. thaliana* treated with 150 mM for 4 days (Renault et al., 2010). Regarding the plasma membrane located GABA transporters described for various species such as *A. thaliana*, wheat and tomato (Fromm, 2020; Xu, Sai, et al., 2021) and the experimental evidence for import of petiole-fed GABA into the GC (Xu, Long, et al., 2021) it is likely that exogenous GABA can be transported into GC. However, it remains to be determined whether the GABA produced in mesophyll under stress is mainly redirected into the mitochondrial TCA cycle as carbon source for respiration or whether it can also reach to GC *via* the apoplast and act as a metabolic signal in the coordination of leaf stomatal regulation. Further studies on GC metabolism supported by *in vivo* GABA imaging (Marvin et al., 2019) might resolve a prospective effect of salt ion-dependent GABA accumulation in GC on stomatal closure under salt stress.

4. The possible role of sulfur in guard cell metabolism under salt stress

Sulfur is essential for plants as it is involved in biosynthesis of the amino acids cysteine and methionine. Moreover, sulfur-containing substances such as reactive oxygen species (ROS) are crucial for abiotic stress resistance and pests and diseases. Thus, sulfur-containing compounds such as glutathione or thioredoxin play a critical role in plant responses to abiotic stresses such as salinity or drought stress (Cao et al., 2014; Fatma et al., 2021). Recently, sulfur dioxide has been considered to be a beneficial bioregulator in animals. However, the positive roles of sulfur in plant adaptation to salt or drought stress are an upcoming research topic. In recent studies there is hint that SO_2 fumigation improves the drought tolerance of foxtail millet seedlings. In that study it was reported that the relative water content in the leaves of drought-stressed seedlings was significantly improved by SO_2 exposure. It was concluded that these responses might be related to decreased stomatal apertures and a reduced leaf transpiration rate, which were induced by SO_2 under drought conditions (Han, Yang, Wu, & Yi, 2019). Another study reports a promoted stomatal behavior by the application of NO plus S (and split N) in the presence of salt stress (Jahan, AlAjmi, Rehman, & Khan, 2020). Therefore, in this paragraph we will provide new ideas of possible role of sulfur in guard cell metabolism under salt stress, even because the topic has not been considered much in the recent literature. We would like to note here, that because of the still limited amount of data, this consideration has a thoroughly speculative momentum, therefore this paragraph provides a starting point for future GC research and the involvement of sulfur.

There is more recent information that increased sulfur fertilization decreases Na^+ and Cl^- concentrations in leaves of *Brassica juncea* L. under NaCl stress (Fatma et al., 2021). These authors found that similarly sulfur uptake was reduced under high NaCl salinity. A possible explanation is the antagonism of sulfate and chloride with a reduction of SO_4^{2-} uptake at increased Cl^- concentration in the soil solution. The antagonism might be enhanced by an increased sulfur fertilization under saline conditions. In addition, a positive influence of sulfur on photosynthetic efficiency was reported under nonsaline conditions and especially under saline conditions, this was characterized by increased chlorophyll *a* and chlorophyll *b* concentrations in leaves, together with increased maximal PSII photochemical efficiency and enhanced net photosynthesis. This is underscored by an increased intercellular CO_2 concentration and an increased stomatal conductance, which also indicates an influence on the regulation of GCs (Fatma et al., 2021).

The cell damage induced by NaCl stress often occurs in association with the accumulation of ROS. Therefore, it is interesting that increased sulfur fertilization was shown to decrease the concentration of H_2O_2 and thiobarbituric acid reactive substances (TBARS), which were associated with a concomitant increase in ascorbate peroxidase, glutathione reductase, dehydroascorbate reductase, and monodehydroascorbate reductase activity (Fig. 1C). Thus, the redox state with reduced glutathione (GSH)/oxidized glutathione (GSSG), was increased under normal conditions and especially under NaCl stress. Here, the higher glutathione reductase activity helped to maintain the GSH level and increase redox state (GSH/GSSG) (Fatma et al., 2021). Together with glutathione, reduced ascorbate (AsA) is an important factor for the detoxification of ROS (Fig. 1C). A monodehydroascorbate reductase plays a key role in sustaining a reduced pool of AsA and maintains its redox state together with dihydroxy-ascorbate reductase under NaCl stress (Fatma et al., 2021). The increase of monodehydroascorbate reductase activity after sulfur treatment under NaCl stress suggests that this enzyme has the potential to reduce NaCl-induced oxidative stress and may increase the AsA pool (Fig. 1C). Thus, the activation of the AsA-GSH cycle by sulfur under NaCl conditions helped to reduce NaCl-induced oxidative stress (H_2O_2 and TBARS) (Fatma et al., 2021). Here is a close link since the AsA-GSH cycle also plays an important role in the detoxification of ROS in GC (Rasouli et al., 2021), it is conceivable that by increasing sulfur availability under NaCl conditions, more GSH is present in GC, thereby lowering the ROS concentration in GC. It has already been shown that ROS can not only cause damage but also act as a signaling molecule (Batool et al., 2018). In GC, ROS have been shown to be an important signal for ABA induced stomatal closure. Thus, lowering the concentration of ROS would support the opening of stomata under NaCl conditions. In turn, it has been shown in GC that SO_4^{2-} activates NADPH oxidases, which generates ROS, leading to stomatal closure in an ABA-dependent manner (Batool et al., 2018) (Fig. 1C). However, our knowledge of the interaction of sulfur metabolism and ABA biosynthesis is scarce. There might be a possible connection because at ABA biosynthesis, the final reaction converting abscisic aldehyde to ABA is catalyzed by abscisic aldehyde oxidase 3 (AAO3), which requires the molybdenum cofactor (Moco) (Cao et al., 2014) and Moco must be sulfurized to act as a cofactor of AAO3.

This sulfurization is catalyzed by Moco sulfurase (ABA3), which requires cysteine as a sulfur donor. The sulfurization step catalyzed by ABA3 is thought to provide an efficient way to regulate the amount of active AAO3 enzymes under various physiological stresses. Indeed, ABA3 transcription is rapidly induced by drought stress, NaCl stress, and ABA treatment. Therefore, AAO3 activity may serve as a link of ABA biosynthesis and sulfur metabolism.

Another indication of a link between ABA biosynthesis and sulfur metabolism is the nature of serine acetyltransferases (SAT) (Fig. 1C). Cysteine is formed by the cysteine synthase complex, depending on SAT. It was shown for *Oryza sativa* L., that gene expression of SATs was upregulated under NaCl stress (Kurt, Filiz, & Aydın, 2021), indicating increased cysteine availability. It is likely that these processes also occur in GC, under saline conditions, this would imply that NaCl stress stimulates the synthesis of cysteine, leading to increased synthesis of ABA. Moreover, assimilation of sulfate into cysteine was shown to induce the transcription of nine-*cis*-epoxy-carotenoid-dioxygenase 3 (NCED3), limiting the synthesis of the substrate for AAO3 (Batool et al., 2018). Therefore, sulfate upregulates the expression of NCED3, which is a key enzyme of ABA synthesis in the cytosol of GC (Malcheska et al., 2017). This process will result in an increased ABA biosynthesis and will further support stomatal closure.

Although our knowledge of the interaction between sulfur metabolism and ABA biosynthesis is scarce, it is emerging that sulfur nutrition may be a way to mitigate negative effects of NaCl conditions through the synthesis of sulfur compounds of the AsA-GSH cycle. It can be hypothesized that adaptation of sulfate uptake and assimilation is one more critical factor for NaCl and drought tolerance of crops. More research on that topic is needed.

5. Conclusion

By far not all questions that arise in relation to salt stress-related deficits of plants have been put in relation to guard cell physiology. We prognosticate potential for further research here. In this review we discuss ions and their interaction in guard cell regulation under saline conditions. Besides, the role of potassium in guard cell physiology is widely known, up to date the understanding of the relevance of a salinity-induced reduction of Mg^{2+} for stomatal opening is rudimentary. Few reports have linked the Mg^{2+}-nutritional status to transpiration, doing so by contemplating a role of mesophyll-derived CO_2- or sugar-signals. In terms of key metabolites, salt stress induces GABA accumulation, GABA negatively regulates tonoplast anion channel activity and thus affects stomatal aperture. High Na^+ concentrations may activate the GABA shunt in guard cells. Moreover we provide an opinion that sulfate may function in guard cell regulation under salinity stress. Fig. 1 provides an overview over the described processes in this review.

References

Altenbuchinger, M., Zacharias, H. U., Solbrig, S., Schäfer, A., Büyüközkan, M., Schultheiß, U. T., et al. (2019). A multi-source data integration approach reveals novel associations between metabolites and renal outcomes in the German Chronic Kidney Disease study. *Scientific Reports*, *9*(1), 1–13.

Batool, S., Uslu, V. V., Rajab, H., Ahmad, N., Waadt, R., Geiger, D., et al. (2018). Sulfate is incorporated into cysteine to trigger ABA production and stomatal closure. *The Plant Cell*, *30*(12), 2973–2987. https://doi.org/10.1105/tpc.18.00612.

Bown, A. W., & Shelp, B. J. (2020). Does the GABA shunt regulate cytosolic GABA? *Trends in Plant Science*, *25*(5), 422–424. https://doi.org/10.1016/j.tplants.2020.03.001.

Cao, M. J., Wang, Z., Zhao, Q., Mao, J. L., Speiser, A., Wirtz, M., et al. (2014). Sulfate availability affects ABA levels and germination response to ABA and salt stress in *Arabidopsis thaliana*. *The Plant Journal*, 77(4), 604–615. https://doi.org/10.1111/tpj.12407.

Che-Othman, M. H., Jacoby, R. P., Millar, A. H., & Taylor, N. L. (2020). Wheat mitochondrial respiration shifts from the tricarboxylic acid cycle to the GABA shunt under salt stress. *The New Phytologist*, *225*(3), 1166–1180. https://doi.org/10.1111/nph.15713.

Che-Othman, M. H., Millar, A. H., & Taylor, N. L. (2017). Connecting salt stress signalling pathways with salinity-induced changes in mitochondrial metabolic processes in C3 plants. *Plant, Cell & Environment*, *40*(12), 2875–2905. https://doi.org/10.1111/pce.13034.

Essa, T. A. (2002). Effect of salinity stress on growth and nutrient composition of three soybean (*Glycine max* L. Merrill) cultivars. *Journal of Agronomy and Crop Science*, *188*(2), 86–93.

Fatma, M., Iqbal, N., Gautam, H., Sehar, Z., Sofo, A., D'Ippolito, I., et al. (2021). Ethylene and sulfur coordinately modulate the antioxidant system and ABA accumulation in mustard. *Plants (Basel)*, *10*(1), 180. https://doi.org/10.3390/plants10010180.

Franzisky, B. L., Geilfus, C.-M., Romo-Pérez, M. L., Fehrle, I., Erban, A., Kopka, J., et al. (2021). Acclimatisation of guard cell metabolism to long-term salinity. *Plant, Cell & Environment*, *44*(3), 870–884. https://doi.org/10.1111/pce.13964.

Fromm, H. (2020). GABA signaling in plants: Targeting the missing pieces of the puzzle. *Journal of Experimental Botany*, *71*(20), 6238–6245. https://doi.org/10.1093/jxb/eraa358.

Gardner, W. K. (2016). Sodium, calcium and magnesium ratios in soils of NW Victoria, Australia may restrict root growth and crop production. *Journal of Plant Nutrition*, *39*, 1205–1215.

Geilfus, C. M. (2018). Chloride: From nutrient to toxicant. *Plant and Cell Physiology*, *59*(5), 877–886.

Geilfus, C. M., Lan, J., & Carpentier, S. (2018). Dawn regulates guard cell proteins in *Arabidopsis thaliana* that function in ATP production from fatty acid beta-oxidation. *Plant Molecular Biology*, *98*(6), 525–543.

Geilfus, C.-M., Mithofer, A., Ludwig-Muller, J., Zörb, C., & Mühling, K.-H. (2015). Chloride-inducible transient apoplastic alkalinizations induce stomata closure by

controlling abscisic acid distribution between leaf apoplast and guard cells in salt-stressed *Vicia faba*. *The New Phytologist, 208*(3), 803–816.
Goss, R., Oroszi, S., & Wilhelm, C. (2007). The importance of grana stacking for xanthophyll cycle-dependent NPQ in the thylakoid membranes of higher plants. *Physiologia Plantarum, 131*(3), 496–507.
Grigore, M. N., Boscaiu, M., Llinares, J., & Vicente, O. (2012). Mitigation of salt stress-induced inhibition of *Plantago crassifolia* reproductive development by supplemental calcium or magnesium. *Notulae Botanicae Horti Agrobotanici Cluj-Napoca, 40*(2), 58–66.
Han, Y., Yang, H., Wu, M., & Yi, H. (2019). Enhanced drought tolerance of foxtail millet seedlings by sulfur dioxide fumigation. *Ecotoxicology and Environmental Safety, 30*(178), 9–16. https://doi.org/10.1016/j.ecoenv.2019.04.006.
Hanstein, S., Wang, X., Qian, X., Friedhoff, P., Fatima, A., Shan, Y., et al. (2011). Changes in cytosolic Mg^{2+} levels can regulate the activity of the plasma membrane H^{+}-ATPase in maize. *Biochemical Journal, 435*(1), 93–101.
Hedrich, R., & Shabala, S. (2018). Stomata in a saline world. *Current Opinion in Plant Biology, 46*, 87–95. https://doi.org/10.1016/j.pbi.2018.07.015.
Hermans, C., Johnson, G. N., Strasser, R. J., & Verbruggen, N. (2004). Physiological characterisation of Mg^{2+} deficiency in sugar beet: Acclimation to low magnesium differentially affects photosystems I and II. *Planta, 220*(2), 344–355.
Höhner, R., Aboukila, A., Kunz, H. H., & Venema, K. (2016). Proton gradients and proton-dependent transport processes in the chloroplast. *Frontiers in Plant Science, 7*, 218.
Inoue, S. I., & Kinoshita, T. (2017). Blue light regulation of stomatal opening and the plasma membrane H^{+}-ATPase. *Plant Physiology, 174*(2), 531–538. https://doi.org/10.1104/pp.17.00166.
Jahan, B., AlAjmi, M. F., Rehman, M. T., & Khan, N. A. (2020). Treatment of nitric oxide supplemented with nitrogen and sulfur regulates photosynthetic performance and stomatal behavior in mustard under salt stress. *Physiologia Plantarum, 168*(2), 490–510. https://doi.org/10.1111/ppl.13056.
Jezek, M., & Blatt, M. R. (2017). The membrane transport system of the guard cell and its integration for stomatal dynamics. *Plant Physiology, 174*(2), 487–519.
Jezek, M., Geilfus, C. M., Bayer, A., & Mühling, K. H. (2015). Photosynthetic capacity, nutrient status, and growth of maize (*Zea mays* L.) upon $MgSO_4$ leaf-application. *Frontiers in Plant Science, 5*, 781.
Karimi, S. M., Freund, M., Wager, B. M., Knoblauch, M., Fromm, J. M., Mueller, H., et al. (2021). Under salt stress guard cell rewire ion transport and abscisic acid signaling. *The New Phytologist, 231*(3), 1040–1055. https://doi.org/10.1111/nph.17376.
Kinnersley, A. M., & Turano, F. J. (2000). Gamma aminobutyric acid (GABA) and plant responses to stress. *Critical Reviews in Plant Sciences, 19*(6), 479–509. https://doi.org/10.1080/07352680091139277.
Kotula, L., Caparros, P. G., Zörb, C., Colmer, T. D., & Flowers, T. J. (2020). Improving crop salt tolerance using transgenic approaches: An update and physiological analysis. *Plant, Cell & Environment, 43*(12), 2932–2956. https://doi.org/10.1111/pce.13865.
Kurt, F., Filiz, E., & Aydın, A. (2021). Genome-wide identification of serine acetyltransferase (SAT) gene family in rice (*Oryza sativa*) and their expressions under salt stress. *Molecular Biology Reports, 48*(9), 6277–6290. https://doi.org/10.1007/s11033-021-06620-6.
Lawson, T. (2009). Guard cell photosynthesis and stomatal function. *The New Phytologist, 181*(1), 13–34.
Lawson, T., & Blatt, M. R. (2014). Stomatal size, speed, and responsiveness impact on photosynthesis and water use efficiency. *Plant Physiology, 164*(4), 1556–1570.
Lawson, T., & Matthews, J. (2020). Guard cell metabolism and stomatal function. *Annual Review of Plant Biology, 71*, 273–302.

Lee, J. D., & Hastie, T. J. (2015). Learning the structure of mixed graphical models. *Journal of Computational and Graphical Statistics*, *24*(1), 230–253.

Loupassaki, M. H., Chartzoulakis, K. S., Digalaki, N. B., & Androulakis, I. I. (2002). Effects of salt stress on concentration of nitrogen, phosphorus, potassium, calcium, magnesium, and sodium in leaves, shoots, and roots of six olive cultivars. *Journal of Plant Nutrition*, *25*(11), 2457–2482.

Malcheska, F., Ahmad, A., Batool, S., Müller, H. M., Ludwig-Müller, J., Kreuzwieser, J., et al. (2017). Drought-enhanced xylem sap sulfate closes stomata by affecting ALMT12 and guard cell ABA synthesis. *Plant Physiology*, *174*(2), 798–814. https://doi.org/10.1104/pp.16.01784.

Marvin, J. S., Shimoda, Y., Magloire, V., Leite, M., Kawashima, T., Jensen, T. P., et al. (2019). A genetically encoded fluorescent sensor for in vivo imaging of GABA. *Nature Methods*, *16*(8), 763–770. https://doi.org/10.1038/s41592-019-0471-2.

McAusland, L., Vialet-Chabrand, S., Davey, P., Baker, N. R., Brendel, O., & Lawson, T. (2016). Effects of kinetics of light-induced stomatal responses on photosynthesis and water-use efficiency. *The New Phytologist*, *211*(4), 1209–1220.

McLachlan, D. H., Lan, J., Geilfus, C. M., Dodd, A. N., Larson, T., Baker, A., et al. (2016). The breakdown of stored triacylglycerols is required during light-induced stomatal opening. *Current Biology*, *26*(5), 707–712.

Mehne-Jakobs, B. (1996). Magnesium deficiency treatment causes reductions in photosynthesis of well-nourished Norway spruce. *Trees*, *10*(5), 293–300.

Messinger, S. M., Buckley, T. N., & Mott, K. A. (2006). Evidence for involvement of photosynthetic processes in the stomatal response to CO_2. *Plant Physiology*, *140*(2), 771–778.

Milne, B. F., Toker, Y., Rubio, A., & Nielsen, S. B. (2015). Unraveling the intrinsic color of chlorophyll. *Angewandte Chemie International Edition*, *54*(7), 2170–2173.

Naeini, M. R., Khoshgoftarmanesh, A. H., Lessani, H., & Fallahi, E. (2005). Effects of sodium chloride-induced salinity on mineral nutrients and soluble sugars in three commercial cultivars of pomegranate. *Journal of Plant Nutrition*, *27*(8), 1319–1326.

Olsen, R. L., Pratt, R. B., Gump, P., Kemper, A., & Tallman, G. (2002). Red light activates a chloroplast-dependent ion uptake mechanism for stomatal opening under reduced CO_2 concentrations in *Vicia* spp. *The New Phytologist*, *153*(3), 497–508.

Ramesh, S. A., Tyerman, S. D., Gilliham, M., & Xu, B. (2017). γ-Aminobutyric acid (GABA) signalling in plants. *Cellular and Molecular Life Sciences*, *74*(9), 1577–1603. https://doi.org/10.1007/s00018-016-2415-7.

Rasouli, F., Kiani-Pouya, A., Li, L., Zhang, H., Chen, Z., Hedrich, R., et al. (2020). Sugar beet (*Beta vulgaris*) guard cell responses to salinity stress: A proteomic analysis. *International Journal of Molecular Sciences*, *21*(7), 2331. https://doi.org/10.3390/ijms21072331.

Rasouli, F., Kiani-Pouya, A., Shabala, L., Li, L., Tahir, A., Yu, M., et al. (2021). Salinity effects on guard cell proteome in *Chenopodium quinoa*. *International Journal of Molecular Sciences*, *22*(1), 428. https://doi.org/10.3390/ijms22010428.

Renault, H., Roussel, V., El Amrani, A., Arzel, M., Renault, D., Bouchereau, A., et al. (2010). The Arabidopsis pop2-1 mutant reveals the involvement of GABA transaminase in salt stress tolerance. *BMC Plant Biology*, *10*(1), 20. https://doi.org/10.1186/1471-2229-10-20.

Roelfsema, M. R. G., Hanstein, S., Felle, H. H., & Hedrich, R. (2002). CO_2 provides an intermediate link in the red light response of GC. *The Plant Journal*, *32*(1), 65–75.

Sanchez, D. H., Siahpoosh, M. R., Roessner, U., Udvardi, M. K., & Kopka, J. (2008). Plant metabolomics reveals conserved and divergent metabolic responses to salinity. *Physiologia Plantarum*, *132*(2), 209–219.

Serrano, E. E., Zeiger, E., & Hagiwara, S. (1988). Red light stimulates an electrogenic proton pump in *Vicia* guard cell protoplasts. *Proceedings of the National Academy of Sciences of the United States of America*, *85*(2), 436–440.

Shahzad, M., Zörb, C., Geilfus, C. M., & Mühling, K. H. (2013). Apoplastic Na^+ in *Vicia faba* leaves rises after short-term salt stress and is remedied by silicon. *Journal of Agronomy and Crop Science, 199*(3), 161–170.
Shelp, B. J., Bown, A. W., & Zarei, A. (2017). 4-Aminobutyrate (GABA): A metabolite and signal with practical significance. *Botany, 95*(11), 1015–1032. https://doi.org/10.1139/cjb-2017-0135.
Signorelli, S., Dans, P. D., Coitiño, E. L., Borsani, O., & Monza, J. (2015). Connecting proline and γ-aminobutyric acid in stressed plants through non-enzymatic reactions. *PLoS One, 10*(3), e0115349. https://doi.org/10.1371/journal.pone.0115349.
Sun, O. J., & Payn, T. W. (1999). Magnesium nutrition and photosynthesis in *Pinus radiata*: Clonal variation and influence of potassium. *Tree Physiology, 19*(8), 535–540.
Tang, N., Li, Y., & Chen, L. S. (2012). Magnesium deficiency–induced impairment of photosynthesis in leaves of fruiting *Citrus reticulata* trees accompanied by up-regulation of antioxidant metabolism to avoid photo-oxidative damage. *Journal of Plant Nutrition and Soil Science, 175*(5), 784–793.
Tränkner, M., Jákli, B., Tavakol, E., Geilfus, C. M., Cakmak, I., Dittert, K., et al. (2016). Magnesium deficiency decreases biomass water-use efficiency and increases leaf water-use efficiency and oxidative stress in barley plants. *Plant and Soil, 406*(1–2), 409–423.
Xing, S. G., Jun, Y. B., Hau, Z. W., & Liang, L. Y. (2007). Higher accumulation of gamma-aminobutyric acid induced by salt stress through stimulating the activity of diamine oxidases in *Glycine max* (L.) Merr. roots. *Plant Physiology and Biochemistry, 45*(8), 560–566. https://doi.org/10.1016/j.plaphy.2007.05.007.
Xu, B., Long, Y., Feng, X., Zhu, X., Sai, N., Chirkova, L., et al. (2021). GABA signalling modulates stomatal opening to enhance plant water use efficiency and drought resilience. *Nature Communications, 12*(1), 1952. https://doi.org/10.1038/s41467-021-21694-3.
Xu, B., Sai, N., & Gilliham, M. (2021). The emerging role of GABA as a transport regulator and physiological signal. *Plant Physiology, 187*(4), 2005–2016. https://doi.org/10.1093/plphys/kiab347.
Yildirim, E., Karlidag, H., & Turan, M. (2009). Mitigation of salt stress in strawberry by foliar K, Ca and Mg nutrient supply. *Plant, Soil and Environment, 55*, 213–221.
Zörb, C., Geilfus, C.-M., and Dietz, K-J. (2019) Salinity and crop yield. Plant Biology 21, 31–38.
Zörb, C., Senbayram, M., & Peiter, E. (2014). Potassium in agriculture—Status and perspectives. *Journal of Plant Physiology, 171*(9), 656–669. https://doi.org/10.1016/j.jplph.2013.08.008.

CHAPTER SIX

Stomata under salt stress—What can mechanistic modeling tell us?

Nguyen B.A. Thu, Anna Amtmann, Michael R. Blatt*, and Thanh-Hao Nguyen†
Laboratory of Plant Physiology and Biophysics, Bower Building, University of Glasgow, Glasgow, United Kingdom
*Corresponding author: e-mail address: michael.blatt@glasgow.ac.uk

Contents

1. Introduction 140
   1.1 The fundamentals of stomatal movement 141
   1.2 Changes in stomatal ion content under salt stress 143
   1.3 Signaling and transport systems involved in regulation of stomatal movement upon salt exposure 145
   1.4 How does Na^+ stress propagate through guard cell physiology? 148
   1.5 Developing Na^+-sensitivity within the OnGuard platform 149
   1.6 Stomata under salt stress—What can we learn from OnGuard modeling? 155
Acknowledgments 156
References 156

Abstract

Stomata are pores that form between pairs of guard cells and are commonly found in the leaf epidermis. The pores allow gaseous exchange between the inner air spaces of the leaf and the atmosphere, opening for CO_2 entry to support photosynthesis and closing to reduce transpirational water loss. Guard cell membrane transport and its coordination play a central part in regulating the pore aperture. We know how guard cells respond to light, CO_2 and drought with sufficient detail to model with quantitative accuracy their mechanics. By contrast, there is surprisingly little detail about the impact of salt stress on guard cells. Here we review the topic and introduce Na^+ transport within the proven OnGuard3 modeling platform to question the mechanics of stomatal responses to salinity. The analysis indicates a critical role for ABA-mediated Ca^{2+} signaling as an early and overriding process leading to stomatal closure, a beneficial behavior for plants in the face of salt stress.

† Senior author.

Advances in Botanical Research, Volume 103
ISSN 0065-2296
https://doi.org/10.1016/bs.abr.2022.02.012

1. Introduction

Salinity is a major environmental factor limiting plant growth and development, and thus threatening crop productivity and food security (DESA, 2019). Sodium is the fifth most abundant metal in the Earth's crust and the second most abundant ion in water (Greenwood & Earnshow, 1997), and its place among the alkali earth metals means that the Na^+ ion competes with K^+ in many biological activities. Research has long sought to understand and overcome the limitations imposed by Na^+ by generating salt tolerant crops. The bulk of global food production relies on plants that exhibit developmental delays and reduced fitness when Na^+ concentrations rise above 20–30 mM. These effects arise as a combination of osmotic stress and the cytotoxicity of elevated Na^+ and Cl^- within the plant tissues (Munns & Tester, 2008).

Halophytes can sustain growth and development in high salinity through unique anatomical characteristics and associated physiological adaptations that allow them to survive in salt rich habitats. Among these characteristics, for example, are leaf succulence, root suberization, salt secreting glands and bladders (see other chapters of this book). Glycophytes show a range of tolerance to moderate salinity, and research using different varieties or mutants have highlighted important genes (e.g., HKT1, Munns & Tester, 2008) and pathways (e.g., SOS, Zhu, 2003) that allow the plants to cope with moderate salt stress. Such knowledge has provided much information, albeit often corelative or inferred through genetic impairment. The conceptual problem, however, is the reverse of this situation: We want to know what are the minimum of characteristics that might be introduced into crops, whether anatomical, biochemical or molecular, to reduce the impacts of salinity. To this end, it becomes essential both to identify the primary targets of Na^+ on salt-sensitive plants and to understand how their physiological impacts of Na^+ may propagate.

In this chapter, we focus on just one aspect of salt stress, its consequences for stomatal function, and hence for gas exchange, photosynthesis and water use by the plant, that profoundly affects growth and crop yields (Maas & Grattan, 1999; Tang, Mu, Shao, Wang, & Brestic, 2015). There exists a great depth of knowledge about stomata gathered over the past half-century (Jezek & Blatt, 2017; Kollist, Nuhkat, & Roelfsema, 2014; Lawson & Blatt, 2014). Yet remarkably little of this knowledge addresses the detail of how salinity impacts on their function. Some insight in this case can be drawn from mechanistic modeling, such as afforded by the OnGuard platform. To date, OnGuard models have incorporated the mechanics of solute transport, H^+ and Ca^{2+} buffering, and malate metabolism and have shown true power in uncovering homeostatic networks of guard cells and their connections to the environment (Blatt, Jezek, Lew, & Hills, 2022; Blatt, Wang, Leonhardt, & Hills, 2014; Hills, Chen, Amtmann, Blatt, & Lew, 2012). Here, we introduce a minimum of transport for Na^+ within OnGuard3 to ask how salinity may be understood in mechanistic terms as it affects the coordination of transport and metabolism within the stomatal guard cells. In turn, we show how these effects propagate to macroscopic changes in stomatal conductance, photosynthesis and water use.

1.1 The fundamentals of stomatal movement

Stomata, the pore structures on the leaf epidermis formed by pairs of guard cells, facilitate CO_2 uptake for photosynthesis in exchange for water loss by evaporation. This function of stomata directly links to plant water use efficiency (WUE), which defines how much water through transpiration is traded off for carbon fixation in photosynthesis. Stomatal movement is precisely and strictly controlled by the manipulation of the guard cell turgor in response to environmental stimuli. At maturity guard cells lack functional plasmodesmatal connections to adjacent epidermal cells (Wille & Lucas, 1984). Thus, the guard cells rely completely on the ion transport systems available on the plasma membrane and tonoplast to adjust their turgor. During stomatal opening, solutes are transported into the guard cells, including the inorganic solutes K^+ and Cl^-, as well as organics such as malate and sugars. During stomatal closure, guard cells export solutes from the vacuole and cytosol to the apoplast. The changes of solute concentration result in changes in cellular water potential, which then adjusts the cell turgor. The latter is translated to precise stomatal movement. Thus, corresponding transport systems consisting of plasma membrane H^+-ATPase, K^+, Cl^-, and Ca^{2+} channels, and cation- and anion-selective channels at the tonoplast must be involved in these processes. In the past three decades molecular functions of the transport systems and the dynamics of their regulation have been uncovered (for comprehensive reviews, see Kim, Böhmer, Hu, Nishimura, & Schroeder, 2010; Jezek & Blatt, 2017).

Stomatal opening requires the activation of the plasma membrane H^+-ATPase, which in turn, activates and energizes downstream channels and transporters mediating solute uptake. Upon blue light perception,

H^+-ATPase is activated via phosphorylation (Assmann, Simoncini, & Schroeder, 1985; Kinoshita et al., 2001; Kinoshita & Shimazaki, 1999; Shimazaki, Iino, & Zeiger, 1986; Shimazaki & Kondo, 1987). The activated H^+-ATPase hydrolyses one ATP, coupling its dephosphorylation with the translocation of a single H^+ to the apoplast to drive H^+ extrusion and generate the electrochemical gradients needed for the influx of ions from the apoplast (Blatt, 1987; Lawson & Blatt, 2014). Potassium ions are the major cation taken up and can enter the cell via inward rectifying K^+ channels (Nakamura et al., 1995; Pilot et al., 2001; Very & Sentenac, 2003) and via H^+-coupled transport (Clint & Blatt, 1989). In contrast, Cl^- uptake against the electrochemical gradient must couple with the H^+ import via the H^+/anion cotransporter (Jezek & Blatt, 2017; Kollist et al., 2014).

Vacuolar ion transport must also participate actively during stomatal opening. The vacuole makes up 80–90% of the cell volume, and thus the majority of solutes for osmotic adjustment must cross both membranes (MacRobbie, 1998). The vacuolar membrane also contains H^+ pumps, including V-type H^+-ATPases (VHATPase) and H^+-pumping pyrophosphatases (VHPPases), to drive H^+ transport and generate a pH gradient, acid within the vacuole (Bak et al., 2013; Darley, Skiera, Northrop, Sanders, & Davies, 1998; Davies, Hunt, & Sanders, 1994; Rienmüller et al., 2012). The pH gradient facilitates solute uptake, including that of Cl^-, NO_3^-, and malate, in part via Na^+, K^+/H^+ (NHX) antiporters, H^+-coupled anion transporters (CLC transporters), and aluminium-activated anion channels (ALMT) (Barragán et al., 2012; De Angeli et al., 2006; De Angeli, Zhang, Meyer, & Martinoia, 2013; Jossier et al., 2010; Venema, Quintero, Pardo, & Donaire, 2002).

Stomata close through a coordination of multiple ion transport systems on both plasma and vacuolar membranes of the guard cells. This coordination depends in part on a rise of cytosolic $[Ca^{2+}]$ ($[Ca^{2+}]_{cyt}$) triggered by the influx of Ca^{2+} through plasma membrane Ca^{2+} channels that generates a Ca^{2+}-induced Ca^{2+} release from the vacuole through vacuolar Ca^{2+} channels (VCa) (Grabov & Blatt, 1998). The rise of $[Ca^{2+}]_{cyt}$ combined with phosphorylation cascades activates the S-type (SLAC) and R-type (ALMT) anion channels to promote anion efflux and inhibit the plasma membrane H^+-ATPase (Geiger et al., 2010; Kinoshita, Nishimura, & Shimazaki, 1995; Meyer et al., 2010). Both pump inhibition and anion channel activation depolarize the membrane to activate the outward-rectifying K^+ channels—in Arabidopsis the GORK channel—facilitating K^+ efflux. The rise of $[Ca^{2+}]_{cyt}$ also inactivates inward-rectifying K^+

channels—in Arabidopsis including KAT1—as a failsafe against further K^+ uptake across the plasma membrane (Grabov & Blatt, 1997). Channels on the vacuolar membrane, including the two-pore K^+ channel 1 (TPK1), participate to the stomatal closure process by facilitating the release of cation from the vacuole to the cytosol (Bihler et al., 2005; Gobert, Isayenkov, Voelker, Czempinski, & Maathuis, 2007).

1.2 Changes in stomatal ion content under salt stress

The general characteristics of cellular ion homeostasis under salt stress has been comprehensively reviewed before (Zhu, 2003). Although guard cells share many similarities in ion transport and homeostasis with other plant cells, the unique properties of guard cell membranes and the reversibility of osmotic solute flux in these cells can be expected to lead to important differences in how guard cells deal with salt stress. Some knowledge comes from early studies using epidermal strips of the glycophytes *Commelina communis* and *Vicia faba*, as well as of the halophyte *Aster tripolium*. In general, guard cells are permeable not only to K^+ but also to other monovalent cations, including Na^+, Li^+, Cs^+, and Rb^+ (Iljin, 1957). The availability of these alkali cations can have a substantial impact on stomatal movement. Like K^+, the cations Na^+, Li^+, and Cs^+ each promote stomatal opening, but much higher concentrations have been reported to be required (Humble & Hsiao, 1969). However, unlike K^+, all three monovalent cations are reported to inhibit stomatal closure in response to the transition from light to dark conditions (Humble & Hsiao, 1969). The effects of Na^+ in suppressing stomatal closure are of physiological importance (Jarvis & Mansfield, 1980; Willmer & Mansfield, 1970; Yu & Assmann, 2016; Zhao, Wang, Wang, Wang, & Zhang, 2008) simply because Na^+ is present at much higher concentration in nature.

As an example, early studies showed that 50 mM NaCl treatments of *C. communis* stomata in epidermal peels prevented closure as would normally occur in darkness, high CO_2, and with abscisic acid (ABA) (Jarvis & Mansfield, 1980). Jarvis and Mansfield also showed that replacement of K^+ by Na^+ at a 2:1 ratio resulted in a sharp loss in sensitivity of stomata to darkness and CO_2. A similar situation occurs in Arabidopsis where $[K^+]$ in guard cells may decline by up to 70% after 1 day of NaCl exposure (Karimi et al., 2021). These observations highlight a physiological 'exchange' of Na^+ with K^+ and, as in other plant cells under salt stress, how maintaining high K^+:Na^+ ratio is of prime importance for normal stomatal behavior.

A different picture emerges in halophytes, despite similarities in overall stomatal function. For example, an inhibitory effect of Na^+ on stomatal opening was observed in *A. tripolium*, a halophyte without salt secreting glands (Perera, Mansfield, & Malloch, 1994; Véry, Robinson, Mansfield, & Sanders, 1998). This halophyte retains K^+ as the dominant cation even with very high NaCl of 300 mM in the soil, and it maintains a much lower Na^+:K^+ ratio, notably in the guard cells compared with other cell types including mesophyll, epidermal, and subsidiary cells (Perera, De Silva, & Mansfield, 1997). These data were interpreted that the higher Na^+:K^+ ratio in the other cell types and their sequestration of Na^+ is a strategy preventing Na^+ from reaching the guard cells and inhibiting stomatal opening.

Salt exposure also affects the cellular Cl^- content. Although Cl^- could be a beneficial osmoticum in some halophytes and can be accumulated up to 1.5 M in shoot tissues under salinity condition, in non-halophytes the ability to exclude Cl^- is thought to be critical for salt tolerance as its extreme accumulation retards growth (Bazihizina, Colmer, Cuin, Mancuso, & Shabala, 2019). These findings accord also with findings that disrupting vacuolar Cl^- transport via the CLCc transporter of Arabidopsis results in a hypersensitivity to NaCl and stomata that fail to close in ABA (Jossier et al., 2010). Of course, the consequences of genetic disruption do not speak directly to a mechanism, but the results nevertheless point to Cl^- homeostasis as a part of stomatal movement and salt tolerance.

What are the Na^+ and Cl^- concentrations in guard cells and the surrounding apoplast of leaf tissues? These values are often poorly defined and may vary significantly between species. For example, with 100 mM NaCl treatments in the soil, Na^+ and Cl^- concentrations were found to situate around 90 and 200 mM, respectively, in salt-sensitive pea but did not exceed 10 and 15 mM in the more salt-tolerant spinach, respectively (Speer & Kaiser, 1991). Most data across species suggest that Na^+ and Cl^- are maintained below 50–80 mM in the cytosol with soil additions of 100–150 mM NaCl (Isayenkov & Maathuis, 2019). Vacuolar levels of Na^+ and Cl^-, however, can exceed 500 mM (Isayenkov & Maathuis, 2019), indicating a major contribution of vacuoles to Na^+ and Cl^- homeostasis in the face of salt stress.

Of course, the phenomenology of NaCl accumulation leaves open questions about the mechanics behind the transport and sequestration of these ions. One aspect of salt stress is clear: changes in $[Ca^{2+}]_{cyt}$ are important for salt adaptation. Indeed, extracellular Ca^{2+} applications can mitigate the most severe effects of salt stress. For example, adding Ca^{2+} at 10 mM can efficiently inhibit NaCl-induced stomatal opening of guard cells, apparently by blocking Na^+ influx, thereby reducing water loss via transpiration (Zhao, Wang, Wang, Wang, & Zhang, 2011). Changes in $[Ca^{2+}]_{cyt}$ within guard cells have yet to be studied in depth in the context of salt stress. However, we may anticipate many similarities to what is known for other plant cell types. In Arabidopsis leaf tissues, NaCl has been shown to induce transients in $[Ca^{2+}]_{cyt}$ that reach 1.8 μM within 10 s, findings that have been interpreted to arise from both intracellular stores and entry from the apoplast (Knight, Trewavas, & Knight, 1997). Long-term NaCl exposure, by contrast, led to Ca^{2+} levels that were marginally reduced in guard cells (Franzisky et al., 2021). Thus, it remains unclear whether—and how—$[Ca^{2+}]_{cyt}$ may change within guard cells under salt stress. The topic demands a systematic and mechanistic analysis if we are to understand the role of Ca^{2+} in this aspect of stomatal function.

1.3 Signaling and transport systems involved in regulation of stomatal movement upon salt exposure

As most glycophytes cannot tolerate the Na^+ toxicity, avoiding that Na^+ accumulates in the leaf is crucial for their survival and recirculation of Na^+ back to the root plays a key role here (Munns & Tester, 2008). The high affinity K^+ transporter 1 (HKT1) and non-selective cation channels mediate Na^+ influx and are assumed to be responsible for Na^+ recirculation (Amtmann & Sanders, 1998; Mäser et al., 2002; Rus et al., 2001). Homologs of HKT1 are found in many species, with an especially high variety in monocots (Almeida, Katschnig, & De Boer, 2013). HKT transporters function mainly in root. However, their possible roles in guard cells have yet to be addressed.

When Na^+ reaches the guard cell apoplast, it may enter the cell either via high affinity Na^+ transporters or via non-selective cation channels (Demidchik, Davenport, & Tester, 2002; Yamaguchi, Hamamoto, & Uozumi, 2013). Inside the guard cell, Na^+ affects the function of the ion channels on both the plasma membrane and tonoplast. Cytosolic Na^+ has been shown to block homologs of the outward-rectifying K^+ channel GORK in a voltage-dependent manner (Thiel & Blatt, 1991), thereby suppressing stomatal closure (Fig. 1). Sequestration of Na^+ to the vacuole, blocks the SV (TPC1) channels on the vacuolar membrane, by shifting the voltage-dependence toward more positive values (Ivashikina & Hedrich, 2005) although, again, a functional role for this channel in general remains unclear (Jezek & Blatt, 2017).

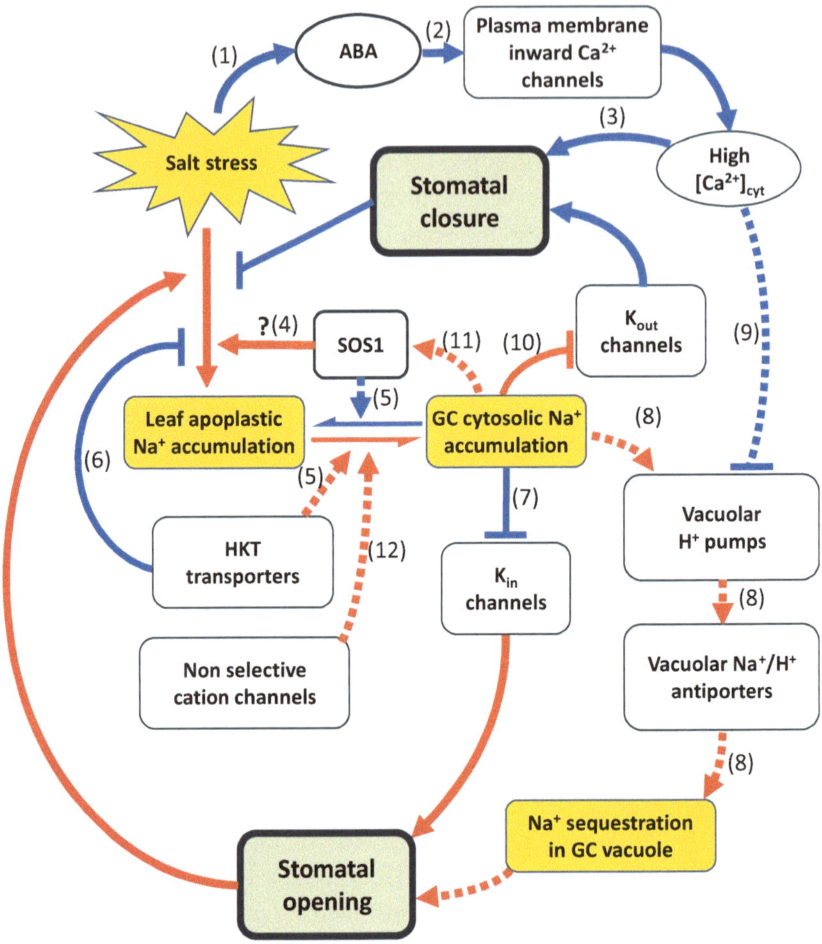

Fig. 1 Important targets of Na^+ in guard cell. Blue connectors represent the pathway responsible for limiting leaf Na^+ accumulation. Red connectors show the pathway promoting Na^+ accumulation. Dashed connectors are the assumptions based on research on other cell types. The perception of salt stress by the root stimulates the ABA synthesis and ABA signaling which then promote stomatal closure via Ca^{2+} signals to avoid Na^+ translocation to the leaf. When Na^+ gets into the guard cell (GC) cytosol, it activates the plasma membrane Na^+/H^+ antiporter SOS1 via phosphorylation cascade and the H^+ pumps and transporters responsible for Na^+ sequestration in vacuole. The cytosolic Na^+ also inhibits the outward-rectifying K^+ channels (K_{out} channels), through which, inhibiting stomatal closure. In the halophyte *A. tripolium*, the properties of the inward-rectifying K^+ channels (K_{in} channels) are changed by Na^+, inhibiting stomatal opening under salt stress. For reference, (1) (Jia, Wang, Zhang, & Zhang, 2002), (2) (Grabov & Blatt, 1998; Hamilton, Hills, Köhler, & Blatt, 2000), (3) (Zhao et al., 2011), (4) (Kronzucker & Britto, 2011), (5) (Hedrich & Shabala, 2018), (6) (Horie et al., 2005), (7) (Véry et al., 1998), (8) (Barkla, Zingarelli, Blumwald, & Smith, 1995; Queirós et al., 2009), (9) (Nakamura, Kasamo, Sakata, & Ohta, 1992), (10) (Thiel & Blatt, 1991), (11) (Qiu, Guo, Dietrich, Schumaker, & Zhu, 2002; Shi, Ishitani, Kim, & Zhu, 2000), (12) (Amtmann & Sanders, 1998).

To avoid Na^+ toxicity, the Na^+/H^+ antiporters in both plasma membrane (SOS1) and vacuolar membrane (NHX) may extrude Na^+ to the apoplast and sequester it to the vacuole, respectively, and both are energized by the activity of H^+ pumps (Ballesteros, Donaire, & Belver, 1996; Nakamura et al., 1992; Qiu et al., 2002; Shi et al., 2000; Yokoi et al., 2002). *SOS1* is highly expressed by salt stress and preferentially in the epidermal cells at the root tip and in the parenchyma cells at the xylem/symplast boundary of roots, suggesting that SOS1 play an important role in regulating long-distance Na^+ transport, including loading Na^+ into the xylem or retrieving it from the xylem (Shi et al., 2000; Shi, Quintero, Pardo, & Zhu, 2002). The latter function is still controversial (Kronzucker & Britto, 2011). As the stomatal expression pattern of SOS1 has not yet been documented, whether SOS1-mediated Na^+ extrusion existing in the guard cells remains to be addressed. In contrast, the vacuolar Na^+/H^+ antiporters NHXs are found in the guard cells, supporting their role in Na^+ sequestration in the vacuole of guard cell during salt stress (Fukuda, Nakamura, Hara, Toki, & Tanaka, 2011; Wang et al., 2007). The presence of cytosolic Na^+ is known to activate VHATPases and VHPPases which create the H^+ gradient across the vacuolar membrane, driving the sequestration of Na^+ to the vacuole in exchange for H^+ export via the NHX antiporters (Fig. 1) (Ballesteros et al., 1996; Nakamura et al., 1992).

Even before Na^+ reaches the guard cells, salt stress may still impact on stomatal function. Roots exposed to Na^+ are stimulated to synthesize ABA in vascular tissue, a common response to osmotic stress, and this ABA is carried via the transpiration stream to elevate ABA in the leaves (Hedrich & Shabala, 2018). ABA can be built up rapidly, for example, within 15 min in roots and 30 min in leaves of *V. faba* (Geilfus, Mithöfer, Ludwig-Müller, Zörb, & Muehling, 2015). In the face of ABA, guard cells undergo a well-established program of responses that lead to stomatal closure (Blatt & Thiel, 1993; Jezek & Blatt, 2017). Indeed, some of the most rapid responses to ABA in guard cells are complete within a few tens of seconds. Notably, gating of the plasma membrane Ca^{2+} channels is enhanced within a few seconds (Hamilton et al., 2000) and the SLAC1 channel responds to the subsequent rise in $[Ca^{2+}]_{cyt}$ and to ABA-triggered phosphorylation cascades (Blatt, 2014; Brandt et al., 2012; Chen, Hills, Lim, & Blatt, 2010; Grabov & Blatt, 1998) to promote stomatal closure. With the stomata closed, transpiration is slowed which must also reduce Na^+ transfer via the transpiration stream to the leaf (Fig. 1).

1.4 How does Na^+ stress propagate through guard cell physiology?

The discussion above makes clear the breadth of impacts imposed by Na^+ as it may relate to stomatal function. It also shows how little we know about these impacts, their interconnections and what might be the most important targets for future bioengineering efforts. One approach to address these questions lies in mechanistic modeling. Here we have used OnGuard, as it is the only platform that incorporates solute transport mechanics, buffering and metabolism, and draws explicitly on the wealth of knowledge available for guard cells (Blatt et al., 2022; Jezek & Blatt, 2017). In its original formulation (Chen et al., 2012; Hills et al., 2012), the OnGuard platform accurately simulated stomatal dynamics in isolation, as if in epidermal peels, and uncovered an emergent metabolic network connecting KAT1 K^+ channel and SLAC1 anion channel activities (Wang et al., 2012). Subsequent releases introduced foliar transpiration of the whole plant (Wang et al., 2017); turgor 'exchange', the opposing mechanical pressure of the surrounding epidermal cells, and the intrinsic visco-elastic properties of the guard cell wall (Jezek, Hills, Blatt, & Lew, 2019); and, with the latest release, OnGuard3, CO_2 exchange between the atmosphere and the leaf, and the feedback of mesophyll photosynthesis (Jezek et al., 2021). Adding CO_2 exchange has yielded new predictions, several since validated, including an unforeseen latency or 'carbon memory' in stomatal kinetics with fluctuating light and the partial pressure of CO_2 (pCO_2) that depends on endomembrane Ca^{2+} stores in the guard cell (Jezek et al., 2021).

Unique to OnGuard, models built on this platform yield the kinetic features of stomata as they arise from the metabolic and transport processes that mediate solute flux with light, the pCO_2 , and relative humidity (RH) as well as water and solute availability. Model outputs include the osmotic solute content, guard cell turgor, volume and stomatal aperture and, hence, stomatal conductance and carbon assimilation. These outputs arise directly from subcellular solute flux and the molecular activities of guard cell transport. All model components are available for analysis and, thus, offer a wealth of information about the connections between whole-plant, cellular and molecular behaviors.

Most important, the OnGuard platform generates real-time kinetics that can be compared directly with experiments. The platform uses an iterative computational cycle with small increments in time to calculate the progress of solute and water flux in and out of the guard cell. As a result, every model 'evolves' over time, constrained only by basic biophysical laws, the equations and their parameters defining each transport and metabolic reaction, and the interactions between these processes that arise from the shared variables, notably that of membrane voltage across each membrane and the common pools of substrates and products, just as they must do in vivo.

OnGuard outputs provide valuable and often unforeseen predictions that emerge from these interactions. Like any modeling platform built on systems of elements, interactions between these elements often give rise to so-called emergent behaviors that could not be anticipated from knowledge of the elements in isolation. This information is then available for study through experiment. Modeling thus becomes a tool complementary to experimentation (Stelling, Sauer, Szallasi, Doyle, & Doyle, 2004; van Riel, 2006), each model representing a hypothesis to be discarded, validated, or refined by comparison with experimental results.

1.5 Developing Na^+-sensitivity within the OnGuard platform

Introducing a new solute to the OnGuard platform requires that the user adds the relevant permeation pathways to each membrane. Practice shows that there must be at least two pathways for the flux of an ion across each membrane to allow for circulation (Chen et al., 2012). Although information about Na^+ transport systems in guard cell remains scarce, much is known for Na^+ transport in other plant cells and can be inferred from this knowledge. Here, we summarize a minimal set of additions to account for Na^+ transport and explore its consequences for stomata.

Members of the HKT family of transporters are plausible candidates for Na^+ influx during membrane hyperpolarization, while the Na^+/H^+ antiporter SOS1 is one candidate for Na^+ extrusion from the cytosol (Hedrich & Shabala, 2018). Therefore, in the model, a Na^+ leak channel and a Na^+/H^+ antiporter were added to the plasma membrane, respectively. Their activities were adjusted as appropriate, constrained by our present knowledge of Na^+ stress effects in guard cells and other plant cell types, as detailed below, and its metabolic consequences, especially for cytosolic pH.

As with the plasma membrane, it was necessary to add at least two Na^+ transporters to the vacuolar membrane within the OnGuard platform. Histochemical staining has shown a strong expression of the vacuolar Na^+/H^+ antiporter AtNHX1 in guard cells (Shi & Zhu, 2002), making this antiporter one of the two tonoplast transporters. This antiporter is responsible for Na^+ sequestration in the vacuole, nominally important in protecting the cell from Na^+ toxicity. The other transporter, in this case mediating Na^+ efflux from the vacuole, could be the tonoplast SV and FV channels (Bonales-Alatorre, Shabala, Chen, & Pottosin, 2013; Ivashikina & Hedrich, 2005) that might comprise a 'leakage' pathway for Na^+ across the vacuolar membrane. The conductance of this leak was set to rise with vacuolar $[Na^+]$, as the SV channel, at least, does not favor Na^+ efflux from the vacuole in most circumstances (Ivashikina & Hedrich, 2005).

Two further additions were introduced to the OnGuard platform in line with published data. First, we added Na^+-dependent block of the outward-rectifying K^+ channels to accommodate cation competition in the channel pore from Na^+ inside the guard cell (Thiel & Blatt, 1991). Second, we added a complementary activation of the tonoplast proton pumps with cytosolic Na^+ (Matsumoto & Chung, 1988; Nakamura et al., 1992; Wang, Lüttge, & Ratajczak, 2001). The increase in pump activity is particularly important during salt stress as it energizes the NHX-mediated Na^+ sequestration to the vacuole. Therefore, Na^+-dependent activation of the VHATPase and VHPPase was added to the model to allow for a 3-fold enhancement of pump activity under salt stress. Those new settings form the basic set of transporters needed for the model to accommodate the Na^+ ion.

As noted above, salt stress also affects the stomatal function via synthesis of the stress-related plant hormone ABA. Although ABA may be synthesized throughout the plant, under salt stress ABA is generally thought to pass to the guard cells from root and vascular tissues, thereby reducing the stomatal aperture (Amjad et al., 2014; Barbieri et al., 2012; Karimi et al., 2021). Central to ABA action in the guard cells is a positive shift of the voltage-dependence of the plasma membrane Ca^{2+} channels, thereby favoring Ca^{2+} influx and an increase of $[Ca^{2+}]_{cyt}$ (Grabov & Blatt, 1998; Hamilton et al., 2000; Köhler & Blatt, 2002). Increasing $[Ca^{2+}]_{cyt}$, in turn, inhibits the plasma membrane H^+-ATPase and inward-rectifying K^+ channels, and it activates several other transporters, including the SLAC1 anion channel, that favor solute loss to close stomata. Therefore, to mimic the action of ABA, we introduced a Na^+-induced shift in gating of the Ca^{2+} channels. Finally, to analyze the effects of Na^+ with and without ABA, we simulated the effects of NaCl additions outside in each case, first without (ABA−) and then with (ABA+) the additional effects on transport of the plant hormone.

Best estimates suggest that, under salt stress, the Na^+ content of the leaf apoplast frequently rises to values between 20 and 30 mM. Therefore, we incorporated in OnGuard3 treatments with 30 mM of NaCl for 4 days and followed the consequences over the diurnal cycle as before (Chen et al., 2012; Wang et al., 2017). In every case, we observed a strong reduction in the dynamic range of apertures and a slowing in both stomatal opening and closure kinetics (Fig. 2). Although the models, with and without ABA, were largely indistinguishable without NaCl outside, their behaviors strongly differed after 4 days under salt stress. The ABA+ model, incorporating Na^+-sensitivity to the plasma membrane Ca^{2+} channel, displayed a reduced stomatal aperture, similar to experimental data (Amjad et al., 2014; Barbieri et al., 2012; Karimi et al., 2021). By contrast, the ABA- model showed an elevated stomatal aperture, resulting in a severe reduction of WUE (Fig. 2). This behavior reflects the observation of stomata in epidermal strips treated with NaCl, in which stomata were seen to open but failed to close in the dark, under high CO_2, and in ABA (Jarvis & Mansfield, 1980). The comparisons, both in model and experiment, indicates a crucial role for long-distance signaling mediated by ABA during salt stress.

A careful analysis of the OnGuard3 simulations shows a number of other differences. Among these, notable is a decrease in pH_{cyt} only for the ABA+ model under salt stress (Fig. 3). The reduction of pH_{vac} is simply the consequence of the addition of Na^+-activated H^+ pumps to the models' vacuolar membrane. The accumulation of H^+ in the cytosol, however, is explained in part by the reduced H^+-ATPase activity and decrease in H^+ efflux across the plasma membrane (Fig. 3). The simulation reflects experimental results that have reported cytosolic acidification occurs during NaCl treatment (Gao, Knight, Trewavas, Sattelmacher, & Plieth, 2004; Okazaki, Kikuyama, Hiramoto, & Iwasaki, 1996) and it adds mechanistic support to the idea of H^+ pump inhibition that has been suggested to depend on rise in $[Ca^{2+}]_{cyt}$ (Goh, Kinoshita, Oku, & Shimazaki, 1996; Kinoshita et al., 1995). Most important, the OnGuard3 simulations predict a critical role for $[Ca^{2+}]_{cyt}$ in connecting salt stress with guard cell pH: without the Na^+ (and ABA −) dependent change of Ca^{2+} channel activity in simulation, inhibition of the H^+-ATPase would not occur and, hence, the acidification of pH_{cyt} would be reduced.

Both models also show substantial changes in the Na^+ content of the cytosol and vacuole, with the most severe consequences for the K^+ contents evident in the ABA − model. Under salt stress, K^+ built up to high levels in both compartments in ABA − model, whereas these effects were moderated in the ABA+ model. Not surprisingly, the Cl^- contents followed a similar pattern with the anion elevated mainly in the vacuole with Na^+ (Fig. 4). Again, the simulations offer some insights to experimental observations,

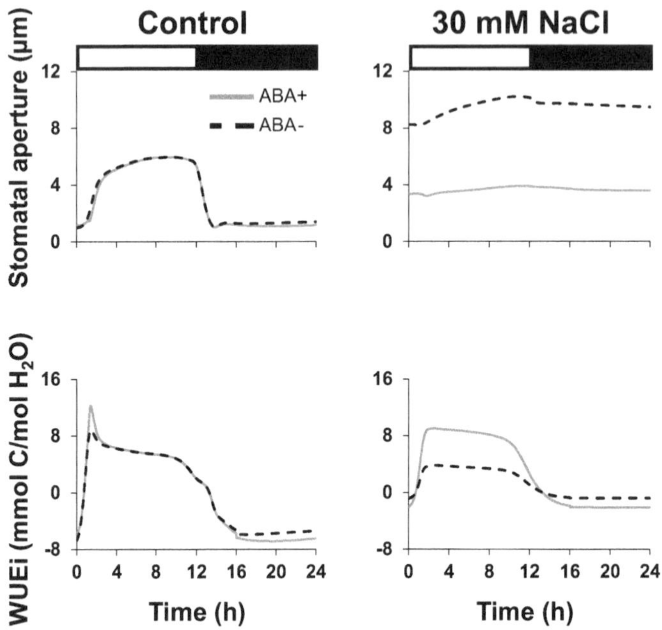

Fig. 2 Stomatal movement kinetics and WUEi of WT and ABA-deficient model under control and dark condition. Output resolved over the standard diurnal cycle (12h light:12h dark, indicated by bars above). Shown are stomatal aperture and WUE (Assimilation/Transpiration ratio) under control condition (left) and day 4 under salt stress (right). Salt stress dramatically reduces the stomatal dynamic and WUE. ABA− model displays larger stomatal aperture, as a result, a severely reduced WUE compared to the ABA+ model.

where the Na^+ sequestration in the vacuole is the common strategy to avoid Na^+ toxicity in the cytosol (Binzel, Hess, Bressan, & Hasegawa, 1988; Hajibagheri, Harvey, & Flowers, 1987; Niu, Bressan, Hasegawa, & Pardo, 1995; Wu et al., 2019). Activation of the Na^+/H^+ antiporter, with pump energization at the vacuolar membrane reduced the Na^+ load on the cytosol in favor of the vacuole (Nakamura et al., 1992; Otoch et al., 2001; Queirós et al., 2009).

Most important, the model analysis predicts that, without Ca^{2+} channel activation by ABA, salt stress leads to extremes in accumulation of K^+ in guard cells, effectively locking the stomata open. A comparison of the two models shows that, under salt stress, $[Ca^{2+}]_{cyt}$ is elevated substantially in the ABA+ model, but not in the ABA− model throughout the day (Fig. 5). The rise of

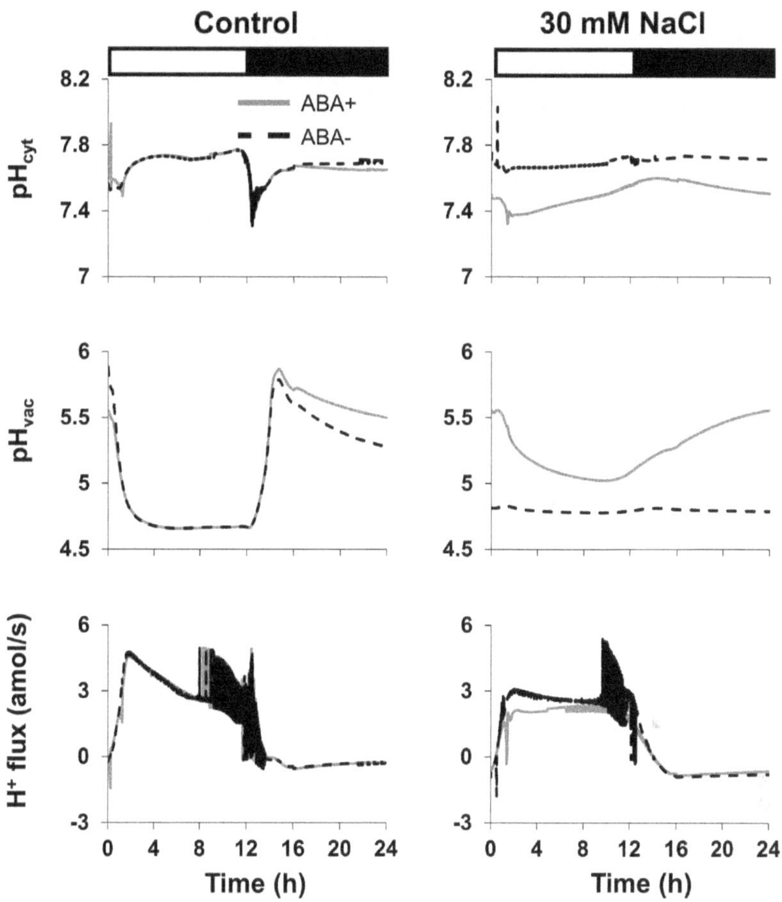

Fig. 3 Salt stress effect on the pH of cytosol and vacuole and the H^+ flux across the plasma membrane. Output resolved over the standard diurnal cycle (12 h light:12 h dark, indicated by bars above). Salt stress causes the acidification of cytosolic and vacuolar pH, and reduces the H^+ efflux across the plasma membrane.

$[Ca^{2+}]_{cyt}$, as noted above, inhibits the H^+ pumps and the inward-rectifying K^+ channels, while promoting anion channel activity for Cl^- and malate efflux. The result is a reduction in stomatal aperture, much as observed in experiments (Amjad et al., 2014; Barbieri et al., 2012; Karimi et al., 2021). In the ABA− model, $[Ca^{2+}]_{cyt}$ remains largely unchanged over the diurnal cycle, accounting for the accumulation in solute content and, of course, a greater stomatal aperture.

Finally, the simulations also help to explain a phenomenon reported four decades ago of stomata from epidermal strips that opened but were

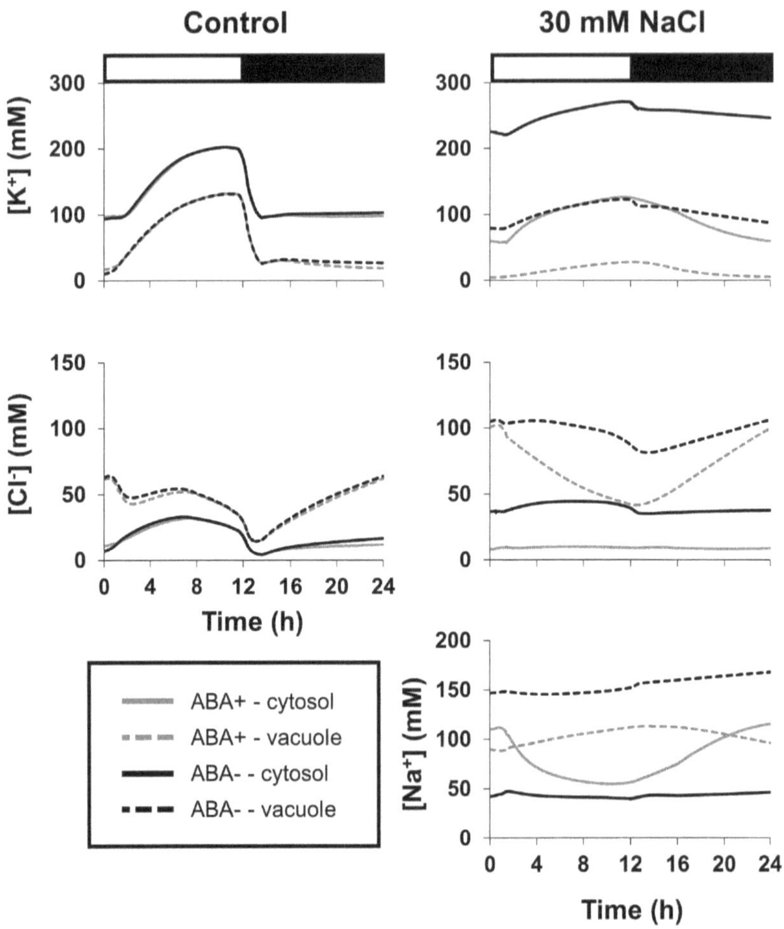

Fig. 4 Changes in solute contents in different guard cell compartments during salt stress. Output resolved over the standard diurnal cycle (12 h light:12 h dark, indicated by bars above). Under salt stress, Na^+ and Cl^- are accumulated at high concentration in guard cell cytosol and vacuole. Due to the competition effect of Na^+, K^+ content decreases in ABA+ model under salt stress. ABA− model, however, can still retain high concentration of K^+ under salt stress.

unable to close in the presence of Na^+ treatment (Jarvis & Mansfield, 1980). In simulation, Na^+ accumulation supressed oscillations in $[Ca^{2+}]_{cyt}$ and plasma membrane voltage at the end of the day (Fig. 5), oscillations which coordinate with the efflux of K^+ and Cl^- to accelerate stomatal closing (Chen et al., 2012; Minguet-Parramona et al., 2016). Thus, under salt stress, stomatal closure is greatly slowed (Fig. 2).

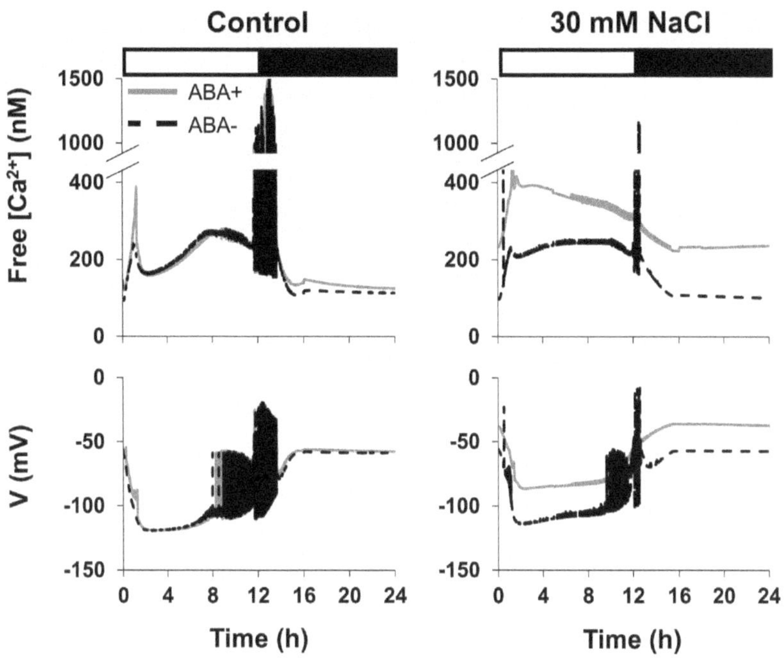

Fig. 5 The effect of salt stress on $[Ca^{2+}]_{cyt}$ and plasma membrane potential. The ABA+ model has higher $[Ca^{2+}]_{cyt}$ throughout the day under salt stress but its oscillation is totally suppressed. Plasma membrane depolarization with salt stress is observed in both models relative to the control and is most evident in the ABA+ model.

1.6 Stomata under salt stress—What can we learn from OnGuard modeling?

The OnGuard platform successfully reproduces the behavior of stomata both in epidermal strips and in the intact plant under salt stress. Required only are the addition of a minimum of four transport pathways for Na^+ at the two membranes, and the documented regulatory effects of cytosolic Na^+ on the tonoplast pumps and ABA-mediated Ca^{2+} channels. Apart from the phenomenology of stomata previously reported under salt stress, the simulations offer key insights that help explain their mechanics. Modeling suggests that under salt stress, stomata fail to close during dark transition because of an uncoupling of solute efflux from $[Ca^{2+}]_{cyt}$ and voltage oscillations that are also important for determining the 'carbon memory' of guard cells when faced with fluctuations in light and CO_2 (Jezek et al., 2021).

The simulations also emphasize the crucial role of ABA-mediated Ca^{2+} signaling during salt stress. The simple addition of a positive-going shift in the voltage-dependence for Ca^{2+} channel gating, the model accurately reproduces stomatal behavior as observed in intact plants under salt stress. This appears the minimum change needed to ensure the behavior of stomata in response to salt stress. The similarity of the ABA – model, which lacks this characteristic, with the behavior of stomata in epidermal strips suggests the more important role of ABA long-distance signaling over the autonomous ABA synthesis of guard cell.

We stress that these are a selection only of the observations and predictions that may be drawn from OnGuard3 and the introduction of Na^{+} transport for salt stress. Indeed, the advantage of the OnGuard platform lies on its flexibility of design and its ability to capture the kinetics of concurrent guard cell mechanics in addition to predicting steady-state stomatal conductances. This unique approach allows users to observe in real time, just as in experiments, and facilitates predictions that benefit from the temporal characteristics of stomatal behaviors. We anticipate that, with minimal expansion, the OnGuard platform will enable explorations of stomatal behaviors under many other environmental challenges.

Acknowledgments

Development of the OnGuard platform was supported with grants BB/L001276/1, BB/L019025/1, BB/M001601/1, and BB/N01832X/1 from the UK Biotechnology and Biological Sciences Research Council. The expansion of OnGuard3 to include salinity was supported by grants BB/T006153/1 and BB/T013508/1 from the UK Biotechnology and Biological Sciences Research Council to MRB.

References

Almeida, P., Katschnig, D., & De Boer, A. H. (2013). HKT transporters - state of the art. *International Journal of Molecular Sciences*, *14*, 20359–20385.

Amjad, M., Akhtar, J., Anwar-ul-Haq, M., Yang, A., Akhtar, S. S., & Jacobsen, S.-E. (2014). Integrating role of ethylene and ABA in tomato plants adaptation to salt stress. *Scientia Horticulturae*, *172*, 109–116.

Amtmann, A., & Sanders, D. (1998). Mechanisms of Na^{+} uptake by plant cells. *Advances in Botanical Research*, *29*, 75–112.

Assmann, S., Simoncini, L., & Schroeder, J. (1985). Blue light activates electrogenic ion pumping in guard cell protoplasts of *Vicia faba*. *Nature*, *318*, 285–287.

Bak, G., Lee, E.-J., Lee, Y., Kato, M., Segami, S., Sze, H., et al. (2013). Rapid structural changes and acidification of guard cell vacuoles during stomatal closure require phosphatidylinositol 3, 5-bisphosphate. *The Plant Cell*, *25*, 2202–2216.

Ballesteros, E., Donaire, J. P., & Belver, A. (1996). Effects of salt stress on H^{+}-ATPase and H^{+}-PPase activities of tonoplast-enriched vesicles isolated from sunflower roots. *Physiologia Plantarum*, *97*, 259–268.

Barbieri, G., Vallone, S., Orsini, F., Paradiso, R., De Pascale, S., Negre-Zakharov, F., et al. (2012). Stomatal density and metabolic determinants mediate salt stress adaptation and water use efficiency in basil (*Ocimum basilicum* L.). *Journal of Plant Physiology*, *169*, 1737–1746.

Barkla, B. J., Zingarelli, L., Blumwald, E., & Smith, J. A. C. (1995). Tonoplast Na^+/H^+ antiport activity and its energization by the vacuolar H^+-ATPase in the halophytic plant *Mesembryanthemum crystallinum* L. *Plant Physiology*, *109*, 549–556.

Barragán, V., Leidi, E. O., Andrés, Z., Rubio, L., De Luca, A., Fernández, J. A., et al. (2012). Ion exchangers NHX1 and NHX2 mediate active potassium uptake into vacuoles to regulate cell turgor and stomatal function in Arabidopsis. *The Plant Cell*, *24*, 1127–1142.

Bazihizina, N., Colmer, T. D., Cuin, T. A., Mancuso, S., & Shabala, S. (2019). Friend or foe? Chloride patterning in halophytes. *Trends in Plant Science*, *24*, 142–151.

Bihler, H., Eing, C., Hebeisen, S., Roller, A., Czempinski, K., & Bertl, A. (2005). TPK1 is a vacuolar ion channel different from the slow-vacuolar cation channel. *Plant Physiology*, *139*, 417–424.

Binzel, M. L., Hess, F. D., Bressan, R. A., & Hasegawa, P. M. (1988). Intracellular compartmentation of ions in salt adapted tobacco cells. *Plant Physiology*, *86*, 607–614.

Blatt, M. R. (1987). Electrical characteristics of stomatal guard cells: the contribution of ATP-dependent,"electrogenic" transport revealed by current-voltage and difference-current-voltage analysis. *The Journal of Membrane Biology*, *98*, 257–274.

Blatt, M. R. (2014). An open-andshut case: ABA and stomata. *The Biochemist*, *36*, 17–22.

Blatt, M. R., Jezek, M., Lew, V. L., & Hills, A. (2022). What can mechanistic models tell us about guard cells, photosynthesis, and water use efficiency? *Trends in Plant Science*, *27*, 166–179.

Blatt, M. R., & Thiel, G. (1993). Hormonal control of ion channel gating. *Annual Review of Plant Biology*, *44*, 543–567.

Blatt, M. R., Wang, Y., Leonhardt, N., & Hills, A. (2014). Exploring emergent properties in cellular homeostasis using OnGuard to model K^+ and other ion transport in guard cells. *Journal of Plant Physiology*, *171*, 770–778.

Bonales-Alatorre, E., Shabala, S., Chen, Z.-H., & Pottosin, I. (2013). Reduced tonoplast fast-activating and slow-activating channel activity is essential for conferring salinity tolerance in a facultative halophyte, quinoa. *Plant Physiology*, *162*, 940–952.

Brandt, B., Brodsky, D. E., Xue, S., Negi, J., Iba, K., Kangasjärvi, J., et al. (2012). Reconstitution of abscisic acid activation of SLAC1 anion channel by CPK6 and OST1 kinases and branched ABI1 PP2C phosphatase action. *Proceedings of the National Academy of Sciences*, *109*, 10593–10598.

Chen, Z.-H., Hills, A., Bätz, U., Amtmann, A., Lew, V. L., & Blatt, M. R. (2012). Systems dynamic modeling of the stomatal guard cell predicts emergent behaviors in transport, signaling, and volume control. *Plant Physiology*, *159*, 1235–1251.

Chen, Z. H., Hills, A., Lim, C. K., & Blatt, M. R. (2010). Dynamic regulation of guard cell anion channels by cytosolic free Ca^{2+} concentration and protein phosphorylation. *The Plant Journal*, *61*, 816–825.

Clint, G. M., & Blatt, M. R. (1989). Mechanisms of fusicoccin action: Evidence for concerted modulations of secondary K^+ transport in a higher plant cell. *Planta*, *178*, 495–508.

Darley, C. P., Skiera, L. A., Northrop, F. D., Sanders, D., & Davies, J. M. (1998). Tonoplast inorganic pyrophosphatase in *Vicia faba* guard cells. *Planta*, *206*, 272–277.

Davies, J. M., Hunt, I., & Sanders, D. (1994). Vacuolar H ($^+$)-pumping ATPase variable transport coupling ratio controlled by pH. *Proceedings of the National Academy of Sciences*, *91*, 8547–8551.

De Angeli, A., Monachello, D., Ephritikhine, G., Frachisse, J., Thomine, S., Gambale, F., et al. (2006). The nitrate/proton antiporter AtCLCa mediates nitrate accumulation in plant vacuoles. *Nature*, *442*, 939–942.

De Angeli, A., Zhang, J., Meyer, S., & Martinoia, E. (2013). AtALMT9 is a malate-activated vacuolar chloride channel required for stomatal opening in Arabidopsis. *Nature Communications*, *4*, 1–10.

Demidchik, V., Davenport, R. J., & Tester, M. (2002). Nonselective cation channels in plants. *Annual Review of Plant Biology*, *53*, 67–107.

Franzisky, B. L., Geilfus, C. M., Romo-Pérez, M. L., Fehrle, I., Erban, A., Kopka, J., et al. (2021). Acclimatisation of guard cell metabolism to long-term salinity. *Plant, Cell & Environment*, *44*, 870–884.

Fukuda, A., Nakamura, A., Hara, N., Toki, S., & Tanaka, Y. (2011). Molecular and functional analyses of rice NHX-type Na^+/H^+ antiporter genes. *Planta*, *233*, 175–188.

Gao, D., Knight, M. R., Trewavas, A. J., Sattelmacher, B., & Plieth, C. (2004). Self-reporting Arabidopsis expressing pH and $[Ca^{2+}]$ indicators unveil ion dynamics in the cytoplasm and in the apoplast under abiotic stress. *Plant Physiology*, *134*, 898–908.

Geiger, D., Scherzer, S., Mumm, P., Marten, I., Ache, P., Matschi, S., et al. (2010). Guard cell anion channel SLAC1 is regulated by CDPK protein kinases with distinct Ca^{2+} affinities. *Proceedings of the National Academy of Sciences*, *107*, 8023–8028.

Geilfus, C. M., Mithöfer, A., Ludwig-Müller, J., Zörb, C., & Muehling, K. H. (2015). Chloride-inducible transient apoplastic alkalinizations induce stomata closure by controlling abscisic acid distribution between leaf apoplast and guard cells in salt-stressed *Vicia faba*. *New Phytologist*, *208*, 803–816.

Gobert, A., Isayenkov, S., Voelker, C., Czempinski, K., & Maathuis, F. J. (2007). The two-pore channel TPK1 gene encodes the vacuolar K^+ conductance and plays a role in K^+ homeostasis. *Proceedings of the National Academy of Sciences*, *104*, 10726–10731.

Goh, C.-H., Kinoshita, T., Oku, T., & Shimazaki, K. (1996). Inhibition of blue light-dependent H^+ pumping by abscisic acid in *Vicia* guard-cell protoplasts. *Plant Physiology*, *111*, 433–440.

Grabov, A., & Blatt, M. R. (1997). Parallel control of the inward-rectifier K^+ channel by cytosolic free Ca^{2+} and pH in *Vicia* guard cells. *Planta*, *201*, 84–95.

Grabov, A., & Blatt, M. R. (1998). Membrane voltage initiates Ca^{2+} waves and potentiates Ca^{2+} increases with abscisic acid in stomatal guard cells. *Proceedings of the National Academy of Sciences*, *95*, 4778–4783.

Greenwood, N., & Earnshow, A. (1997). *Lithium, sodium, potassium, rubidium, caesium and francium* (2nd ed.). Oxford: Elsevier Butterworth-Heinemann.

Hajibagheri, M., Harvey, D., & Flowers, T. (1987). Quntitative ion distribution within root cells of salt-sensitive and salt-tolerant maize varieties. *New Phytologist*, *105*, 367–379.

Hamilton, D. W., Hills, A., Köhler, B., & Blatt, M. R. (2000). Ca^{2+} channels at the plasma membrane of stomatal guard cells are activated by hyperpolarization and abscisic acid. *Proceedings of the National Academy of Sciences*, *97*, 4967–4972.

Hedrich, R., & Shabala, S. (2018). Stomata in a saline world. *Current Opinion in Plant Biology*, *46*, 87–95.

Hills, A., Chen, Z.-H., Amtmann, A., Blatt, M. R., & Lew, V. L. (2012). OnGuard, a computational platform for quantitative kinetic modeling of guard cell physiology. *Plant Physiology*, *159*, 1026–1042.

Horie, T., Motoda, J., Kubo, M., Yang, H., Yoda, K., Horie, R., et al. (2005). Enhanced salt tolerance mediated by AtHKT1 transporter-induced Na^+ unloading from xylem vessels to xylem parenchyma cells. *The Plant Journal*, *44*, 928–938.

Humble, G., & Hsiao, T. C. (1969). Specific requirement of potassium for light-activated opening of stomata in epidermal strips. *Plant Physiology*, *44*, 230–234.

Iljin, W. (1957). Drought resistance in plants and physiological processes. *Annual Review of Plant Physiology*, *8*, 257–274.

Isayenkov, S. V., & Maathuis, F. J. (2019). Plant salinity stress: Many unanswered questions remain. *Frontiers in Plant Science*, *10*, 80.

Ivashikina, N., & Hedrich, R. (2005). K^+ currents through SV-type vacuolar channels are sensitive to elevated luminal sodium levels. *The Plant Journal*, *41*, 606–614.

Jarvis, R., & Mansfield, T. (1980). Reduced stomatal responses to light, carbon dioxide and abscisic acid in the presence of sodium ions. *Plant, Cell & Environment*, *3*, 279–283.
Jezek, M., & Blatt, M. R. (2017). The membrane transport system of the guard cell and its integration for stomatal dynamics. *Plant Physiology*, *174*, 487–519.
Jezek, M., Hills, A., Blatt, M. R., & Lew, V. L. (2019). A constraint-relaxation-recovery mechanism for stomatal dynamics. *Plant, Cell & Environment*, *42*, 2399–2410.
Jezek, M., Silva-Alvim, F. A., Hills, A., Donald, N., Ishka, M. R., Shadbolt, J., et al. (2021). Guard cell endomembrane Ca^{2+}-ATPases underpin a 'carbon memory'of photosynthetic assimilation that impacts on water-use efficiency. *Nature Plants*, 7, 1301–1313.
Jia, W., Wang, Y., Zhang, S., & Zhang, J. (2002). Salt-stress-induced ABA accumulation is more sensitively triggered in roots than in shoots. *Journal of Experimental Botany*, *53*, 2201–2206.
Jossier, M., Kroniewicz, L., Dalmas, F., Le Thiec, D., Ephritikhine, G., Thomine, S., et al. (2010). The Arabidopsis vacuolar anion transporter, AtCLCc, is involved in the regulation of stomatal movements and contributes to salt tolerance. *The Plant Journal*, *64*, 563–576.
Karimi, S. M., Freund, M., Wager, B. M., Knoblauch, M., Fromm, J., Müller, H., et al. (2021). Under salt stress guard cells rewire ion transport and abscisic acid (ABA) signaling. *The New Phytologist*, *231*, 1040–1055.
Kim, T.-H., Böhmer, M., Hu, H., Nishimura, N., & Schroeder, J. I. (2010). Guard cell signal transduction network: Advances in understanding abscisic acid, CO_2, and Ca^{2+} signaling. *Annual Review of Plant Biology*, *61*, 561–591.
Kinoshita, T., Doi, M., Suetsugu, N., Kagawa, T., Wada, M., & Shimazaki, K. (2001). Phot1 and phot2 mediate blue light regulation of stomatal opening. *Nature*, *414*, 656–660.
Kinoshita, T., Nishimura, M., & Shimazaki, K. (1995). Cytosolic concentration of Ca^{2+} regulates the plasma membrane H^+-ATPase in guard cells of fava bean. *The Plant Cell*, 7, 1333–1342.
Kinoshita, T., & Shimazaki, K. I. (1999). Blue light activates the plasma membrane H^+-ATPase by phosphorylation of the C-terminus in stomatal guard cells. *The EMBO Journal*, *18*, 5548–5558.
Knight, H., Trewavas, A. J., & Knight, M. R. (1997). Calcium signalling in Arabidopsis thaliana responding to drought and salinity. *The Plant Journal*, *12*, 1067–1078.
Köhler, B., & Blatt, M. R. (2002). Protein phosphorylation activates the guard cell Ca^{2+} channel and is a prerequisite for gating by abscisic acid. *The Plant Journal*, *32*, 185–194.
Kollist, H., Nuhkat, M., & Roelfsema, M. R. G. (2014). Closing gaps: Linking elements that control stomatal movement. *New Phytologist*, *203*, 44–62.
Kronzucker, H. J., & Britto, D. T. (2011). Sodium transport in plants: A critical review. *New Phytologist*, *189*, 54–81.
Lawson, T., & Blatt, M. R. (2014). Stomatal size, speed, and responsiveness impact on photosynthesis and water use efficiency. *Plant Physiology*, *164*, 1556–1570.
Maas, E. V., & Grattan, S. (1999). Crop yields as affected by salinity. *Agricultural Drainage*, *38*, 55–108.
MacRobbie, E. (1998). Signal transduction and ion channels in guard cells. *Philosophical Transactions of the Royal Society, B: Biological Sciences*, *353*, 1475–1488.
Mäser, P., Eckelman, B., Vaidyanathan, R., Horie, T., Fairbairn, D. J., Kubo, M., et al. (2002). Altered shoot/root Na^+ distribution and bifurcating salt sensitivity in Arabidopsis by genetic disruption of the Na^+ transporter AtHKT1. *FEBS Letters*, *531*, 157–161.
Matsumoto, H., & Chung, G. C. (1988). Increase in proton-transport activity of tonoplast vesicles as an adaptive response of barley roots to NaCl stress. *Plant and Cell Physiology*, *29*, 1133–1140.

Meyer, S., Mumm, P., Imes, D., Endler, A., Weder, B., Al-Rasheid, K. A., et al. (2010). AtALMT12 represents an R-type anion channel required for stomatal movement in Arabidopsis guard cells. *The Plant Journal*, *63*, 1054–1062.

Minguet-Parramona, C., Wang, Y., Hills, A., Vialet-Chabrand, S., Griffiths, H., Rogers, S., et al. (2016). An optimal frequency in Ca^{2+} oscillations for stomatal closure is an emergent property of ion transport in guard cells. *Plant Physiology*, *170*, 33–42.

Munns, R., & Tester, M. (2008). Mechanisms of salinity tolerance. *Annual Review of Plant Biology*, *59*, 651–681.

Nakamura, Y., Kasamo, K., Sakata, M., & Ohta, E. (1992). Stimulation of the extrusion of protons and H^+-ATPase activities with the decline in pyrophosphatase activity of the tonoplast in intact mung bean roots under high-NaCl stress and its relation to external levels of Ca^{2+} ions. *Plant and Cell Physiology*, *33*, 139–149.

Nakamura, R. L., McKendree, W. L., Jr., Hirsch, R. E., Sedbrook, J. C., Gaber, R. F., & Sussman, M. R. (1995). Expression of an Arabidopsis potassium channel gene in guard cells. *Plant Physiology*, *109*, 371–374.

Niu, X., Bressan, R. A., Hasegawa, P. M., & Pardo, J. M. (1995). Ion homeostasis in NaCl stress environments. *Plant Physiology*, *109*, 735.

Okazaki, Y., Kikuyama, M., Hiramoto, Y., & Iwasaki, N. (1996). Short-term regulation of cytosolic Ca^{2+}, cytosolic pH and vacuolar pH under NaCl stress in the charophyte alga *Nitellopsis obtusa*. *Plant, Cell & Environment*, *19*, 569–576.

Otoch, M.d. L. O., Sobreira, A. C. M., de Aragão, M. E. F., Orellano, E. G., Lima, M.d. G. S., & de Melo, D. F. (2001). Salt modulation of vacuolar H^+-ATPase and H^+-pyrophosphatase activities in *Vigna unguiculata*. *Journal of Plant Physiology*, *158*, 545–551.

Perera, L., De Silva, D., & Mansfield, T. (1997). Avoidance of sodium accumulation by the stomatal guard cells of the halophyte *Aster tripolium*. *Journal of Experimental Botany*, *48*, 707–711.

Perera, L., Mansfield, T., & Malloch, A. (1994). Stomatal responses to sodium ions in *Aster tripolium*: A new hypothesis to explain salinity regulation in above-ground tissues. *Plant, Cell & Environment*, *17*, 335–340.

Pilot, G., Lacombe, B. T., Gaymard, F., Chérel, I., Boucherez, J., Thibaud, J.-B., et al. (2001). Guard cell inward K^+ channel activity inarabidopsis involves expression of the twin channel subunits KAT1 and KAT2. *Journal of Biological Chemistry*, *276*, 3215–3221.

Qiu, Q.-S., Guo, Y., Dietrich, M. A., Schumaker, K. S., & Zhu, J.-K. (2002). Regulation of SOS1, a plasma membrane Na^+/H^+ exchanger in *Arabidopsis thaliana*, by SOS2 and SOS3. *Proceedings of the National Academy of Sciences*, *99*, 8436–8441.

Queirós, F., Fontes, N., Silva, P., Almeida, D., Maeshima, M., Gerós, H., et al. (2009). Activity of tonoplast proton pumps and Na^+/H^+ exchange in potato cell cultures is modulated by salt. *Journal of Experimental Botany*, *60*, 1363–1374.

Rienmüller, F., Dreyer, I., Schönknecht, G., Schulz, A., Schumacher, K., Nagy, R., et al. (2012). Luminal and cytosolic pH feedback on proton pump activity and ATP affinity of V-type ATPase from Arabidopsis. *Journal of Biological Chemistry*, *287*, 8986–8993.

Rus, A., Yokoi, S., Sharkhuu, A., Reddy, M., Lee, B.-H., Matsumoto, T. K., et al. (2001). AtHKT1 is a salt tolerance determinant that controls Na^+ entry into plant roots. *Proceedings of the National Academy of Sciences*, *98*, 14150–14155.

Shi, H., Ishitani, M., Kim, C., & Zhu, J.-K. (2000). The Arabidopsis thaliana salt tolerance gene *SOS1* encodes a putative Na^+/H^+ antiporter. *Proceedings of the National Academy of Sciences*, *97*, 6896–6901.

Shi, H., Quintero, F. J., Pardo, J. M., & Zhu, J.-K. (2002). The putative plasma membrane Na^+/H^+ antiporter SOS1 controls long-distance Na^+ transport in plants. *The Plant Cell*, *14*, 465–477.

Shi, H., & Zhu, J.-K. (2002). Regulation of expression of the vacuolar Na^+/H^+ antiporter gene AtNHX1 by salt stress and abscisic acid. *Plant Molecular Biology*, *50*, 543–550.
Shimazaki, K., Iino, M., & Zeiger, E. (1986). Blue light-dependent proton extrusion by guard-cell protoplasts of *Vicia faba*. *Nature*, *319*, 324–326.
Shimazaki, K.-I., & Kondo, N. (1987). Plasma membrane H^+-ATPase in guard-cell protoplasts from *Vicia faba* L. *Plant and Cell Physiology*, *28*, 893–900.
Speer, M., & Kaiser, W. M. (1991). Ion relations of symplastic and apoplastic space in leaves from *Spinacia oleracea* L. and *Pisum sativum* L. under salinity. *Plant Physiology*, *97*, 990–997.
Stelling, J., Sauer, U., Szallasi, Z., Doyle, F. J., & Doyle, J. (2004). Robustness of cellular functions. *Cell*, *118*, 675–685.
Tang, X., Mu, X., Shao, H., Wang, H., & Brestic, M. (2015). Global plant-responding mechanisms to salt stress: Physiological and molecular levels and implications in biotechnology. *Critical Reviews in Biotechnology*, *35*, 425–437.
Thiel, G., & Blatt, M. R. (1991). The mechanism of ion permeation through K^+ channels of stomatal guard cells: Voltage-dependent block by Na^+. *Journal of Plant Physiology*, *138*, 326–334.
UNDESA. (2019). *World Population Prospects 2019*. United Nations. Department of Economic and Social Affairs. World population prospects 2019.
van Riel, N. A. W. (2006). Dynamic modelling and analysis of biochemical networks: Mechanism-based models and model-based experiments. *Briefings in Bioinformatics*, *7*, 364–374.
Venema, K., Quintero, F. J., Pardo, J. M., & Donaire, J. P. (2002). The Arabidopsis Na^+/H^+ exchanger AtNHX1 catalyzes low affinity Na^+ and K^+ transport in reconstituted liposomes. *Journal of Biological Chemistry*, *277*, 2413–2418.
Véry, A. A., Robinson, M. F., Mansfield, T. A., & Sanders, D. (1998). Guard cell cation channels are involved in Na^+–induced stomatal closure in a halophyte. *The Plant Journal*, *14*, 509–521.
Very, A.-A., & Sentenac, H. (2003). Molecular mechanisms and regulation of K^+ transport in higher plants. *Annual Review of Plant Biology*, *54*, 575–603.
Wang, Y., Hills, A., Vialet-Chabrand, S. R., Papanatsiou, M., Griffiths, H., Rogers, S., et al. (2017). Unexpected connections between humidity and ion transport discovered using a model to bridge guard cell-to-leaf scales. *The Plant Cell*, *29*, 2921–2939.
Wang, W., Li, Y., Zhang, Y., Yang, C., Zheng, N., & Xie, Q. (2007). Comparative expression analysis of three genes from the Arabidopsis vacuolar Na^+/H^+ antiporter (AtNHX) family in relation to abiotic stresses. *Chinese Science Bulletin*, *52*, 1754–1763.
Wang, B., Lüttge, U., & Ratajczak, R. (2001). Effects of salt treatment and osmotic stress on V-ATPase and V-PPase in leaves of the halophyte *Suaeda salsa*. *Journal of Experimental Botany*, *52*, 2355–2365.
Wang, Y., Papanatsiou, M., Eisenach, C., Karnik, R., Williams, M., Hills, A., et al. (2012). Systems dynamic modeling of a guard cell cl^- channel mutant uncovers an emergent homeostatic network regulating stomatal transpiration. *Plant Physiology*, *160*, 1956–1967.
Wille, A. C., & Lucas, W. J. (1984). Ultrastructural and histochemical studies on guard cells. *Planta*, *160*, 129–142.
Willmer, C., & Mansfield, T. (1970). Further observations of cation-stimulated stomatal opening in isolated epidermis. *New Phytologist*, *69*, 639–645.
Wu, H., Shabala, L., Zhou, M., Su, N., Wu, Q., Ul-Haq, T., et al. (2019). Root vacuolar Na^+ sequestration but not exclusion from uptake correlates with barley salt tolerance. *The Plant Journal*, *100*, 55–67.
Yamaguchi, T., Hamamoto, S., & Uozumi, N. (2013). Sodium transport system in plant cells. *Frontiers in Plant Science*, *4*, 410.

Yokoi, S., Quintero, F. J., Cubero, B., Ruiz, M. T., Bressan, R. A., Hasegawa, P. M., et al. (2002). Differential expression and function of *Arabidopsis thaliana* NHX Na^+/H^+ antiporters in the salt stress response. *The Plant Journal, 30*, 529–539.

Yu, Y., & Assmann, S. M. (2016). The effect of NaCl on stomatal opening in Arabidopsis wild type and agb1 heterotrimeric G-protein mutant plants. *Plant Signaling & Behavior, 11*, e1085275.

Zhao, X., Wang, Y.-L., Wang, Y.-J., Wang, X.-L., & Zhang, X. (2008). Extracellular Ca^{2+} regulating stomatal movement and plasma membrane K^+ channels in guard cells of *Vicia faba* under salt stress. *Acta Agronomica Sinica, 34*, 1970–1976.

Zhao, X., Wang, Y.-J., Wang, Y.-L., Wang, X.-L., & Zhang, X. (2011). Extracellular Ca^{2+} alleviates NaCl-induced stomatal opening through a pathway involving H_2O_2-blocked Na^+ influx in *Vicia* guard cells. *Journal of Plant Physiology, 168*, 903–910.

Zhu, J.-K. (2003). Regulation of ion homeostasis under salt stress. *Current Opinion in Plant Biology, 6*, 441–445.

CHAPTER SEVEN

Signaling molecules and transcriptional reprogramming for stomata operation under salt stress

Jahanzeb Jafar[a], Heba Hassan[a,b], Sergey Shabala[c,d,*], and Bo Ouyang[a,*]

[a]Key Laboratory of Horticultural Plant Biology, Huazhong Agricultural University, Wuhan, China
[b]Genetics Department, Faculty of Agriculture, Ain Shams University, Cairo, Egypt
[c]International Research Centre for Environmental Membrane Biology, Foshan University, Foshan, Guangdong, China
[d]Tasmanian Institute of Agriculture, College of Science and Engineering, University of Tasmania, Hobart, TAS, Australia
*Corresponding authors: e-mail address: sergey.shabala@utas.edu.au; bouy@mail.hzau.edu.cn

Contents

1. Introduction 164
2. Signaling molecules involved in stomatal closure 165
   2.1 ABA-mediated stomatal movement 165
   2.2 Other signaling molecules 168
3. Transcription factors involved in the regulation of stomatal structure under salinity stress 172
   3.1 MYB 172
   3.2 bZIP 177
   3.3 NAC 178
   3.4 WRKY 180
   3.5 Other TFs 181
4. Concluding remarks 182
Acknowledgments 183
References 183

Abstract

Abiotic stresses such as salinity and other unfavorable conditions affect more than half of the crop production in the world. Salinity ranks as the second limiting factor after drought. By 2050, over 50% of the world arable land will most likely be affected by salinity. Salinity stress reduces water availability and affects stomatal movement in plants. Under salt stress, guard cells are reprogrammed at the transcriptional level to conserve water by more efficient stomatal operation. The intertwined signaling network consisting of signal molecules (such as ABA, JA, Ca^{2+}, NO, H_2S, ROS) and various transcription

Advances in Botanical Research, Volume 103
ISSN 0065-2296
https://doi.org/10.1016/bs.abr.2022.02.013

factors (such as MYB, NAC, bZIP, WRKY family members) fine-tune the stomatal aperture, optimizing leaf gas exchange. Many of these transcription factors are essential for activating the basic components of signaling pathways, from stress receptors to protein kinases and membrane effectors, mediating stomata movements. Among these signaling molecules, abscisic acid (ABA) is considered as the core member. Most of the other molecules are ABA-dependent or work as secondary messengers to increase ABA production. Active molecules such as reactive oxygen/nitrogen species are thought to cause oxidative stress in plants, but in controlled quantity can also act as signals in plant defense mechanisms. This review focuses on the signal molecules and transcription factors involved in regulating stomatal opening to optimize the performance of plants under saline conditions.

Abbreviations

ABA	abscisic acid
CPKs	calcium-dependent protein kinase
ET	ethylene
GORK	guard cell outward rectifying K^+channel
H_2S	hydrogen sulphide
JA	jasmonic acid
K_{in}	potassium influx
NO	nitric oxide
NOS	nitric oxide synthases
NR	nitrite reductase
QUAC1	quick-activating anion channel 1
ROS	reactive oxygen species
SA	salicylic acid
SLAC1	slow anion channel-associated 1

1. Introduction

By 2050, the human population will exceed 10 billion, and the demand for staple crops and livestock will increase by 60% to match the growth (Rawat et al., 2022; Springmann et al., 2018). Meanwhile, crop productivity is strongly affected by abiotic stresses such as salinity (Razzaq et al., 2021). Salinity issue prevails in semiarid and arid regions where irrigation is regularly practiced, due to the fact that the rate of evapotranspiration is greater than that of irrigation. Annually, salinity causes over $27.3 billion loss to agribusiness (Qadir et al., 2014).

Salinization is spreading rapidly, and it is dominant in irrigated areas. By 2050, 50% of arable land is likely to be lost due to salinity (Alam et al., 2021).

Saline water irrigation regimes (de Lima-Neto et al., 2016) and water scarcity (Tri & Tuyet, 2016) are major sources of soil salinity. Compost fertilizers can also cause soil salinity, as both food and municipal organic waste contain significant amounts of NaCl (Gondek, Weindorf, Thiel, & Kleinheinz, 2020). Salinity reduces plant growth and productivity by imposing osmotic, ionic, and oxidative stresses (Zhao et al., 2020). Stomatal traits have been found to be associated with crop yield under salinity stress (Liu et al., 2017).

Stomata are tiny pores on the leaf surface surrounded by a pair of guard cells, which balance CO_2 assimilation and transpiration water loss, and are responsible for approximately 95% of leaf gas exchange (Raven, 2002; Vatén & Bergmann, 2012). They also play key roles in nutrient uptake and evaporative cooling of leaf tissue (Morison, Stewart, & Howell, 2003). Thus, optimization of stomatal opening and closure is instrumental in maximizing plant performance under adverse stress conditions (Ding, Kang, Du, Hao, & Zhang, 2014).

2. Signaling molecules involved in stomatal closure

A variety of phytohormones, including abscisic acid (ABA), jasmonic acid (JA), ethylene (ET) and salicylic acid (SA), form a complex signaling network to combat stress conditions (Sierla, Waszczak, Vahisalu, & Kangasjärvi, 2016). Some phytohormones, such as ABA or JA, regulate stomatal closure (Förster et al., 2019), whereas auxin and cytokinin mediate stomatal opening, although the exact mechanism involved is unclear (Song, She, He, Huang, & Song, 2006). The action mode of ET is ambiguous because it can act as either a positive or a negative regulator, depending on tissues and conditions (Daszkowska-Golec & Szarejko, 2013). In addition, various simple molecules also participate in stomatal regulation, such as Ca^{2+}, NO, H_2S, and reactive oxygen species (ROS) (Fig. 1).

2.1 ABA-mediated stomatal movement

Abscisic acid (ABA) is known as the "stress hormone" that regulates stress responses in plants. Environmental stress induces ABA accumulation, which represents one of the fastest responses to abiotic stress (Yang, Li, Liu, Xie, & Sui, 2020). ABA plays an important role in stress-related signal transduction networks *via* increasing the binding of ABA to its receptor and initiating signal transduction for cellular responses to stresses (Delgado, Mora-Poblete, Ahmar, Chen, & Figueroa, 2021; Sah, Reddy, & Li, 2016). ABA concentration controls stomatal behavior and water loss in leaf to adapt and survive stress

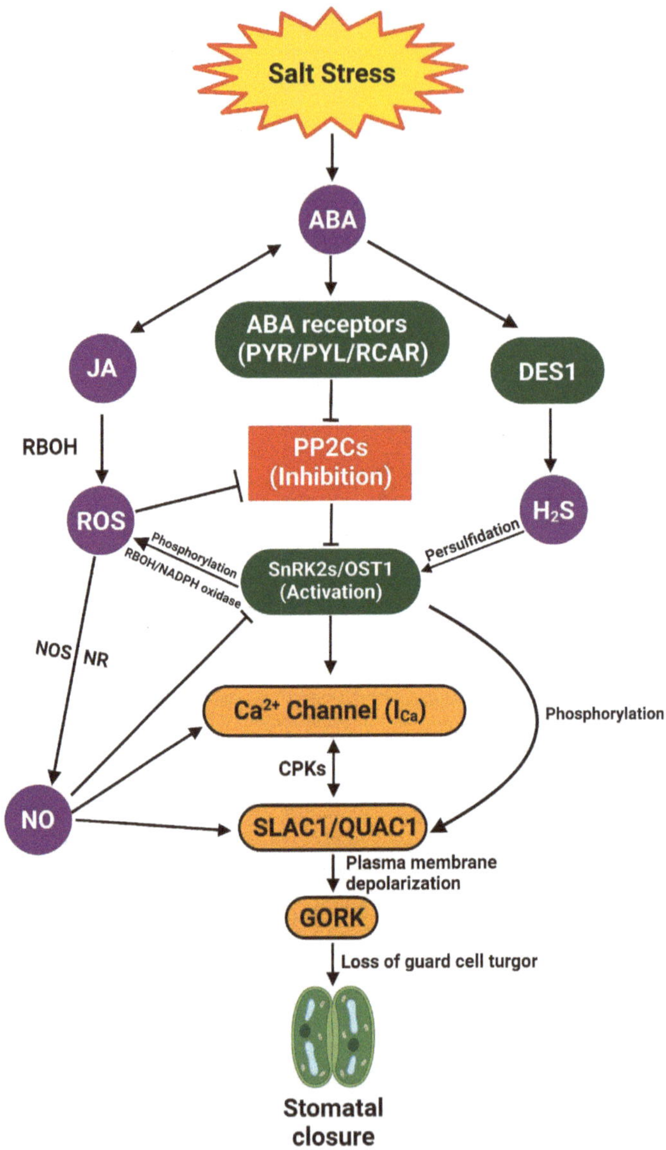

Fig. 1 A summarized model of signaling pathways and their cross-talks in regulation of stomatal closure. The positive regulators are shown in green boxes, while the negative regulators are shown in red. In the guard cells, the ABA receptors (PYR/PYL/RCAR) suppress the activity of PP2C phosphatase (Park et al., 2009). This results in the activation of SnRK2.6 (OST1) kinase. SLAC1 (Slow Anion Channel-associated 1) and QUAC1 (Quick-activating Anion Channel 1) (Lee et al., 2009; Sah et al., 2016), the anion channels involved

conditions (Niu et al., 2018). Under salinity stress, ABA is significantly increased, triggering changes in gene expression and stomatal closure (Liotenberg, North, & Marion-Poll, 1999). This increase is likely to be due to the osmotic component of salt stress, rather than being specific to NaCl (Davies, Wilkinson, & Loveys, 2002). In addition to hormone signals, stomatal regulation also involves hydraulic signals during salinity or drought stress (Rodriguez-Dominguez et al., 2016). Guard cells directly perceive the change of air relative humidity, while roots perceive the changes of water potential, which in turn allocate ABA to stomata (Ko & Helariutta, 2017). The leaf water potential in plants decreases with the increase of salinity-related ions (Wilkinson & Davies, 2010). Some studies have shown that the sensitivity of stomata to low water potential is due to the increase in ABA level. ABA works both chemically and hydraulically by regulating stomatal closure as well as restricting hydraulic conductivity of leaf (Dodd, 2013). This decrease in hydraulic conductivity in turn leads to the decrease of stomatal conductance in leaves, thus allowing ABA to mediate the indirect closure of stomata (Wang et al., 2018).

The PYR/PYL/RCAR protein family was identified as the receptor of ABA (Ma et al., 2009). These receptors trigger the ABA signaling cascade by phosphorylating downstream proteins in cells. ABA binds directly to PYR/PYL/RCAR receptors to inhibit PP2C (Type-2C Protein Phosphatase) (Park et al., 2009). The inhibition of PP2C becomes the cause of SnRK2 (SNF1-Related Protein kinase 2) activation, but in the absence of ABA, it is inhibited by PP2C controlled dephosphorylation (Umezawa et al., 2009; Vlad et al., 2009).

in the core signaling process are activated *via* phosphorylation by SnRK2.2/2.3/2.6 protein kinases (Boudsocq et al., 2004). The, signaling cross-talk between JA (Jasmonic Acid) and ABA causes the generation of ROS (Reactive Oxygen Species) and NO (Nitric Oxide). NADPH oxidase (RBOH) is the primary ROS source. The generation of NO appears to be dependent on the activity of two enzymes NOS (Nitric Oxide Synthases) and NR (Nitrite Reductase) (Gayatri, Agurla, & Raghavendra, 2013). Ca^{2+} influx from apoplast causes ROS and NO to elevate guard cell $[Ca^{2+}]$cyt, while I_{Ca} channels mediate Ca^{2+} influx from apoplast. CPKs (Calcium-dependent Protein Kinase) detect elevated $[Ca^{2+}]$cyt and thus activate the S-type anion channels (Ghorbel et al., 2021). On the other hand, H_2S controls ABA signaling through persulfidation of SnRK2.6 (OST1) (Chen et al., 2020). DES1 (L-cysteine DeSulfhydrase 1), produces H_2S (Hydrogen Sulfide) in the cytoplasm using cysteine as substrate (Zhang et al., 2021). ABA induces an increase in the cytosolic Ca^{2+}, leading to membrane depolarization, deactivation of K^+ influx channels, and activation of guard cell outward rectifying K^+ channel (GORK), eventually resulting in K^+ efflux from guard cells (Miedema & Assmann, 1996). The outward flow of K^+ and other anions reduces turgor in guard cells, leading to stomatal closure (MacRobbie, 1998).

Recently, Takahashi et al. (2020) reported M3Ks (Raf-like MAP3 Kinase) as another essential component for the activation of SnRK2 kinases. The M3Ks phosphorylate SnRK2.6 (OST1) at a specific site (Ser171), which is necessary for the reactivation of SnRK2 kinases. The activated SnRK2s, most importantly SnRK2.6 (OST1), cause the phosphorylation of SLAC1 (Slow Anion Channel-associated 1), some transcription factors and other different targets to trigger stomatal closure (Geiger et al., 2009; Lee, Lan, Buchanan, & Luan, 2009). When SnRK2 protein kinases are activated by ABA, a downstream signaling pathway of ABA is initiated. In *Arabidopsis*, three types of SnRK2 kinase genes, *SnRK2.2*, *SnRK2.3* and *SnRK2.6* (*OST1*), are activated in the presence of ABA (Boudsocq, Barbier-Brygoo, & Laurière, 2004). These three protein kinases have common functions to mediate ABA response, *i.e.*, to phosphorylate different substrate proteins (*e.g.*, SnRK2.6 (OST1) phosphorylates SLAC1 at serine 120) (Geiger et al., 2009). Moreover, OST1 binds and activates QUAC1 (Quick-activating Anion Channel 1) by phosphorylation, leading to stomatal closure (Fernando & Schroeder, 2016; Sah et al., 2016).

Interestingly, salinity-induced elevation of ABA content in the xylem sap (Albacete et al., 2008; Kefu, Munns, & King, 1991) is often not observed in halophytes (Reef, Schmitz, Rogers, Ball, & Lovelock, 2012). The meta-analysis of the published literature revealed that under control conditions, the ABA content in leaves of halophytes is usually much lower than that of glycophytes (Hedrich & Shabala, 2018). Salinity stress results in a two- to threefold increase in leaf ABA content in glycophytes, but the ABA levels in halophytes remain fairly stable. This prompted a suggestion that modulation of stomata aperture by altering ABA levels operates over a much lower concentration range in halophytes than in glycophytes, calling for a need for direct experimental validation of this hypothesis (Hedrich & Shabala, 2018).

2.2 Other signaling molecules

Besides ABA as a major signal modulating stomata movement, there are many other signals involved in the regulation of stomata aperture, such as jasmonic acid (JA), hydrogen sulfide (H_2S), and nitric oxide (NO).

JA biosynthesis pathway has been extensively studied in tomato and *Arabidopsis* since its first discovery (Vick & Zimmerman, 1984). Under non-stress conditions, the level of JA-Ile (Jasmonoyl-Isoleucine, a JA derivative) is very low, so various transcription factors could not activate JA responsive genes. These transcription factors are repressed by a series of transcription repressors, such as the jasmonate zinc-finger inflorescence meristem domain (JAZ) proteins (Zhou & Memelink, 2016). Under stress conditions, JAT1 (Jasmonate-inducible Alkaloid Transporter 1) transports JA-Ile into the nucleus, forming a coreceptor complex SCFCOI1-JAZ, which participates in the ubiquitination of JAZ and its degradation in the 26S proteasome (Li et al., 2018). COI1 is a member of F-box proteins, and it acts as a receptor of jasmonate. JA-Ile is the only form of jasmonate with which the coreceptor complex can bind, while other forms of jasmonate such as JA, MeJA (Methyl Jasmonate) or 12-OPDA (12-OxoPhytodienoic Acid) do not have this binding ability (Li et al., 2018). COI1 controls the substrate selectivity of the SCF (SKP1-Cullin1-F-box protein)-type ubiquitin (E3) ligases (Wasternack & Feussner, 2018). After binding of SCF/COI1 with JA-Ile, ubiquitination of JAZ family proteins starts. Two domains (ZIM and Jas) of JAZ proteins mediate the interaction of JAZ and other proteins. The JAZ–COI1 relationship is mediated by the Jas domain, which is responsible for its dimerization and association with NINJA (Novel Interactor of JAZ), and NINJA connects the TPL (TOPLESS) transcription suppressor with JA signal transduction (Hamant & Haswell, 2017; Kosová et al., 2015). After JAZ degradation, transcription factors are derepressed and bind immediately with G-box regulatory element, and then RNA Pol II, MED25 and GTF (General Transcription Factor) are recruited, which results in the expression of JA responsive genes (Ali & Baek, 2020).

JA or MeJA also mediates stomata closure in response to drought and salinity by reducing transpiration rate (Chen et al., 2011; Munemasa, Mori, & Murata, 2011). Like ABA, JA activates S-type anion channels (Munemasa, Mori, et al., 2011), causing chloride efflux and, hence, reduction in cell turgor (Chen et al., 2011).

MeJA also interacts with other signaling molecules like ABA, ROS (Reactive Oxygen Species), NO and Ca^{2+} (Ghorbel, Brini, Sharma, & Landi, 2021). The stomatal movement mediated by JA is often dependent on the presence of ABA (de Ollas & Dodd, 2016). A recent finding also suggests that MeJA may be involved in the regulation of ABA receptor complex (Ghorbel et al., 2021). The stimulation of OST1-mediated stomatal closure is controlled by ABA and JA (Yin et al., 2016). In MeJA and ABA upstream signaling cross-talk, components like PP2A and RCN1 (Reticulocalbin 1) are involved in the production of ROS and NO. MeJA and ABA signaling contain TGG1 (Myrosinase 1), TGG2 (Myrosinase 2), which mediate ROS production in guard cells.

Similar to ABA, MeJA triggers the inward Ca^{2+} current through Ca-permeable channels in the plasma membrane of the guard cells, increasing the cytosolic free calcium (Munemasa, Hossain, Nakamura, Mori, & Murata, 2011). A separate study suggests that blocking the plasma membrane $[Ca^{2+}]_{cyt}$ will inhibit MeJA-induced stomatal closure (Munemasa, Mori, et al., 2011; Suhita, Kolla, Vavasseur, & Raghavendra, 2003). MeJA-induced signaling is highly dependent upon $[Ca^{2+}]_{cyt}$ that act as a central response to the abiotic stresses (Carroll & Peralta, 1998). In *Arabidopsis*, CPK6 (Calcium-dependent Protein Kinase 6) acts as a central regulator of MeJA signaling. The knockdown mutant of CPK6 fails to activate the MeJA induced I_{ca}, which decreases the level of $[Ca^{2+}]_{cyt}$ and fails to regulate stomatal closure (Munemasa, Hossain, et al., 2011). CPK6 is also required for the activation of SLAC.

Hydrogen sulfide (H_2S) is another signaling molecule involved in plant responses to various biotic or abiotic stressors (Li, Jia, Wang, Cao, & Wen, 2014). Many enzymes, including DES1 (L-Cysteine Desulfhydrase 1), LCD (L-Cysteine Desulfhydrase), DCD (D-Cysteine Desulfhydrase), SiR (Sulfite Reductase) and CAS (Cyanoalanine Synthase), have been discovered in plants, which produce hydrogen sulfide (Corpas & Palma, 2020; Liu & Xue, 2021; Pandey & Gautam, 2020; Xuan, Li, Wang, & Wang, 2020).

Stomatal closure mediated by H_2S follows the ABA dependent signaling pathway by the persulfidation of DES1, which in turn positively regulates SnRK2.6 (OST1) and leads to stomatal closure (Chen et al., 2020). The H_2S level in the guard cells also increases with the persulfidation of LCD1 stimulated by ABA. This ultimately leads to the persulfidation of the NADPH oxidase RBOHD that enhances ROS production. The ABA signaling is fine-tuned by a negative feedback loop mediated by persulfide-oxidation of RBOHD and DES1, which results in ABA desensitivity (Shen et al., 2020). Recently in *Arabidopsis*, it was discovered that LCD generated H_2S works downstream of phospholipase D (PLDδ) for inducing stomatal closure (Liu et al., 2021). But the effects of H_2S in stomatal closure are inhibited by glibenclamide (an ATP-binding cassette (ABC) transporter inhibitor) and scavenging of L-Cys desulfhydrase and cystathionine γ lyase (responsible for endogenous production of H_2S). These factors also cause partial inhibition of ABA-dependent stomatal closure, indicating that H_2S plays a role in ABA-mediated stomatal closure, which is likely related to the role of ABC transporter in guard cells (García-Mata & Lamattina, 2010). According to Ma et al. (2019), H_2S is highly likely to be involved in salinity related closing of stomata in the downstream signaling pathway of H_2O_2 in faba bean (*Vicia faba*).

In addition to ABA, ET has a partnership with H_2S-mediated stomatal closure, but this relationship is rather complex. The exogenous application of ACC (1-AminoCyclopropane-1-Carboxylic acid, the precursor of ET) enhances the production of H_2S and D/L-Cysteine desulfhydrase (D/L-CDes) in *Arabidopsis* leaves, causing stomatal closure (Hou, Wang, Liu, Hou, & Liu, 2013). In the same study, the *Arabidopsis* D/L-CDes overexpression plants show increased stomatal closure in the presence of ethylene as compared to the control plants while no effects were exhibit by the knockdown plants *Atl-cdes* and *Atd-cdes*, suggesting the role of H_2S in the ethylene induced stomatal closure.

Although nitric oxide (NO) is not classified as a hormone, it plays a very important signaling role in various development and adaptive processes of plants, such as seed germination, photosynthesis, senescence, stomatal movement and flowering regulation (Bhardwaj et al., 2021). The beneficial effects of NO in plants can only be observed at lower concentrations (Lau, Hamdan, Pua, Saidi, & Tan, 2021; Sami et al., 2018). In *Arabidopsis*, the exogenous application of ABA and GSNO (Nitrosoglutathione; NO donor) causes the S-nitrosylation of OST1/SnRK2.6 (Wang et al., 2015). This process negatively regulates ABA-dependent stomatal closure. It was also observed that the knockdown mutants of GSNO reductase (*gsnor1–3*) accumulated massive amount of NO, causing the S-nitrosylation of OST1 at Cys137 residue level and disabling ABA induced closing of stomata. Another group of researchers reported the negative interaction of NO in ABA mediated stomatal closure. The nitration of ABA receptor proteins at tyrosine residue causes decreased activity of receptors and they undergo proteasome degradation (Castillo et al., 2015). Another study revealed the positive role of NO in ABA regulation of anion channels, as ABA inactivates K^{+}_{in} and activates anion channels in the guard cells. But this inhibition of K^{+}_{in} ends when NIA1 and NIA2 are defected, causing the deactivation of anion channels (Chen et al., 2016). In the *nia1nia2* double mutant of *Arabidopsis*, the expression of genes encoding ABA receptors, OST1, CPKs and transcription factors are impaired, suggesting that NO regulates ABA-induced stomatal closure *via* influencing these factors (Zhao, Cai, Wang, & Chen, 2016).

Recent studies have found that L-NAME (NG-Nitro-L-arginine Methyl Ester, Hydrochloride, a nitric oxide synthase inhibitor) affects MeJA-induced stomatal closure in *Arabidopsis*. A notable reduction in size and numbers of stomata can be observed in the presence of L-arginine as well as sodium nitrite. The oxidation of L-arginine and sodium nitrite reduction (nitrate/nitrite) in the guard cells are very important for NO production and stomatal closure induced by MeJA (Yastreb, Kolupaev, Kokorev, Horielova, & Dmitriev, 2018). Similarly, the NO-mediated stomatal closure as well as $[Ca^{2+}]_{cyt}$ and $K^{+}{}_{in}$ suppression in guard cells are negatively regulated in LOX2 (catalyst for synthesizing oxylipins) knockdown mutants. These findings indicate that NO plays a role in stomatal closure by regulating the levels of Ca^{2+} and K^{+} in guard cells (Sun & Yu, 2015).

Stomatal response to salinity and drought is similar in nature (Chaves, Flexas, & Pinheiro, 2009). Increased salinity causes a significant decrease in stomatal conductance (Clough & Sim, 1989). High salinity and increased level of CO_2 are expected to coexist (Hoque et al., 2016). Severe salt stress and increased CO_2 alone or in combination can reduce stomatal conductance (Nguyen, Stanton, Schmitz, Farquhar, & Ball, 2015). High salinity stress may cause oxidative damage in plant tissue, similar to the severe drought effect (Sanoubar et al., 2016). However, greater CO_2 levels may reduce oxidative stress damage caused by salinity by lowering ROS levels and increasing ABA, thus improving plant development in high salinity environment (Pérez-López et al., 2009).

3. Transcription factors involved in the regulation of stomatal structure under salinity stress

Transcriptome analysis is an effective strategy to identify the expression of candidate genes and understand the response mechanism of abiotic stress (Guo et al., 2020). Stomatal closure is induced by sodium ions in the apoplast surrounding the guard cells and causes a reduction in photosynthesis rate and water uptake (Kerstiens, Tych, Robinson, & Mansfield, 2002). We have already discussed the role of hormones in stomatal closure, such as the role of ABA, in avoiding water loss (Fatma et al., 2021). At the molecular level, it is known that the regulation of some transcription factor families in plants will program stomatal closure. Families including MYB, bZIP, NAC, WRKY, bHLH, and many others can induce stomatal closure directly or through hormone pathway. These TF families are described in detail below (Table 1).

3.1 MYB

The 198 members of the MYB superfamily constitute the largest transcription factor group in *Arabidopsis* (Chen et al., 2006). The first MYB TF involved in the stomatal movement, namely *AtMYB60*, was identified in

Table 1 Transcription factors involved in the stomatal movement from different family.

#	TF	Plant species	Pathway	Stomata status	References
NAC					
1	NAC019	*Arabidopsis thaliana*	JA	Opening	Lindemose, O'Shea, Jensen, & Skriver, 2013; Wang et al., 2018
2	NAC72	*A. thaliana*	ABA	Closure	Yu et al., 2021
3	SNAC1	*Oryza sativa*	ABA	Closure	Cominelli, Galbiati, & Tonelli, 2010; You et al., 2013
4	NAC11	*A. thaliana*	ABA	Closure	Yu et al., 2021
5	ONAC022	*O. sativa*	ABA	Closure	Hong, Zhang, Huang, Li, & Song, 2016
MYB					
6	MYB30	*Diplotaxis tenuifolia*	ABA	—	Franzoni, Cocetta, Trivellini, & Ferrante, 2019
7	MYB85	*Gossypium arboreum*	ABA	Reduced stomatal density	Butt, Yang, Gong, et al., 2017; Hussain et al., 2021
8	MYB96	*A. thaliana*	—	Stomatal opening	Baldoni, Genga, & Cominelli, 2015
9	MYB1	*Hordeum vulgare*	—	Stomatal conductance	Manna et al., 2021
10	MYB5	*Nicotiana tabacum*	—	Decreased stomatal aperture	Wei et al., 2017
11	MYB60	*A. thaliana*	Light	Opening	Jung et al., 2008; Saibo, Lourenço, & Oliveira, 2009; Cominelli et al., 2010; Butt, Yang, Gong, et al., 2017

Continued

Table 1 Transcription factors involved in the stomatal movement from different family.—cont'd

#	TF	Plant species	Pathway	Stomata status	References
12	MYB2	*O. sativa*	JA and ABA	Closure	
13	MYB61	*A. thaliana*	Dark/light and ABA	Closure	Jung et al., 2008; Saibo et al., 2009; Cominelli et al., 2010; Butt, Yang, Gong, et al., 2017; Yu et al., 2021
14	MYB61	*Citrus sinensis*	ABA	Closure	Yu et al., 2021
15	MYB44	*A. thaliana*	ABA	Closure	Cominelli et al., 2010; Butt, Yang, Gong, et al., 2017
16	MYB15	*A. thaliana*	ABA	Closure	Cominelli et al., 2010; Butt, Yang, Gong, et al., 2017
17	MYB109	*A. thaliana*	—	Closure	So et al., 2020
18	MYB5	*G. barbadense*	ABA	Closure	Butt, Yang, Gong, et al., 2017
19	MYB96	*A. thaliana*	ABA	Closure	Butt, Yang, Gong, et al., 2017
WRKY					
20	WRKY53	*A. thaliana*	—	Opening	Li, Pang, Lu, & Jin, 2020; Hussain et al., 2021
21	WRKY41	*G. herbaceum*	—	Closure	Hussain et al., 2021
22	WRKY46	*A. thaliana*	Light	Opening	Ding, Yan, et al., 2014
23	WRKY68	*G. herbaceum*	—	Closure	Hussain et al., 2021
24	WRKY65	*A. thaliana*	ABA pathway	Closure	Huo et al., 2021
25	WRKY81	*Solanum lycopersicum*	—	Closure	Hussain et al., 2021

Table 1 Transcription factors involved in the stomatal movement from different family.—cont'd

#	TF	Plant species	Pathway	Stomata status	References
26	WRKY17	*N. benthamiana*	ABA	Closure	Yan et al., 2014
27	WRKY70	*A. thaliana*	ABA	A negative regulator of stomatal closure	Li et al., 2013
28	WRKY54	*A. thaliana*	ABA	A negative regulator of stomatal closure	Li et al., 2013
29	*WRKY36*	*A. thaliana*	ABA	Closure	Li et al., 2020
30	*WRKY20*	*A. thaliana*	ABA	Stomatal density	Ma et al., 2019
31	*WRKY146*	*A. thaliana*	—	Closure	Ma et al., 2017
AREB					
32	AREB1	*A. thaliana*	ABA	Closure	Fujita et al., 2005; Saibo et al., 2009
33	AREB2	*A. thaliana*	ABA	Closure	Saibo et al., 2009
bHLH					
34	bHLH35	*Populus euphratica*	ABA	Increase stomatal pore size	Yu et al., 2021; Zhao et al., 2020
35	bHLH130	*N. tabacum*	ABA	Closure	Zhao et al., 2020
36	bHLH38	*Myrothamnus flabellifolia*	ABA	Closure	Qiu et al., 2020
37	bHLH148	*O. sativa*	JA	Closure	de Ollas & Dodd, 2016
ABI					
38	ABI1	*A. thaliana*	ABA	Closure	Sun et al., 2019
39	ABI2	*A. thaliana*	ABA	Closure	Sun et al., 2019
40	ABF3	*Populus euphratica*	ABA	Closure	Saibo et al., 2009; Yu et al., 2021

Arabidopsis, which belongs to the R2R3-MYB family (Cominelli et al., 2005). *AtMYB60* has a distinct expression pattern in guard cells (Galbiati et al., 2008). Its expression is negatively regulated by signals that promote stomatal closure, such as ABA treatment (Cominelli et al., 2005). *MYB62L* is another novel R2R3 TF in *Arabidopsis*, which helps to reduce stomatal density and stomatal opening under drought, salt and cold conditions (Butt, Yang, Chen, et al., 2017; Yang et al., 2016). *MYB62L* is also found to be induced by ABA treatment (Butt, Yang, Chen, et al., 2017).

Other *Arabidopsis R2R3MYB* genes involved in transcriptional reprogramming of stomata are *AtMYB15* and *AtMYB44* (Jung et al., 2008). The expression of these genes was upregulated in guard cells under the conditions of ABA and abiotic stresses. The results of transgenic lines overexpressing *AtMYB15* revealed that it plays a role in regulating stomatal closure, with increased sensitivity to ABA and salinity tolerance. *AtMYB15* also controls the expression of a variety of genes that are upregulated in response to abiotic stress (Ding et al., 2009). Similarly, compared to wild type, the overexpression of *AtMYB44* shows more tolerance under salinity and drought conditions, while its knockout mutant *atmyb44* shows negative stomatal regulation. *AtMYB44* inhibits the expression of genes encoding PP2C, which have previously been identified as negative regulators of the ABA signaling pathway. The overexpression lines have a stronger response to ABA and ABA-induced stomatal closure, which is consistent with the role of *AtMYB44* in the regulation of PP2C genes (Jung et al., 2008).

Another study has revealed the role of *AtMYB96*, another R2R3 MYB TF, in abiotic stress tolerance. The activation-tagged overexpression line, *myb96-ox*, demonstrates dwarf growth with changed morphological changes in leaf. This line shows increased tolerance to osmotic stress, while the T-DNA insertional knockout mutant, *myb96-1*, shows the opposite (Seo et al., 2009). Under normal conditions, the stomatal movement in the knockout and overexpression plants was not obvious. However, in the presence of ABA, the stomatal size of the overexpression lines was reduced, as compared to wild type (Seo et al., 2009). The study also revealed *AtRD22*, the direct target of *AtMYB96*, which is a popular dehydration responsive gene involved in abiotic stress tolerance (Yamaguchi-Shinozaki & Shinozaki, 1993).

Other R2R3-type of MYB TFs involved in the stomatal movement are MYB-FLP (Four Lips, also named *AtMYB124*) and its paralog *AtMYB88*. These transcription factors are required to control division of the cell in stomatal background. The *flp-1* mutant shows more stomatal clusters, which is the result of extra cell division. The double mutant *flp-1myb88* displays even bigger clusters of stomata than the single *flp-1* mutant (Xie, Lee, et al., 2010). *FLP* and *AtMYB88* are also essential for abiotic stress tolerance such as salinity and drought. The impact of the drought and salt stress are more severe in the double mutant. The double mutant transpires water far more quickly than the wild type. To further explore the role of these TFs in abiotic stress tolerance, ABA was introduced, which resulted in 30% stomatal closure in the double mutant while 70% closed in the wild type. This indicates the ABA-dependent activity of *AtMYB124/AtMYB88* (Xie, Li, Wang, Sack, & Grotewold, 2010).

Similarly, the rice MYB gene, *OSMYB48*, contributes to plant tolerance to drought and salinity (Xiong et al., 2014). The research also showed that plants overexpressing *OsMYB48* were highly sensitive to ABA under abiotic stress. Compared to wild type, the higher accumulation of ABA and low water loss rate were observed in transgenic plants. These facts indicate that *OsMYB48* might play a role in conferring abiotic stress tolerance *via* stomatal movement (Xiong et al., 2014).

3.2 bZIP

As one of the largest and most diverse transcription factor family, bZIP is also well-known for their ability to respond to different abiotic stresses, such as drought, salinity and extreme temperature (Hartmann et al., 2015; Xiang, Tang, Du, Ye, & Xiong, 2008). These stress conditions result in osmotic stress in plants. In the process of osmotic stress, ABA is found to be crucial in signal transduction, thereby inducing the expression of DRGs (Drought Responsive Genes) (Lee & Luan, 2012). A study has revealed that *Arabidopsis* bZIP TFs such as *AREB1*, *AREB2*, *ABF3*, *ABF1*, and *ABF4* play a part in regulating the expression of ABA-dependent genes (Choi, Hong, Ha, Kang, & Kim, 2000; Saxena et al., 2013). ABF3, AREB1 and AREB2, are considered to be associated with ABA-dependent stomatal closure under drought or osmotic stress (Yoshida et al., 2010). In rice, bZIPs, *i.e.*, *OsbZIP2*, *OsbZIP23*, *OsbZIP42*, *OsbZIP46*, *TRAB1*, *OsABF1*, and *OsABI5* are also involved in ABA signaling and osmotic stress responses (Joo, Lee, & Song, 2019; Yoon et al., 2017; Zou, Guan, Zhang, & Chen, 2008).

Under abiotic stress conditions and ABA treatment, transgenic potato plants overexpressing *CaBZ1* (belonging to the S group of bZIP TFs) had lower water loss rate and faster stomatal closure than wild type plants. Overexpression of *CaBZ1* improved osmotic stress tolerance by increasing the expression of ABA- and stress-related genes (such as *CYP707A1*, CBF, and NAC-like genes) (Moon et al., 2015).

Soybean *GmbZIP1* belongs to the AREB subfamily of the bZIP TFs family, and it also positively regulates stomatal closure. Compared with wild type plants, the overexpression lines of *GmbZIP1* show increased tolerance to various abiotic stresses such as salinity, drought and cold. The transgenic plants showed increased sensitivity to ABA, which triggers stomatal closure and enhances plant tolerance to stress (Gao et al., 2011).

In contrast, another study on soybean bZIP TF, namely *GmbZIP15*, revealed its negative role in salinity and drought stress tolerance. Compared to wild type, transgenic plants overexpressing *GmbZIP15* are more sensitive to abiotic stress conditions such as salinity and drought, with increased stomatal aperture and lower expression levels of stress-responsive genes (Zhang et al., 2020). *GmbZIP19* is another soybean bZIP TF whose overexpression was also found to be negatively regulating stomatal closure with increased sensitivity to salinity and drought stress conditions (He et al., 2020).

Overexpression of *ZmABP9* gene (Maize Auxin Binding protein 9) in cotton plants leads to stomatal closure, decreased ROS level, higher chlorophyll contents and better germination under salt, drought and oxidative stress (Wang et al., 2017). In *Arabidopsis*, *ABP2*, a maize bZIP TF, increases the tolerance to drought and salinity. The overexpression plants of *ABP2* show increased stomatal closure (ABA-dependent) and increased stress-related gene expression, which enhances the plant tolerance to drought and salinity (Kim, Kang, Cho, Park, & Kim, 2004).

Similarly, transferring the *Brassica napus* gene *BnaABF2* to *Arabidopsis* induced expression of LEAs genes (*RAB18, RD29B, KIN2*) and decreased stomata size, which confers better tolerance to salinity and drought stress (Zhao, Hu, Li, & Yao, 2016).

3.3 NAC

The name of NAC was derived from three proteins, namely no apical meristem (NAM), ATAF1–2, and CUC2. The NAC TF family participates in salinity and osmotic stress tolerance by mediating different types of signaling pathways. According to genome-wide identification studies, *Arabidopsis* contains 117, rice has 151, and maize has 152 NAC TFs (Ooka et al., 2003; Shiriga et al., 2014).

In *Arabidopsis* and *Rosa indica*, the overexpression of *RhNAC3* increased seed germination and reprogrammed stomatal closure in response to osmotic stress and ABA treatment (Xue-Xuan et al., 2010). The expression of rice NAC gene *OsNAC6* increases under salt, osmotic, cold stress, which is also induced by ABA (Ooka et al., 2003). This statement was endorsed by Ohnishi et al. (2005) but with the addition of MeJA to the list. These facts suggest that *OsNAC6* is involved in the positive regulation of stomata. Similarly, overexpression of the stress-responsive *SNAC1* gene increased drought and salt tolerance in rice *via* ABA-dependent stomatal closure (Hu et al., 2006). *SNAC1* improved osmotic tolerance in transgenic rice plants during the reproductive growth in field under severe stress without decreasing yield (Nakashima, Takasaki, Mizoi, Shinozaki, & Yamaguchi-Shinozaki, 2012). Another study on *SNAC1* revealed its involvement in the activation of the *OsSRO1c* gene, leading to osmotic stress tolerance by stomatal closure (You et al., 2013).

Similarly, *OsNAC10* overexpression plants increase tolerance to abiotic stresses in field conditions. Microarray analysis revealed that this TF could be induced by ABA and abiotic stress conditions such as salinity and drought (Jeong et al., 2010). During drought and osmotic stress conditions, overexpression of *ONAC002*, *ONAC009*, or *ONAC122* in rice results in the increase of stomatal movement and root diameter (Jeong et al., 2013). Another stress-responsive TF, *ONAC022*, enhances rice tolerance to salinity and drought stress conditions. The study suggests that the overexpression plants under abiotic stress conditions have shown a significantly higher number of closed stomata as compared to wild type plants (Hong et al., 2016).

Overexpression of *OsNAC5* in rice results in transgenic plants which follow the ABA-dependent pathway. Transgenic plants showed increased expression of ABA biosynthesis-related genes and endogenous ABA levels to confer salinity and drought tolerance (Jeong et al., 2013; Rehman et al., 2021). On the other hand, the expression of *OsNAC4* is enhanced in the presence of JA, also resulting in salt stress tolerance (Rehman et al., 2021).

In another study, sorghum *SbNAC2* TF showed enhanced tolerance to abiotic stress conditions by increasing ROS scavenging and stress-related gene expression. It was also revealed that the expression of *SbNAC2* is strongly induced by abiotic stresses and ABA treatment. In addition, the stomatal aperture of *SbNAC2*-overexpressing plants was lower than that of the wild type. These facts indicate that stomatal closure is regulated by the ABA signaling pathway (Jin et al., 2021). Similarly, under salt and drought stress, overexpression of grapevine stress-responsive gene *VvNAC17* in *Arabidopsis* plants resulted in a higher germination rate and length of the root as compared to the wild type (Ju et al., 2020). By exogenous application of ABA to the overexpression lines, the stomatal opening was inhibited, and the density of stomata was reduced, showing an improved survival rate due to dehydration stress as compared to wild type plants. Furthermore, when the *VvNAC17*-overexpresing lines were exposed to ABA and drought, the expression of marker genes such as *COR15A*, *COR47*, *ABI5*, *AREB1*, *P5CS*, *RD22*, and *RD29A* was upregulated, which revealed the ABA-dependent role of this TF in abiotic stress (Ju et al., 2020).

3.4 WRKY

WRKY is an important TF family in vascular plants. It plays a vital role in responding to various biotic and abiotic stresses, which are regulated by different plant hormonal signaling transduction (Jiang et al., 2017). In total, 74 WRKY family members have been identified in *Arabidopsis*, while 161 in maize, 100 in rice and 197 in soybean plants (Fan et al., 2015; He et al., 2016; Leng & Zhao, 2020; Rushton, Somssich, Ringler, & Shen, 2010).

In *Arabidopsis*, *GhWRKY6*-overexpressing plants have been found to be highly sensitive to salt and drought stress and the stomatal closure is negatively regulated, whereas the loss-of-function mutant *wrky6* shows low sensitivity to stress and displays stomatal closure as compared to wild type. These facts indicate that *GhWRKY6* plays a negative role in stomatal regulation and abiotic stress tolerance (Li et al., 2019). Another cotton WRKY TF, *GhWRKY41*, was found to enhance tolerance to salinity and drought stress when overexpressed in tobacco. The *GhWRKY41*-overexpressing plants can also regulate stomatal closure under osmotic stress (Chu et al., 2016). In addition, when cotton *GhWRKY68* is overexpressed in tobacco, the plants show improved tolerance to salinity and drought stress, with more frequent of stomatal closure (Jia et al., 2015).

In *Arabidopsis*, *WRKY46*, has been found to be negatively regulating abiotic stress tolerance. Overexpression plants of *WRKY46* demonstrate high sensitivity to salt and drought stress with an increased rate of water loss. *WRKY46* was also found to be involved in the negative regulation of ROS accumulation in guard cells, indicating its role in inhibiting stomatal closure (Ding, Yan, et al., 2014). Similarly, osmoregulation of H_2O_2 and stomatal movement were found in plants overexpressing *AtWRKY53*, which confers osmotic stress tolerance in *Arabidopsis*. Two *Arabidopsis* WRKY genes, *WRKY54/70*, were found to negatively regulate stomatal closure. It is suggested that both genes cooperate with each other to negatively regulate the production of salicylic acid, which is found to be involved in stomatal closure (Li et al., 2013).

Similarly, overexpression of *GsWRKY20* in soybean shows increased sensitivity to ABA, resulting in stomatal closure in the early stages of plant growth and inducing tolerance against dehydration or osmotic stress (Luo et al., 2013). It has also been reported that overexpression of *WRKY146* in *Arabidopsis* can enhance drought tolerance by inducing stomatal closure and reducing transpiration rate (Ma et al., 2017).

A maize WRKY gene family member, *ZmWRKY58*, when expressed in rice, shows increased tolerance to osmotic stress by maintaining the integrity of cell membrane and the closure of stomata because of the involvement of ABA and calcium-dependent signaling cascade (Cai et al., 2014; Leng & Zhao, 2020). In tomato, the *SlWRKY8*-overexpressing plants show enhanced tolerance to salinity and drought stress by reducing stomatal opening, increasing proline accumulation and the activity of ROS scavengers (Gao et al., 2020). Another tomato WRKY gene, *SlWRKY81,* can increase osmotic tolerance in plants by increasing ABA contents, which leads to stomatal closure and ROS accumulation in guard cells (Ahammed et al., 2020). Ahammed et al. (2021) also reported that the activity of *SlWRKY81* in stomatal closure is induced by reducing the accumulation of NO in guard cells under osmotic stress.

3.5 Other TFs

The bHLH (basic helix-loop-helix proteins) is a superfamily of transcription factors involving the regulation of various abiotic stress conditions and plant growth and development. In *Arabidopsis*, 167 bHLH members have been identified while rice has 177 members (Qiu et al., 2020). An *Arabidopsis* bHLH, *AtbHLH68*, with high expression in root tissue, functions in an ABA-dependent pathway. Overexpression of *AtBHLH68* increases plant sensitivity to ABA and ABA biosynthesis. Therefore, *AtBHLH68* regulates the ABA signaling pathway, thus increasing the tolerance of plants to osmotic stress (Le Hir et al., 2017). These factors indicate the involvement of stomatal regulation as part of the stress tolerance mechanism. Similarly, in *Arabidopsis*, *AtbHLH112* is another ABA-dependent bHLH gene which improves the tolerance of overexpressed plants by reducing stomatal aperture under salt and osmotic stress (Liu et al., 2015). Tolerance is also achieved by increasing proline contents and antioxidant enzyme activity, and decreasing the damage effects of H_2O_2 (Liu et al., 2015). Another study also showed that salt and osmotic tolerance are reduced in *AtBHLH112*-knockout plants (Chen et al., 2017). Overexpression of apple *MdbHLH130* in tobacco, positively regulates stomatal closure and ROS scavenging, and enhances tolerance to abiotic stress (Zhao et al., 2020). Similarly, overexpression of wheat *TabHLH1* in tobacco shows enhanced tolerance to osmotic stress. Under salinity and drought stress, *TabHLH1* transgenic plants showed positive regulation of stomatal closure, and reduced ROS accumulation. It is suggested that TabHLH1 follows an ABA-dependent pathway to enhance plant tolerance to abiotic stress conditions (Yang et al., 2016).

Homeobox-leucine zipper (HD-Zip) TF superfamily has also been found to be related to abiotic stress tolerance in plants (Hussain, Kayani, & Amjad, 2011). In *Arabidopsis*, the HD-Zip gene *ATHB6* is expressed in developing leaves, roots and carpels, and it is upregulated by osmotic stress and has been found to be involved in ABA-dependent stomatal closure (Söderman, Hjellström, Fahleson, & Engström, 1999). *AtHB7*, another member of HD-Zip in *Arabidopsis*, was found to be involved in stomatal regulation and abiotic stress responses (Ré, Capella, Bonaventure, & Chan, 2014). Overexpression of the maize *Zmhdz10* in rice plants resulted in an improved tolerance to osmotic and salt stress. Similarly, *Arabidopsis* plants overexpressing *Zmhdz10* show enhanced salt and drought stress tolerance by altering the expression of ABA and stress-responsive genes (*P5CS1*, *RD22*, *RD29B*, and *ABI1*). Under abiotic stress, *Zmhdz10* activates the ABA-dependent pathway, resulting in tolerance to these stressors (Zhao et al., 2014), indicating its possible role in the regulation of stomatal aperture. In rice, the stress-related gene, *OsHOX22*, which belongs to the homeodomain TF family, is upregulated under abiotic stress. Compared with wild-type plants, the plants of the T-DNA mutant *oshox22-1* showed a reduced stomatal density (Zhang et al., 2012), indicating the possible mechanism of *OsHOX22* in the regulation of osmotic stress tolerance in plants. The overexpression of OsTF1L, a rice HD-Zip TF, regulates stomatal closure under osmotic stress conditions. Results revealed that compared to nontransgenic plants, the tolerance to osmotic stress is increased, with better reproductive development and higher yield (Bang et al., 2019).

In addition, nuclear factor Y (NF-Y) is a ubiquitous transcription factor, such as NFYA5, which plays an important role in controlling stomatal aperture through ABA response in guard cells (Li et al., 2008). Some other TFs involved in the regulation of stomatal movement are listed in Table 1.

4. Concluding remarks

To overcome the future threats to global food security, we need to address the situation by making plants tolerant enough to survive harsh conditions like salinity and drought. The cross-talks between TFs and phytohormones form a complex signaling network to regulate stomatal closure and defend the plants from dehydration. However, it is very important to understand the mechanism of these signaling pathways and transcription factors involved in stomatal regulation. Signaling molecules such as ABA, JA, ROS, NO, H_2S, and Ca^{2+} are all involved in stomatal movements. ABA is considered as a primary stress-related hormone, therefore, the study on ABA is very extensive. Other molecules are considered to have a secondary signaling role and could be ABA-dependent or independent. But most of the components involved in regulating these secondary messengers are unknown. To better understand how this complex signaling network works, we need to uncover the mysterious components involved in the regulation of stomatal closing. On the other hand, due to the invention of revolutionary gene-editing technology, scientists are now discovering various TFs in the gene families, such as MYB, NAC, WRKY, bZIP, which are involved in the regulation of biotic and abiotic stresses. Since there are many missing links within some signaling pathways in stomatal closure, the mechanism of most signaling molecules is still unclear. Therefore, more work needs to be done to reveal the mechanism lying in the signaling pathway regulating stomatal movement.

Acknowledgments

This work was supported by the National Natural Science Foundation of China (31972416, U1906205), the National Key Research and Development Program of China (2018YFD1000800), and the Innovative Research Group Project of Natural Science Foundation of Hubei Province, China (2019CFA017).

References

Ahammed, G. J., Li, X., Mao, Q., Wan, H., Zhou, G., & Cheng, Y. (2021). The *SlWRKY81* transcription factor inhibits stomatal closure by attenuating nitric oxide accumulation in the guard cells of tomato under drought. *Physiologia Plantarum*, *172*(2), 885–895.

Ahammed, G. J., Li, X., Yang, Y., Liu, C., Zhou, G., Wan, H., et al. (2020). Tomato *WRKY81* acts as a negative regulator for drought tolerance by modulating guard cell H_2O_2-mediated stomatal closure. *Environmental and Experimental Botany*, *171*, 103960.

Alam, H., Khattak, J. Z., Ksiksi, T. S., Saleem, M. H., Fahad, S., Sohail, H., et al. (2021). Negative impact of long-term exposure of salinity and drought stress on native *Tetraena mandavillei* L. *Physiologia Plantarum*, *172*(2), 1336–1351.

Albacete, A., Ghanem, M. E., Martínez-Andújar, C., Acosta, M., Sánchez-Bravo, J., Martínez, V., et al. (2008). Hormonal changes in relation to biomass partitioning and shoot growth impairment in salinized tomato (*Solanum lycopersicum* L.) plants. *Journal of Experimental Botany*, *59*(15), 4119–4131.

Ali, M., & Baek, K. H. (2020). Jasmonic acid signaling pathway in response to abiotic stresses in plants. *International Journal of Molecular Sciences*, *21*(2), 621.

Baldoni, E., Genga, A., & Cominelli, E. (2015). Plant MYB transcription factors: Their role in drought response mechanisms. *International Journal of Molecular Sciences*, *16*(7), 15811–15851.

Bang, S. W., Lee, D. K., Jung, H., Chung, P. J., Kim, Y. S., Choi, Y. D., et al. (2019). Overexpression of *OsTF1L*, a rice HD-Zip transcription factor, promotes lignin biosynthesis and stomatal closure that improves drought tolerance. *Plant Biotechnology Journal*, *17*(1), 118–131.

Bhardwaj, S., Kapoor, D., Singh, S., Gautam, V., Dhanjal, D. S., Jan, S., et al. (2021). Nitric oxide: A ubiquitous signal molecule for enhancing plant tolerance to salinity stress and their molecular mechanisms. *Journal of Plant Growth Regulation*, 1–13.

Boudsocq, M., Barbier-Brygoo, H., & Laurière, C. (2004). Identification of nine sucrose nonfermenting 1-related protein kinases 2 activated by hyperosmotic and saline stresses in Arabidopsis thaliana. *Journal of Biological Chemistry*, *279*(40), 41758–41766.

Butt, H. I., Yang, Z., Chen, E., Zhao, G., Gong, Q., Yang, Z., et al. (2017). Functional characterization of cotton *GaMYB62L*, a novel R2R3 TF in transgenic *Arabidopsis*. *PLoS One*, *12*(1), e0170578.

Butt, H. I., Yang, Z., Gong, Q., Chen, E., Wang, X., Zhao, G., et al. (2017). *GaMYB85*, an R2R3 MYB gene, in transgenic *Arabidopsis* plays an important role in drought tolerance. *BMC Plant Biology*, *17*(1), 1–17.

Cai, R., Zhao, Y., Wang, Y., Lin, Y., Peng, X., Li, Q., et al. (2014). Overexpression of a maize *WRKY58* gene enhances drought and salt tolerance in transgenic rice. *Plant Cell, Tissue and Organ Culture*, *119*(3), 565–577.

Carroll, R. C., & Peralta, E. G. (1998). The m3 muscarinic acetylcholine receptor differentially regulates calcium influx and release through modulation of monovalent cation channels. *The EMBO Journal*, *17*(11), 3036–3044.

Castillo, M. C., Lozano-Juste, J., González-Guzmán, M., Rodriguez, L., Rodriguez, P. L., & León, J. (2015). Inactivation of PYR/PYL/RCAR ABA receptors by tyrosine nitration may enable rapid inhibition of ABA signaling by nitric oxide in plants. *Science Signaling*, *8*(392), ra89.

Chaves, M. M., Flexas, J., & Pinheiro, C. (2009). Photosynthesis under drought and salt stress: Regulation mechanisms from whole plant to cell. *Annals of Botany*, *103*(4), 551–560.

Chen, H. C., Hsieh-Feng, V., Liao, P. C., Cheng, W. H., Liu, L. Y., Yang, Y. W., et al. (2017). The function of *OsbHLH068* is partially redundant with its homolog, *AtbHLH112*, in the regulation of the salt stress response but has opposite functions to control flowering in *Arabidopsis*. *Plant Molecular Biology*, *94*(4), 531–548.

Chen, S., Jia, H., Wang, X., Shi, C., Wang, X., Ma, P., et al. (2020). Hydrogen sulfide positively regulates abscisic acid signaling through persulfidation of SnRK2.6 in guard cells. *Molecular Plant*, *13*(5), 732–744.

Chen, Y., Pang, Q., Dai, S., Wang, Y., Chen, S., & Yan, X. (2011). Proteomic identification of differentially expressed proteins in *Arabidopsis* in response to methyl jasmonate. *Journal of Plant Physiology*, *168*(10), 995–1008.

Chen, Y., Xiaoyuan, Y., Kun, H., Meihua, L., Jigang, L., Zhaofeng, G., et al. (2006). The MYB transcription factor superfamily of Arabidopsis: Expression analysis and phylogenetic comparison with the rice MYB family. *Plant Molecular Biology*, *60*(1), 107–124.

Chen, Z. H., Wang, Y., Wang, J. W., Babla, M., Zhao, C., García-Mata, C., et al. (2016). Nitrate reductase mutation alters potassium nutrition as well as nitric oxide-mediated control of guard cell ion channels in *Arabidopsis*. *New Phytologist*, *209*(4), 1456–1469.

Choi, H. I., Hong, J. H., Ha, J. O., Kang, J. Y., & Kim, S. Y. (2000). ABFs, a family of ABA-responsive element binding factors. *Journal of Biological Chemistry*, *275*(3), 1723–1730.

Chu, X., Wang, C., Chen, X., Lu, W., Li, H., Wang, X., et al. (2016). Correction: The cotton WRKY gene *GhWRKY41* positively regulates salt and drought stress tolerance in transgenic Nicotiana benthamiana. *PLoS One*, *11*(6), e0157026.

Clough, B., & Sim, R. G. (1989). Changes in gas exchange characteristics and water use efficiency of mangroves in response to salinity and vapour pressure deficit. *Oecologia*, *79*(1), 38–44.

Cominelli, E., Galbiati, M., & Tonelli, C. (2010). Transcription factors controlling stomatal movements and drought tolerance. *Transcription*, *1*(1), 41–45.

Cominelli, E., Galbiati, M., Vavasseur, A., Conti, L., Sala, T., Vuylsteke, M., et al. (2005). A guard-cell-specific MYB transcription factor regulates stomatal movements and plant drought tolerance. *Current Biology*, *15*(13), 1196–1200.

Corpas, F. J., & Palma, J. M. (2020). H_2S signaling in plants and applications in agriculture. *Journal of Advanced Research*, *24*, 131–137.

Daszkowska-Golec, A., & Szarejko, I. (2013). Open or close the gate–stomata action under the control of phytohormones in drought stress conditions. *Frontiers in Plant Science*, *4*, 138.

Davies, W. J., Wilkinson, S., & Loveys, B. (2002). Stomatal control by chemical signalling and the exploitation of this mechanism to increase water use efficiency in agriculture. *New Phytologist*, *153*(3), 449–460.

Delgado, C., Mora-Poblete, F., Ahmar, S., Chen, J. T., & Figueroa, C. R. (2021). Jasmonates and plant salt stress: Molecular players, physiological effects, and improving tolerance by using genome-associated tools. *International Journal of Molecular Sciences*, *22*(6), 3082.

de Lima-Neto, A. J., Cavalcante, L. F., Mesquita, F. D. O., Souto, A. G. D. L., dos Santos, G. P., dos Santos, J. Z., et al. (2016). Papaya seedlings irrigation with saline water in soil with bovine biofertilizer. *Chilean Journal of Agricultural Research*, *76*(2), 236–242.

de Ollas, C., & Dodd, I. C. (2016). Physiological impacts of ABA-JA interactions under water-limitation. *Plant Molecular Biology*, *91*(6), 641–650.

Ding, R., Kang, S., Du, T., Hao, X., & Zhang, Y. (2014). Scaling up stomatal conductance from leaf to canopy using a dual-leaf model for estimating crop evapotranspiration. *PLoS One*, *9*(4), e95584.

Ding, Z., Li, S., An, X., Liu, X., Qin, H., & Wang, D. (2009). Transgenic expression of *MYB15* confers enhanced sensitivity to abscisic acid and improved drought tolerance in *Arabidopsis thaliana*. *Journal of Genetics and Genomics*, *36*(1), 17–29.

Ding, Z. J., Yan, J. Y., Xu, X. Y., Yu, D. Q., Li, G. X., Zhang, S. Q., et al. (2014). Transcription factor *WRKY46* regulates osmotic stress responses and stomatal movement independently in *Arabidopsis*. *The Plant Journal*, *79*(1), 13–27.

Dodd, I. C. (2013). Abscisic acid and stomatal closure: A hydraulic conductance conundrum? *New Phytologist*, *197*(1), 6–8.

Fan, X., Guo, Q., Xu, P., Gong, Y., Shu, H., Yang, Y., et al. (2015). Transcriptome-wide identification of salt-responsive members of the WRKY gene family in *Gossypium aridum*. *PLoS One*, *10*(5), e0126148.

Fatma, M., Iqbal, N., Gautam, H., Sehar, Z., Sofo, A., D'Ippolito, I., et al. (2021). Ethylene and sulfur coordinately modulate the antioxidant system and ABA accumulation in mustard plants under salt stress. *Plants (Basel, Switzerland)*, *10*(1), 180.

Fernando, V. D., & Schroeder, D. F. (2016). Role of ABA in Arabidopsis salt, drought, and desiccation tolerance. In *Abiotic and biotic stress in plants—Recent advances and future perspectives*, IntechOpen.

Förster, S., Schmidt, L. K., Kopic, E., Anschütz, U., Huang, S., Schlücking, K., et al. (2019). Wounding-induced stomatal closure requires jasmonate-mediated activation of GORK K^+ channels by a Ca^{2+} sensor-kinase CBL1-CIPK5 complex. *Developmental Cell*, *48*(1), 87–99.

Franzoni, G., Cocetta, G., Trivellini, A., & Ferrante, A. (2019). Transcriptional regulation in rocket leaves as affected by salinity. *Plants*, *9*(1), 20.

Fujita, Y., Fujita, M., Satoh, R., Maruyama, K., Parvez, M. M., Seki, M., et al. (2005). AREB1 is a transcription activator of novel ABRE-dependent ABA signaling that enhances drought stress tolerance in *Arabidopsis*. *The Plant Cell*, *17*(12), 3470–3488.

Galbiati, M., Simoni, L., Pavesi, G., Cominelli, E., Francia, P., Vavasseur, A., et al. (2008). Gene trap lines identify *Arabidopsis* genes expressed in stomatal guard cells. *The Plant Journal*, *53*(5), 750–762.

Gao, S. Q., Chen, M., Xu, Z. S., Zhao, C. P., Li, L., Xu, H. J., et al. (2011). The soybean *GmbZIP1* transcription factor enhances multiple abiotic stress tolerances in transgenic plants. *Plant Molecular Biology*, *75*(6), 537–553.

Gao, Y. F., Liu, J. K., Yang, F. M., Zhang, G. Y., Wang, D., Zhang, L., et al. (2020). The WRKY transcription factor *WRKY8* promotes resistance to pathogen infection and mediates drought and salt stress tolerance in *Solanum lycopersicum*. *Physiologia Plantarum*, *168*(1), 98–117.

García-Mata, C., & Lamattina, L. (2010). Hydrogen sulphide, a novel gasotransmitter involved in guard cell signalling. *New Phytologist*, *188*(4), 977–984.

Gayatri, G., Agurla, S., & Raghavendra, A. S. (2013). Nitric oxide in guard cells as an important secondary messenger during stomatal closure. *Frontiers in Plant Science*, *4*, 425.

Geiger, D., Scherzer, S., Mumm, P., Stange, A., Marten, I., Bauer, H., et al. (2009). Activity of guard cell anion channel SLAC1 is controlled by drought-stress signaling kinase-phosphatase pair. *Proceedings of the National Academy of Sciences of the United States of America*, *106*(50), 21425–21430.

Ghorbel, M., Brini, F., Sharma, A., & Landi, M. (2021). Role of jasmonic acid in plants: The molecular point of view. *Plant Cell Reports*, *40*(8), 1471–1494.

Gondek, M., Weindorf, D., Thiel, C., & Kleinheinz, G. (2020). Soluble salts in compost and their effects on soil and plants: A review. *Compost Science & Utilization*, *28*, 59–75.

Guo, X., Wang, Q., Liu, Y., Zhang, X., Zhang, L., & Fan, S. (2020). Screening of salt stress responsive genes in *Brachypodium distachyon* (L.) Beauv. by transcriptome analysis. *Plants*, *9*(11), 1522.

Hamant, O., & Haswell, E. S. (2017). Life behind the wall: Sensing mechanical cues in plants. *BMC Biology*, *15*(1), 1–9.

Hartmann, L., Pedrotti, L., Weiste, C., Fekete, A., Schierstaedt, J., Göttler, J., et al. (2015). Crosstalk between two bZIP signaling pathways orchestrates salt-induced metabolic reprogramming in *Arabidopsis* roots. *The Plant Cell*, *27*(8), 2244–2260.

He, Q., Cai, H., Bai, M., Zhang, M., Chen, F., Huang, Y., et al. (2020). A soybean bZIP transcription factor *GmbZIP19* confers multiple biotic and abiotic stress responses in plant. *International Journal of Molecular Sciences*, *21*(13), 4701.

He, G. H., Xu, J. Y., Wang, Y. X., Liu, J. M., Li, P. S., Chen, M., et al. (2016). Drought-responsive WRKY transcription factor genes *TaWRKY1* and *TaWRKY33* from wheat confer drought and/or heat resistance in *Arabidopsis*. *BMC Plant Biology*, *16*(1), 1–16.

Hedrich, R., & Shabala, S. (2018). Stomata in a saline world. *Current Opinion in Plant Biology*, *46*, 87–95.

Hong, Y., Zhang, H., Huang, L., Li, D., & Song, F. (2016). Overexpression of a stress-responsive NAC transcription factor gene *ONAC022* improves drought and salt tolerance in rice. *Frontiers in Plant Science*, *7*, 4.

Hoque, M. A., Scheelbeek, P. F. D., Vineis, P., Khan, A. E., Ahmed, K. M., & Butler, A. P. (2016). Drinking water vulnerability to climate change and alternatives for adaptation in coastal South and South East Asia. *Climatic Change*, *136*(2), 247–263.

Hou, Z., Wang, L., Liu, J., Hou, L., & Liu, X. (2013). Hydrogen sulfide regulates ethylene-induced stomatal closure in *Arabidopsis thaliana*. *Journal of Integrative Plant Biology*, *55*(3), 277–289.

Hu, H., Dai, M., Yao, J., Xiao, B., Li, X., Zhang, Q., et al. (2006). Overexpressing a *NAM, ATAF,* and *CUC* (NAC) transcription factor enhances drought resistance and salt tolerance in rice. *Proceedings of the National Academy of Sciences of the United States of America, 103*(35), 12987–12992.

Huo, T., Wang, C. T., Yu, T. F., Wang, D. M., Li, M., Zhao, D., et al. (2021). Overexpression of *ZmWRKY65* transcription factor from maize confers stress resistances in transgenic *Arabidopsis*. *Scientific Reports, 11*(1), 1–15.

Hussain, Q., Asim, M., Zhang, R., Khan, R., Farooq, S., & Wu, J. (2021). Transcription factors interact with ABA through gene expression and signaling pathways to mitigate drought and salinity stress. *Biomolecules, 11*(8), 1159.

Hussain, S. S., Kayani, M. A., & Amjad, M. (2011). Transcription factors as tools to engineer enhanced drought stress tolerance in plants. *Biotechnology Progress, 27*(2), 297–306.

Jeong, J. S., Kim, Y. S., Baek, K. H., Jung, H., Ha, S. H., Do Choi, Y., et al. (2010). Root-specific expression of *OsNAC10* improves drought tolerance and grain yield in rice under field drought conditions. *Plant Physiology, 153*(1), 185–197.

Jeong, J. S., Kim, Y. S., Redillas, M. C., Jang, G., Jung, H., Bang, S. W., et al. (2013). *OsNAC5* overexpression enlarges root diameter in rice plants leading to enhanced drought tolerance and increased grain yield in the field. *Plant Biotechnology Journal, 11*(1), 101–114.

Jia, H., Wang, C., Wang, F., Liu, S., Li, G., & Guo, X. (2015). *GhWRKY68* reduces resistance to salt and drought in transgenic *Nicotiana benthamiana*. *PLoS One, 10*(3), e0120646.

Jiang, J., Ma, S., Ye, N., Jiang, M., Cao, J., & Zhang, J. (2017). WRKY transcription factors in plant responses to stresses. *Journal of Integrative Plant Biology, 59*(2), 86–101.

Jin, X., Long, Y., Xiong, S., Yang, Z., Chen, W., Hawar, A., et al. (2021). *SbNAC2* enhances abiotic stress tolerance by upregulating ROS scavenging activities and inducing stress-response genes in sorghum. *Environmental and Experimental Botany, 192*, 104664.

Joo, J., Lee, Y. H., & Song, S. I. (2019). *OsbZIP42* is a positive regulator of ABA signaling and confers drought tolerance to rice. *Planta, 249*(5), 1521–1533.

Ju, Y. L., Yue, X. F., Min, Z., Wang, X. H., Fang, Y. L., & Zhang, J. X. (2020). *VvNAC17*, a novel stress-responsive grapevine (*Vitis vinifera L.*) NAC transcription factor, increases sensitivity to abscisic acid and enhances salinity, freezing, and drought tolerance in transgenic *Arabidopsis*. *Plant Physiology and Biochemistry, 146*, 98–111.

Jung, C., Seo, J. S., Han, S. W., Koo, Y. J., Kim, C. H., Song, S. I., et al. (2008). Overexpression of *AtMYB44* enhances stomatal closure to confer abiotic stress tolerance in transgenic *Arabidopsis*. *Plant Physiology, 146*(2), 623–635.

Kefu, Z., Munns, R., & King, R. W. (1991). Abscisic acid levels in NaCl-treated barley, cotton and saltbush. *Functional Plant Biology, 18*(1), 17–24.

Kerstiens, G., Tych, W., Robinson, M. F., & Mansfield, T. A. (2002). Sodium-related partial stomatal closure and salt tolerance of *Aster* tripolium. *New Phytologist, 153*(3), 509–515.

Kim, S., Kang, J. Y., Cho, D. I., Park, J. H., & Kim, S. Y. (2004). *ABF2*, an ABRE-binding bZIP factor, is an essential component of glucose signaling and its overexpression affects multiple stress tolerance. *The Plant Journal, 40*(1), 75–87.

Ko, D., & Helariutta, Y. (2017). Shoot–root communication in flowering plants. *Current Biology, 27*(17), R973–R978.

Kosová, K., Vítámvás, P., Urban, M. O., Klíma, M., Roy, A., & Prášil, I. T. (2015). Biological networks underlying abiotic stress tolerance in temperate crops—A proteomic perspective. *International Journal of Molecular Sciences, 16*(9), 20913–20942.

Lau, S. E., Hamdan, M. F., Pua, T. L., Saidi, N. B., & Tan, B. C. (2021). Plant nitric oxide signaling under drought stress. *Plants (Basel, Switzerland), 10*(2), 360.

Le Hir, R., Castelain, M., Chakraborti, D., Moritz, T., Dinant, S., & Bellini, C. (2017). *AtbHLH68* transcription factor contributes to the regulation of ABA homeostasis and drought stress tolerance in *Arabidopsis thaliana*. *Physiologia Plantarum, 160*(3), 312–327.

Lee, S. C., Lan, W., Buchanan, B. B., & Luan, S. (2009). A protein kinase-phosphatase pair interacts with an ion channel to regulate ABA signaling in plant guard cells. *Proceedings of the National Academy of Sciences of the United States of America*, *106*(50), 21419–21424.

Lee, S. C., & Luan, S. (2012). ABA signal transduction at the crossroad of biotic and abiotic stress responses. *Plant, Cell & Environment*, *35*(1), 53–60.

Leng, P., & Zhao, J. (2020). Transcription factors as molecular switches to regulate drought adaptation in maize. *Theoretical and Applied Genetics*, *133*(5), 1455–1465.

Li, J., Besseau, S., Törönen, P., Sipari, N., Kollist, H., Holm, L., et al. (2013). Defense-related transcription factors *WRKY70* and *WRKY54* modulate osmotic stress tolerance by regulating stomatal aperture in *Arabidopsis*. *New Phytologist*, *200*(2), 457–472.

Li, J., Jia, H., Wang, J., Cao, Q., & Wen, Z. (2014). Hydrogen sulfide is involved in maintaining ion homeostasis via regulating plasma membrane Na^+/H^+ antiporter system in the hydrogen peroxide-dependent manner in salt-stress *Arabidopsis* thaliana root. *Protoplasma*, *251*, 899–912.

Li, Z., Li, L., Zhou, K., Zhang, Y., Han, X., Din, Y., et al. (2019). *GhWRKY6* acts as a negative regulator in both transgenic *Arabidopsis* and cotton during drought and salt stress. *Frontiers in Genetics*, *10*, 392.

Li, W. X., Oono, Y., Zhu, J., He, X. J., Wu, J. M., Iida, K., et al. (2008). The *Arabidopsis NFYA5* transcription factor is regulated transcriptionally and posttranscriptionally to promote drought resistance. *The Plant Cell*, *20*(8), 2238–2251.

Li, W., Pang, S., Lu, Z., & Jin, B. (2020). Function and mechanism of WRKY transcription factors in abiotic stress responses of plants. *Plants (Basel, Switzerland)*, *9*(11), 1515.

Li, J., Zhang, K., Meng, Y., Hu, J., Ding, M., Bian, J., et al. (2018). Jasmonic acid/ethylene signaling coordinates hydroxycinnamic acid amides biosynthesis through *ORA59* transcription factor. *The Plant Journal*, *95*(3), 444–457.

Lindemose, S., O'Shea, C., Jensen, M. K., & Skriver, K. (2013). Structure, function and networks of transcription factors involved in abiotic stress responses. *International Journal of Molecular Sciences*, *14*(3), 5842–5878.

Liotenberg, S., North, H., & Marion-Poll, A. (1999). Molecular biology and regulation of abscisic acid biosynthesis in plants. *Plant Physiology and Biochemistry*, *37*(5), 341–350.

Liu, X., Fan, Y., Mak, M., Babla, M., Holford, P., Wang, F., et al. (2017). QTLs for stomatal and photosynthetic traits related to salinity tolerance in barley. *BMC Genomics*, *18*(1), 1–13.

Liu, Y., Ji, X., Nie, X., Qu, M., Zheng, L., Tan, Z., et al. (2015). *Arabidopsis AtbHLH112* regulates the expression of genes involved in abiotic stress tolerance by binding to their E-box and GCG-box motifs. *New Phytologist*, *207*(3), 692–709.

Liu, H., & Xue, S. (2021). Interplay between hydrogen sulfide and other signaling molecules in the regulation of guard cell signaling and abiotic/biotic stress response. *Plant Communications*, *2*, 100179.

Liu, Q., Zhou, Y., Li, H., Liu, R., Wang, W., Wu, W., et al. (2021). Osmotic stress-triggered stomatal closure requires phospholipase Dδ and hydrogen sulfide in *Arabidopsis thaliana*. *Biochemical and Biophysical Research Communications*, *534*, 914–920.

Luo, X., Bai, X., Sun, X., Zhu, D., Liu, B., Ji, W., et al. (2013). Expression of wild soybean *WRKY20* in *Arabidopsis* enhances drought tolerance and regulates ABA signalling. *Journal of Experimental Botany*, *64*(8), 2155–2169.

Ma, J., Gao, X., Liu, Q., Shao, Y., Zhang, D., Jiang, L., et al. (2017). Overexpression of *TaWRKY146* increases drought tolerance through inducing stomatal closure in *Arabidopsis thaliana*. *Frontiers in Plant Science*, *8*, 2036.

Ma, Y., Szostkiewicz, I., Korte, A., Moes, D., Yang, Y., Christmann, A., et al. (2009). Regulators of PP2C phosphatase activity function as abscisic acid sensors. *Science*, *324*(5930), 1064–1068.

Ma, Q., Xia, Z., Cai, Z., Li, L., Cheng, Y., Liu, J., et al. (2019). *GmWRKY16* enhances drought and salt tolerance through an ABA-mediated pathway in *Arabidopsis thaliana*. *Frontiers in Plant Science*, *9*, 1979.

MacRobbie, E. A. C. (1998). Signal transduction and ion channels in guard cells. *Philosophical Transactions of the Royal Society of London. Series B: Biological Sciences*, *353*(1374), 1475–1488.

Manna, M., Thakur, T., Chirom, O., Mandlik, R., Deshmukh, R., & Salvi, P. (2021). Transcription factors as key molecular target to strengthen the drought stress tolerance in plants. *Physiologia Plantarum*, *172*(2), 847–868.

Miedema, H., & Assmann, S. M. (1996). A membrane-delimited effect of internal pH on the K^+ outward rectifier of *Vicia faba* guard cells. *The Journal of Membrane Biology*, *154*(3), 227–237.

Moon, S. J., Han, S. Y., Kim, D. Y., Yoon, I. S., Shin, D., Byun, M. O., et al. (2015). Ectopic expression of a hot pepper bZIP-like transcription factor in potato enhances drought tolerance without decreasing tuber yield. *Plant Molecular Biology*, *89*(4), 421–431.

Morison, J. I., Stewart, B. A., & Howell, T. (2003). Plant water use, stomatal control. In *Encyclopedia of water science* (pp. 680–685). New York: Marcel Dekker.

Munemasa, S., Hossain, M. A., Nakamura, Y., Mori, I. C., & Murata, Y. (2011). The *Arabidopsis* calcium-dependent protein kinase, CPK6, functions as a positive regulator of methyl jasmonate signaling in guard cells. *Plant Physiology*, *155*(1), 553–561.

Munemasa, S., Mori, I. C., & Murata, Y. (2011). Methyl jasmonate signaling and signal cross-talk between methyl jasmonate and abscisic acid in guard cells. *Plant Signaling & Behavior*, *6*(7), 939–941.

Nakashima, K., Takasaki, H., Mizoi, J., Shinozaki, K., & Yamaguchi-Shinozaki, K. (2012). NAC transcription factors in plant abiotic stress responses. *Biochimica et Biophysica Acta. Gene Regulatory Mechanisms*, *1819*(2), 97–103.

Nguyen, H. T., Stanton, D. E., Schmitz, N., Farquhar, G. D., & Ball, M. C. (2015). Growth responses of the mangrove *Avicennia marina* to salinity: Development and function of shoot hydraulic systems require saline conditions. *Annals of Botany*, *115*(3), 397–407.

Niu, M., Xie, J., Chen, C., Cao, H., Sun, J., Kong, Q., et al. (2018). An early ABA-induced stomatal closure, Na^+ sequestration in leaf vein and K^+ retention in mesophyll confer salt tissue tolerance in Cucurbita species. *Journal of Experimental Botany*, *69*(20), 4945–4960.

Ohnishi, T., Sugahara, S., Yamada, T., Kikuchi, K., Yoshiba, Y., Hirano, H. Y., et al. (2005). *OsNAC6*, a member of the NAC gene family, is induced by various stresses in rice. *Genes and Genetic Systems*, *80*(2), 135–139.

Ooka, H., Satoh, K., Doi, K., Nagata, T., Otomo, Y., Murakami, K., et al. (2003). Comprehensive analysis of NAC family genes in Oryza sativa and *Arabidopsis thaliana*. *DNA Research*, *10*(6), 239–247.

Pandey, A. K., & Gautam, A. (2020). Stress responsive gene regulation in relation to hydrogen sulfide in plants under abiotic stress. *Physiologia Plantarum*, *168*(2), 511–525.

Park, S. Y., Fung, P., Nishimura, N., Jensen, D. R., Fujii, H., Zhao, Y., et al. (2009). Abscisic acid inhibits type 2C protein phosphatases via the PYR/PYL family of START proteins. *Science*, *324*(5930), 1068–1071.

Pérez-López, U., Robredo, A., Lacuesta, M., Sgherri, C., Muñoz-Rueda, A., Navari-Izzo, F., et al. (2009). The oxidative stress caused by salinity in two barley cultivars is mitigated by elevated CO_2. *Physiologia Plantarum*, *135*(1), 29–42.

Qadir, M., Quillérou, E., Nangia, V., Murtaza, G., Singh, M., Thomas, R. J., et al. (2014). Economics of salt-induced land degradation and restoration. *Natural Resources Forum*, *38*(2), 282–295.

Qiu, J. R., Huang, Z., Xiang, X. Y., Xu, W. X., Wang, J. T., Chen, J., et al. (2020). *MfbHLH38*, a *Myrothamnus flabellifolia* bHLH transcription factor, confers tolerance to drought and salinity stresses in *Arabidopsis*. *BMC Plant Biology*, *20*(1), 1–14.

Raven, J. (2002). Selection pressures on stomatal evolution. *New Phytologist, 153*, 371–386.

Rawat, N., Wungrampha, S., Singla-Pareek, S. L., Yu, M., Shabala, S., & Pareek, A. (2022). Rewilding staple crops for the lost halophytism: Towards sustainability and profitability of agricultural production systems. *Molecular Plant, 15*, 45–64.

Razzaq, A., Wani, S. H., Saleem, F., Yu, M., Zhou, M., & Shabala, S. (2021). Rewilding crops for climate resilience: Economic analysis and de novo domestication strategies. *Journal of Experimental Botany, 72*, 6123–6139.

Ré, D. A., Capella, M., Bonaventure, G., & Chan, R. L. (2014). *Arabidopsis AtHB7* and *AtHB12* evolved divergently to fine tune processes associated with growth and responses to water stress. *BMC Plant Biology, 14*(1), 1–14.

Reef, R., Schmitz, N., Rogers, B. A., Ball, M. C., & Lovelock, C. E. (2012). Differential responses of the mangrove Avicennia marina to salinity and abscisic acid. *Functional Plant Biology, 39*(12), 1038–1046.

Rehman, A., Azhar, M. T., Hinze, L., Qayyum, A., Li, H., Peng, Z., et al. (2021). Insight into abscisic acid perception and signaling to increase plant tolerance to abiotic stress. *Journal of Plant Interactions, 16*(1), 222–237.

Rodriguez-Dominguez, C. M., Buckley, T. N., Egea, G., de Cires, A., Hernandez-Santana, V., Martorell, S., et al. (2016). Most stomatal closure in woody species under moderate drought can be explained by stomatal responses to leaf turgor. *Plant, Cell & Environment, 39*(9), 2014–2026.

Rushton, P. J., Somssich, I. E., Ringler, P., & Shen, Q. J. (2010). WRKY transcription factors. *Trends in Plant Science, 15*(5), 247–258.

Sah, S. K., Reddy, K. R., & Li, J. (2016). Abscisic acid and abiotic stress tolerance in crop plants. *Frontiers in Plant Science, 7*, 571.

Saibo, N. J., Lourenço, T., & Oliveira, M. M. (2009). Transcription factors and regulation of photosynthetic and related metabolism under environmental stresses. *Annals of Botany, 103*(4), 609–623.

Sami, F., Faizan, M., Faraz, A., Siddiqui, H., Yusuf, M., & Hayat, S. (2018). Nitric oxide-mediated integrative alterations in plant metabolism to confer abiotic stress tolerance, NO cross-talk with phytohormones and NO-mediated post translational modifications in modulating diverse plant stress. *Nitric Oxide, 73*, 22–38.

Sanoubar, R., Cellini, A., Veroni, A. M., Spinelli, F., Masia, A., Vittori Antisari, L., et al. (2016). Salinity thresholds and genotypic variability of cabbage (*Brassica oleracea* L.) grown under saline stress. *Journal of the Science of Food and Agriculture, 96*(1), 319–330.

Saxena, S. C., Salvi, P., Kaur, H., Verma, P., Petla, B. P., Rao, V., et al. (2013). Differentially expressed myo-inositol monophosphatase gene (*CaIMP*) in chickpea (*Cicer arietinum* L.) encodes a lithium-sensitive phosphatase enzyme with broad substrate specificity and improves seed germination and seedling growth under abiotic stresses. *Journal of Experimental Botany, 64*(18), 5623–5639.

Seo, P. J., Xiang, F., Qiao, M., Park, J. Y., Lee, Y. N., Kim, S. G., et al. (2009). The *MYB96* transcription factor mediates abscisic acid signaling during drought stress response in *Arabidopsis*. *Plant Physiology, 151*(1), 275–289.

Shen, J., Zhang, J., Zhou, M., Zhou, H., Cui, B., Gotor, C., et al. (2020). Persulfidation-based modification of cysteine desulfhydrase and the NADPH oxidase RBOHD controls guard cell abscisic acid signaling. *The Plant Cell, 32*(4), 1000–1017.

Shiriga, K., Sharma, R., Kumar, K., Yadav, S. K., Hossain, F., & Thirunavukkarasu, N. (2014). Genome-wide identification and expression pattern of drought-responsive members of the NAC family in maize. *Meta Gene, 2*, 407–417.

Sierla, M., Waszczak, C., Vahisalu, T., & Kangasjärvi, J. (2016). Reactive oxygen species in the regulation of stomatal movements. *Plant Physiology, 171*(3), 1569–1580.

So, W. M., Huque, A. M., Shin, H. Y., Kim, S. Y., Shin, J. S., Cui, M., et al. (2020). *AtMYB109* negatively regulates stomatal closure under osmotic stress in *Arabidopsis thaliana*. *Journal of Plant Physiology, 255*, 153292.

Söderman, E., Hjellström, M., Fahleson, J., & Engström, P. (1999). The HD-Zip gene *AtHB6* in Arabidopsis is expressed in developing leaves, roots and carpels and up-regulated by water deficit conditions. *Plant Molecular Biology*, *40*(6), 1073–1083.
Song, X. G., She, X. P., He, J. M., Huang, C., & Song, T. S. (2006). Cytokinin- and auxin-induced stomatal opening involves a decrease in levels of hydrogen peroxide in guard cells of *Vicia faba*. *Functional Plant Biology*, *33*(6), 573–583.
Springmann, M., Clark, M., Mason-D'Croz, D., Wiebe, K., Bodirsky, B. L., Lassaletta, L., et al. (2018). Options for keeping the food system within environmental limits. *Nature*, *562*(7728), 519–525.
Suhita, D., Kolla, V. A., Vavasseur, A., & Raghavendra, A. S. (2003). Different signaling pathways involved during the suppression of stomatal opening by methyl jasmonate or abscisic acid. *Plant Science*, *164*(4), 481–488.
Sun, L. R., Yue, C. M., & Hao, F. S. (2019). Update on roles of nitric oxide in regulating stomatal closure. *Plant Signaling & Behavior*, *14*(10), e1649569.
Sun, Y., & Yu, D. (2015). Activated expression of *AtWRKY53* negatively regulates drought tolerance by mediating stomatal movement. *Plant Cell Reports*, *34*(8), 1295–1306.
Takahashi, Y., Zhang, J., Hsu, P. K., Ceciliato, P. H., Zhang, L., Dubeaux, G., et al. (2020). MAP3Kinase-dependent SnRK2-kinase activation is required for abscisic acid signal transduction and rapid osmotic stress response. *Nature Communications*, *11*(1), 1–12.
Tri, D. Q., & Tuyet, Q. T. T. (2016). Effect of climate change on the salinity intrusion: Case study Ca River Basin, Vietnam. *Journal of Climate Change*, *2*(1), 91–101.
Umezawa, T., Sugiyama, N., Mizoguchi, M., Hayashi, S., Myouga, F., Yamaguchi-Shinozaki, K., et al. (2009). Type 2C protein phosphatases directly regulate abscisic acid-activated protein kinases in *Arabidopsis*. *Proceedings of the National Academy of Sciences of the United States of America*, *106*(41), 17588–17593.
Vatén, A., & Bergmann, D. C. (2012). Mechanisms of stomatal development: An evolutionary view. *EvoDevo*, *3*(1), 1–9.
Vick, B. A., & Zimmerman, D. C. (1984). Biosynthesis of jasmonic acid by several plants species. *Plant Physiology*, *75*(2), 458–461.
Vlad, F., Rubio, S., Rodrigues, A., Sirichandra, C., Belin, C., Robert, N., et al. (2009). Protein phosphatases 2C regulate the activation of the Snf1-related kinase OST1 by abscisic acid in Arabidopsis. *The Plant Cell*, *21*(10), 3170–3184.
Wang, P., Du, Y., Hou, Y. J., Zhao, Y., Hsu, C. C., Yuan, F., et al. (2015). Nitric oxide negatively regulates abscisic acid signaling in guard cells by S-nitrosylation of OST1. *Proceedings of the National Academy of Sciences of the United States of America*, *112*(2), 613–618.
Wang, J., Lian, W., Cao, Y., Wang, X., Wang, G., Qi, C., et al. (2018). Overexpression of *BoNAC019*, a NAC transcription factor from *Brassica oleracea*, negatively regulates the dehydration response and anthocyanin biosynthesis in *Arabidopsis*. *Scientific Reports*, *8*(1), 1–15.
Wang, C., Lu, G., Hao, Y., Guo, H., Guo, Y., Zhao, J., et al. (2017). *ABP9*, a maize bZIP transcription factor, enhances tolerance to salt and drought in transgenic cotton. *Planta*, *246*(3), 453–469.
Wasternack, C., & Feussner, I. (2018). The oxylipin pathways: Biochemistry and function. *Annual Review of Plant Biology*, *69*, 363–386.
Wei, Q., Luo, Q., Wang, R., Zhang, F., He, Y., Zhang, Y., et al. (2017). A wheat R2R3-type MYB transcription factor *TaODORANT1* positively regulates drought and salt stress responses in transgenic tobacco plants. *Frontiers in Plant Science*, *8*, 1374.
Wilkinson, S., & Davies, W. J. (2010). Drought, ozone, ABA and ethylene: New insights from cell to plant to community. *Plant, Cell & Environment*, *33*(4), 510–525.
Xiang, Y., Tang, N., Du, H., Ye, H., & Xiong, L. (2008). Characterization of *OsbZIP23* as a key player of the basic leucine zipper transcription factor family for conferring abscisic acid sensitivity and salinity and drought tolerance in rice. *Plant Physiology*, *148*(4), 1938–1952.

Xie, Z., Lee, E., Lucas, J. R., Morohashi, K., Li, D., Murray, J. A., et al. (2010). Regulation of cell proliferation in the stomatal lineage by the *Arabidopsis* MYB FOUR LIPS via direct targeting of core cell cycle genes. *The Plant Cell, 22*(7), 2306–2321.

Xie, Z., Li, D., Wang, L., Sack, F. D., & Grotewold, E. (2010). Role of the stomatal development regulators *FLP/MYB88* in abiotic stress responses. *The Plant Journal, 64*(5), 731–739.

Xiong, H., Li, J., Liu, P., Duan, J., Zhao, Y., Guo, X., et al. (2014). Overexpression of *OsMYB48-1*, a novel MYB-related transcription factor, enhances drought and salinity tolerance in rice. *PLoS One, 9*(3), e92913.

Xuan, L., Li, J., Wang, X., & Wang, C. (2020). Cross-talk between hydrogen sulfide and other signal molecules regulates plant growth and development. *International Journal of Molecular Sciences, 21*(13), 4593.

Xue-Xuan, X., Hong-Bo, S., Yuan-Yuan, M., Gang, X., Jun-Na, S., Dong-Gang, G., et al. (2010). Biotechnological implications from abscisic acid (ABA) roles in cold stress and leaf senescence as an important signal for improving plant sustainable survival under abiotic-stressed conditions. *Critical Reviews in Biotechnology, 30*(3), 222–230.

Yamaguchi-Shinozaki, K., & Shinozaki, K. (1993). The plant hormone abscisic acid mediates the drought-induced expression but not the seed-specific expression of *RD22*, a gene responsive to dehydration stress in *Arabidopsis thaliana. Molecular and General Genetics, 238*(1), 17–25.

Yan, H., Jia, H., Chen, X., Hao, L., An, H., & Guo, X. (2014). The cotton WRKY transcription factor *GhWRKY17* functions in drought and salt stress in transgenic *Nicotiana benthamiana* through ABA signaling and the modulation of reactive oxygen species production. *Plant and Cell Physiology, 55*(12), 2060–2076.

Yang, Z., Li, J. L., Liu, L. N., Xie, Q., & Sui, N. (2020). Photosynthetic regulation under salt stress and salt-tolerance mechanism of sweet sorghum. *Frontiers in Plant Science, 10*, 1722.

Yang, T., Yao, S., Hao, L., Zhao, Y., Lu, W., & Xiao, K. (2016). Wheat bHLH-type transcription factor gene *TabHLH1* is crucial in mediating osmotic stresses tolerance through modulating largely the ABA-associated pathway. *Plant Cell Reports, 35*(11), 2309–2323.

Yastreb, T. O., Kolupaev, Y. E., Kokorev, A. I., Horielova, E. I., & Dmitriev, A. P. (2018). Methyl jasmonate and nitric oxide in regulation of the stomatal apparatus of *Arabidopsis* thaliana. *Cytology and Genetics, 52*(6), 400–405.

Yin, Y., Adachi, Y., Nakamura, Y., Munemasa, S., Mori, I. C., & Murata, Y. (2016). Involvement of OST1 protein kinase and PYR/PYL/RCAR receptors in methyl jasmonate-induced stomatal closure in *Arabidopsis* guard cells. *Plant and Cell Physiology, 57*(8), 1779–1790.

Yoon, S., Lee, D. K., Yu, I. J., Kim, Y. S., Do Choi, Y., & Kim, J. K. (2017). Overexpression of the *OsbZIP66* transcription factor enhances drought tolerance of rice plants. *Plant Biotechnology Reports, 11*(1), 53.

Yoshida, T., Fujita, Y., Sayama, H., Kidokoro, S., Maruyama, K., Mizoi, J., et al. (2010). *AREB1, AREB2*, and *ABF3* are master transcription factors that cooperatively regulate ABRE-dependent ABA signaling involved in drought stress tolerance and require ABA for full activation. *The Plant Journal, 61*, 672–685.

You, J., Zong, W., Li, X., Ning, J., Hu, H., Li, X., et al. (2013). The *SNAC1*-targeted gene *OsSRO1c* modulates stomatal closure and oxidative stress tolerance by regulating hydrogen peroxide in rice. *Journal of Experimental Botany, 64*(2), 569–583.

Yu, M., Liu, J., Du, B., Zhang, M., Wang, A., & Zhang, L. (2021). NAC transcription factor *PwNAC11* activates *ERD1* by interaction with *ABF3* and *DREB2A* to enhance drought tolerance in transgenic Arabidopsis. *International Journal of Molecular Sciences, 22*(13), 6952.

Zhang, S., Haider, I., Kohlen, W., Jiang, L., Bouwmeester, H., Meijer, A. H., et al. (2012). Function of the *HD-Zip I* gene *Oshox22* in ABA-mediated drought and salt tolerances in rice. *Plant Molecular Biology, 80*(6), 571–585.

Zhang, M., Liu, Y., Cai, H., Guo, M., Chai, M., She, Z., et al. (2020). The bZIP transcription factor *GmbZIP15* negatively regulates salt-and drought-stress responses in soybean. *International Journal of Molecular Sciences, 21*(20), 7778.

Zhang, J., Zhou, H., Zhou, M., Ge, Z., Zhang, F., Foyer, C. H., et al. (2021). The coordination of guard-cell autonomous ABA synthesis and *DES1* function in situ regulates plant water deficit responses. *Journal of Advanced Research, 27*, 191–197.

Zhao, C., Cai, S., Wang, Y., & Chen, Z. H. (2016). Loss of nitrate reductases NIA1 and NIA2 impairs stomatal closure by altering genes of core ABA signaling components in *Arabidopsis*. *Plant Signaling & Behavior, 11*(6), 1456–1469.

Zhao, Q., Fan, Z., Qiu, L., Che, Q., Wang, T., Li, Y., et al. (2020). *MdbHLH130*, an apple bHLH transcription factor, confers water stress resistance by regulating stomatal closure and ROS homeostasis in transgenic tobacco. *Frontiers in Plant Science, 11*, 543696.

Zhao, B. Y., Hu, Y. F., Li, J. J., & Yao, X. (2016). *BnaABF2*, a bZIP transcription factor from rapeseed (*Brassica napus* L.), enhances drought and salt tolerance in transgenic *Arabidopsis*. *Botanical Studies, 57*(1), 1–12.

Zhao, Y., Ma, Q., Jin, X., Peng, X., Liu, J., Deng, L., et al. (2014). A novel maize homeodomain–leucine zipper (HD-Zip) I gene, *Zmhdz10*, positively regulates drought and salt tolerance in both rice and *Arabidopsis*. *Plant and Cell Physiology, 55*(6), 1142–1156.

Zhou, M., & Memelink, J. (2016). Jasmonate-responsive transcription factors regulating plant secondary metabolism. *Biotechnology Advances, 34*(4), 441–449.

Zou, M., Guan, Y., Zhang, F., & Chen, F. (2008). A bZIP transcription factor, OsABI5, is involved in rice fertility and stress tolerance. *Plant Molecular Biology, 66*(6), 675–683.

CHAPTER EIGHT

Root-borne signals and their control of guard cell operation under saline conditions: The role of root signals in stomata regulation

Mengliang Niu[a], Zhilong Bie[b,*], and Yuan Huang[b,*]
[a]School of Life Sciences, Jianghan University, Wuhan, PR China
[b]Key Laboratory of Horticultural Plant Biology, Ministry of Education/College of Horticulture and Forestry Sciences, Huazhong Agricultural University, Wuhan, PR China
*Corresponding authors: e-mail address: biezl@mail.hzau.edu.cn; huangyuan@mail.hzau.edu.cn

Contents

1. Introduction 196
 1.1 Soil salinization as a major problem 196
 1.2 Plant adaptive responses to salinity 196
2. How do plants perceive stress and trigger signals in the root? 198
 2.1 GIPC sphingolipids sense salt and trigger Ca^{2+} influx 199
 2.2 Ca^{2+} and reactive oxygen species (ROS) signaling play as the initial signal under osmotic stress 199
3. Root to shoot signaling 199
 3.1 Think of plants as complete systems 199
 3.2 Physical signals 200
 3.3 Chemical signals 203
4. Local control of stomatal opening and closing under salinity 210
 4.1 Guard cell and aquaporins 210
 4.2 Functions of H_2O_2 210
 4.3 ABA and guard cell 211
 4.4 Sensitivity of stomata 211
5. Conclusions and prospects 212
 5.1 How plants integrate different signals participate in stomatal regulation? 212
 5.2 The plant vascular system: Plant signaling highways 213
 5.3 Grafting as a tool to study root-sourced signal 213
 5.4 How root itself perceive salt stress or osmotic stress? 214
Acknowledgments 215
References 215

Advances in Botanical Research, Volume 103
ISSN 0065-2296
https://doi.org/10.1016/bs.abr.2022.02.014

Abstract

Salinity is a serious problem all over the world, and is an important factor limiting agriculture production. When water availability becomes limiting for plant physiological processes under salinity, stomatal closure is its early response buffering the drop of xylem water potential and catastrophic hydraulic failure. Stomatal movement is controlled by multifaceted signaling network. Root-borne signal is considered one of these crucial factors. However, the mechanism behind the roots and shoots 'signals' that control the stomatal closure, in order to cope with osmotic stress remains to be poorly understood. In this review we discuss the role of root-to-shoot signal under salt stress in regulating stomatal movement.

1. Introduction

1.1 Soil salinization as a major problem

Soil salinity is a major environmental stress, affecting crop production worldwide. Around 950 million hectares of agricultural land worldwide are affected by soil salinization (Shabala, Wu, & Bose, 2015). Salt stress can be divided into osmotic stress and ion toxicity according to the timescale of their operation. Roots first sense salinity and affect water absorption by reducing transpiration; early osmotic stress results in a rapid stomatal closure and root water absorption difficulties which lead to a suppressed photosynthetic rate (Liu et al., 2013). In the second stage, salt ion toxicity caused by ion imbalance inhibits the activities of various enzymes in plants (Neto, Prisco, Eneas, de Abreu, & Gomes, 2006). As the time of stress is prolonged, salt stress also induces excess accumulation of many reactive oxygen species (ROS), which can cause irreversible damage to the plasma membrane.

1.2 Plant adaptive responses to salinity

1.2.1 Plant osmotic regulation

The accumulation and compartmentalization of toxic ions to increase osmotic tolerance are essential for plant adaptation to salt tolerance. Studies have shown that plants are involved in osmotic regulation through absorption or synthesis of diverse macromolecules like carbohydrates (sucrose, sorbitol, mannitol, glycerol, arabinitol, pinitol) (Hare, Cress, & van Staden, 1999); nitrogen compounds (proteins, betaine, glutamate, aspartate, glycine, proline, choline), and organic acids (malate and oxalate) (Kinnersley & Turano, 2000).

1.2.2 Plant ionic relations

Regulation of Na^+ uptake, transport and sequestration is complex but essential for plants to cope with excess salt ions. Salt exclusion is one of the most important strategies to prevent the entry of salts into the vascular system. Na^+ efflux in the root is the first and most important line of defense salinity, preventing the accumulation of large amounts of Na^+ to the shoot. Most plants do exclude about 98% of the salt in the soil solution, allowing only 2% to be transported in the xylem to the shoot (Munns et al., 2012). Efflux occurs through Na^+/H^+ antiporters on the plasma membrane mediated by *SOS1* (salt overly sensitive 1). The SOS pathway is one of the first proven key pathways with Na^+ regulation. Ca^{2+} is first increased by Na^+ level, which binds to SOS3, leading to activation of the kinase SOS2, which in turn, activates *SOS1* (Zhu, 2003). When salt ions enter the root system, the plant has another mechanism to prevent salt ions transport to the shoot. High-affinity K^+ transporter-1 (HKT1) encodes a Na^+ transporter that mainly controls Na^+ delivery from root-to-shoot by withdrawal of Na^+ from the xylem sap. HKT1 transporters are involved in retrieving Na^+ from the transpiration stream by mediating Na^+ into the xylem parenchyma cells in the roots (Sun et al., 2018).

1.2.3 Stomata are key components of plant response to salinity

Several studies have found that stomata regulation play an important role in water deficit tolerance. Stomata regulate transpirational water loss in plants and control carbon dioxide assimilation rate (Duan et al., 2015). Thus, stomatal conductance is taken as a reliable indicator for assessing plant salt tolerance. Stomata are microscopic pores delineated by a pair of guard cells. In some plants guard cells are surrounded by specialized subsidiary cells to form the stomatal complexes (Maurel, Verdoucq, & Rodrigues, 2016). Gas exchanges occur largely through stomata. Stomatal regulation is one of the key mechanisms that participate in regulating CO_2 assimilation versus evaporative water loss. Physiological drought induced by salinity puts a major strain on photosynthesis of plants. The dilemma is whether to close stomata and counteract the lack of electron receptors, or to open stomata and tolerate water desorption by transpiration (Christmann, Weiler, Steudle, & Grill, 2007). The amount of leaf transpiration depends on real-time soil moisture. In addition, transpiration is important for plants to maintain leaf temperature at high temperatures and drive xylem-mediated mass flow of nutrients from the soil into the shoot (Medina & Gilbert, 2016).

1.2.4 Root is the first organ to sense salinity

Root is the first organ to be exposed to an increased salinity, and their responses can affect the overall plant response to it. Early signaling events are crucial for the adaptation of plants at saline environments (Shabala et al., 2015). Studies have shown that reduced water availability in the root environment due to salt and osmotic stress can reduce the hydraulic conductivity of roots—the ability of roots to transport water from the soil to the stem through a water potential gradient. The endodermis plays an important role as an apoplast barrier in plant root structure, which is regulated by hormonal signals. Endodermal structures are plastic under many abiotic stress conditions, and they act on suberin through ethylene and ABA hormone pathways (Barberon et al., 2016). These facts raise the question whether the endodermis functioning as a communication center during root responses to the external environment as well as an important apoplastic barrier is a coincidence.

1.2.5 Significance of rapid stomatal closure under salt stress

The opening or closing of stomata is initiated within 1–2 min after salt stress and completed within 5 min or even less time (Kollist et al., 2019). It has now been accepted that stomatal conductance can respond to water deficit through root-to-shoot chemical signaling. The limited water supply in the soil is sensed by the roots and passed on to the stems in order to adjust the plant's transpiration in time. But how the root signal regulates the stomata through long distance is unclear. Hence, we are focusing mainly on three related questions in this review: (i) how plants perceive salt stress (or osmotic stress), (ii) how plants transmit stress signals to their leaves, and (iii) how stomata respond to stress signals.

2. How do plants perceive stress and trigger signals in the root?

A cellular or tissue-based sensor refers to the protein or molecule that can respond to the environmental fluctuation, and then encode this information into a cascade of biophysical or biochemical events aimed to optimize an organism's performance under altered conditions (Shabala et al., 2015). However, our current knowledge of how salt stress is sensed by plant tissues is severely limited. Until now, only a few putative sensors have been identified. Because the sensors are essential for plants, the mutant with a loss of function is lethal; this stops further research on sensors function (Zhu, 2016). Besides that, demonstrating a protein as a sensor for stress signal is very difficult, and direct experimental evidence is absent (Zhu, 2016).

2.1 GIPC sphingolipids sense salt and trigger Ca^{2+} influx

The *Arabidopsis* OSCA1 (reduced hyperosmolality-induced calcium increase 1) is a putative sensor for a hyperosmotic stress. Various environmental factors such as osmotic stress, salinity, cold, and heat as well as oxidative stress, can cause increases in the cytosolic Ca^{2+} concentration in cells, which can be detected by genetically encoded aequorin or other calcium specific reporters. *Arabidopsis osca1* mutants display a reduced calcium spike compared to WT plants under osmotic stress (using sorbitol and mannitol). Osmotic and salt stresses in roots are sensed by their sensors MOCA1, a putative osmotic sensor, encoding a glucuronosyl transferase located in cell surface GIPC sphingolipids. MOCA1 is necessary for salt-induced cell surface potential depolarization or Na^+/H^+ antiporter activation, and then regulation of growth. Na^+ binds to GIPCs to gate Ca^{2+} influx channels. This mechanism might imply that plasma-membrane lipids are involved in adaption to various salt levels (Jiang et al., 2019).

2.2 Ca^{2+} and reactive oxygen species (ROS) signaling play as the initial signal under osmotic stress

Although the relationship between the Ca^{2+}-triggered SOS pathway and ROS signaling has so far remained elusive, it is currently thought that these two components act in tandem, interacting and amplifying each other during salt sensing. This so-called "Ca^{2+}–ROS hub" plays a crucial role in the activation of several downstream processes critical for salt acclimation. Such as vacuolar sequestration of Na^+ via the vacuolar Na^+/H^+ antiporter (NHX), the increased activity of the vacuolar H^+-ATPase (V-ATPase) via establishing an electrochemical H^+ gradient across the tonoplast (Niu et al., 2018) and extruding Na^+ into the apoplast through SOS1 Na^+/H^+ exchanger activity. In parallel, these salt-activated signaling pathways also relay information from root to shoot tissues, triggering appropriate adaptive responses at the shoot level, even before Na^+ ions can reach the shoot tissues (Steinhorst & Kudla, 2019).

3. Root to shoot signaling

3.1 Think of plants as complete systems

Local stress can cause global changes in gene expression, respiration and photosynthesis processes, transpiration, and plant hormone production. To trigger this stress response at the whole plant level, it is necessary to produce and transmit long-distance stress signals from a single organ to the whole plant. These signals include different physical (electrical, hydraulic) and chemicals (ABA, salicylic acid, and methyl jasmonate) signals, which are defined as systemic signals.

Among these various signals, the chemical signals were thought not to be the original cause of the systemic signal. Chemical signals travel relatively slowly; their rate of propagation can be varied from several $mm\,s^{-1}$ to about $mm\,s^{-1}$ (Huber & Bauerle, 2016). However, the transfer rate can be increased if chemical elements are transported via phloem (several $mm\,s^{-1}$) or xylem (several $cm\,s^{-1}$) (Bauer et al., 2013). For tall woody plants, the transport of chemical signals from root to crown via transpiration flows may take more than 24h and can be as fast as $2\,m\,h^{-1}$ (Longman, Zimmermann, Brown, & Tyree, 1972). This low speed obviously cannot meet the requirements of rapid response of plants to environmental changes. Based on this, hydraulic signal is considered as the main signal of early plant response to water deficit.

3.2 Physical signals

3.2.1 Hydraulic signals

3.2.1.1 Hydraulic signals and AQPs

Stomatal closure in response to soil water deficit is a confirmed example of the long-distance root-to-shoot signaling (Kudoyarova, Kholodova, & Veselov, 2013). Water movement in the xylem is controlled by the hydrostatic pressure gradient. The driving force of water in xylem is produced by leaf transpiration tension, and tension is formed in xylem during transpiration (Shabala, White, Djordjevic, Ruan, & Mathesius, 2016). Xylem pressure can change rapidly in response to salinity (Wegner et al., 2011), changing light conditions (Shabala et al., 2009), and changes of nutrient availability in the rhizosphere (Wegner, Sattelmacher, Läuchli, & Zimmermann, 1999). Hydraulic signals can be transmitted quickly over long distances. The xylem of a plant can be regarded as an incompressible tube, under which hydraulic signals can travel at the speed of sound. This high-speed information transfer provides an effective way to communicate water status between roots and leaves to optimize plant photosynthesis efficiency under water stress (Christmann et al., 2007).

3.2.1.2 Perception of hydraulic signals

A turgor sensing mechanism relies on serine-threonine protein kinases associated with cell walls (Peyronnet, Tran, Girault, & Frachisse, 2014). Several membrane-localized receptor-like kinases (RLK) sense the strain and transduction of cell wall (Monshausen & Haswell, 2013). The RLK family has domains binding at the cell wall; this component is thought to mediate carbohydrate which plays the role of hydraulic sensor. Another type of turgor sensor named wall-associated protein kinases (WAK) has similar function as RLK (Christmann, Grill, & Huang, 2013). Turgor pressure changes in xylem caused by fluctuation in root external conditions can be directly translated into changes in stomatal aperture by decoding mechanical sensory channels located in the stem (Kurusu, Kuchitsu, Nakano, Nakayama, & Iida, 2013).

Aquaporin is one of the intrinsic membrane proteins present in the plasma membrane and most inner part of plant cells (Maurel et al., 2016). The activity of aquaporins in plants has significant influence on stomatal function. By facilitating embolism refilling, aquaporins may thus indirectly promote stomatal opening (Secchi & Zwieniecki, 2014). As aquaporins determine the water permeability of bundle sheath cells (another type of hydraulic constriction for leaf water supply), their down-regulation by chemical signal (such as ABA) provides an efficient mechanism for stomatal regulation (Sade et al., 2014). Membrane aquaporins (AQPs) play an important role in plant dehydration. At the early stage of water deficit, the root hydraulic conductivity decreases rapidly due to the decrease in AQP activity. Post-transcriptional modification of water channels affects AQP activity and thus changes the membrane permeability to water.

Many studies have indicated that hydraulic signal interacts with other type of systemic signals. Although the question of how hydraulic waves are linked to electric, ROS, and ABA remain to be answered, it was proposed that different mechano-sensitive channels could sense hydraulic waves and convert them into Ca^{2+} signals (Basu & Haswell, 2017).

3.2.2 Electrical signals

Within the last decade, the research focus on plant electrophysiology has shifted from the short-distance to long-distance or systemic signaling as signaling pathway has been formed in the phloem, allowing plants to communicate over long distances (Galle, Lautner, Flexas, & Fromm, 2015). Contrast to chemical signals electrical signals can transmit signal over long distances with higher speed: most of the action potentials (APs) in plant investigated so far revealed velocities ranging from 0.005 to $0.2\,m\,s^{-1}$ (Fromm & Lautner, 2007). Although this speed is slower than the rate of hydraulic signal, but is still very fast.

Electrical signal and hydraulic signal travel independent from each other and trigger distinct responses. Re-irrigation of maize plants after soil drying evokes both hydraulic signal and electrical signal, followed by a two-phase response of the net CO_2 uptake rate and stomatal conductance of leaves (Christmann et al., 2007). In the first phase, a rapid decrease in net CO_2 uptake rate and stomatal conductance occurred, while in the second phase both parameters gradually increased. Hydraulic signals trigger the first phase, a reduction in stomatal aperture, and electrical signals trigger an increase in a gas exchange. Experiments with grape have also shown that re-irrigation after drought induces an electrical signal, which are characterized by continuous depolarization with an amplitude of more than 100 mV (Gil et al., 2014). Electrical signals can travel short distances across plasmodesmata, and such transmission has been demonstrated in many species such as *Nitella* (Spanswick & Costerton, 1967). While plasmodesmata relies in the signaling network between neighboring cells, long-distance signals can only be quickly bridged by low-resistance connections that extend throughout the plant, such as the elongated sieve tubes.

Electrical signals are instantaneous changes in the potential gradients across the plasma membrane. The propagation of this signal can strongly influence physiological processes in non-stressed parts of plants. Electrical signals can also be considered as important long-distance system signal in higher plants, as it can interact with other types of signals. It is hypothesized that electrical potential propagation interacts with the transmission of ROS and Ca^{2+} waves (Hilleary & Gilroy, 2018) or with hydraulic signal (Vodeneev, Akinchits, & Sukhov, 2015). Therefore, electrical signals may play a very important role in inducing systemic physiological responses of plants under local stress. The electrical signals response in plants is linked with the activation of potential-dependent Ca^{2+} channels, and then with the Ca^{2+} concentration increasing in the cytoplasm (Demidchik, Shabala, Isayenkov, Cuin, & Pottosin, 2018). The increased Ca^{2+} concentration activates Ca^{2+}-dependent anion channels and H^+-ATPase in the plasma membrane. This process may depend on the function of voltage-dependent slowly activating anion channel 1 (SLAC1) (Cuin, Dreyer, & Michard, 2018). In some studies, electrical signals were induced by the hydraulic signal, which propagates through the xylem. This can be used as direct evidence that electrical signals participate in long-distance regulation. Electrical signals increase the production of ABA in non-stressed plants parts. The probable mechanisms of the electrical signal-induced increase of ABA production may be linked with the elevated Ca^{2+} in cytoplasm and the subsequent stimulation of ROS production (Cuin et al., 2018).

Actually, every cell can be considered as the source of electrical signals. In the state of water deficit in plants, the root was thought to be the original source of electrical signals. Salvador-Recatala (2016) used KCl and NaCl solutions to evoke depolarization *Arabidopsis thaliana* root cells. Similar system potentials were also administered by salts in *Vicia faba* and *Hordeum vulgare* to evoke depolarization (Zimmermann, Maischak, Mithofer, Boland, & Felle, 2009). Osmotically derived APs along with the stem have been found in sunflower, which indicates that not only salt stimuli but osmotically active solutions induce electrical signals propagating along plant stem (Stolarz & Dziubinska, 2017).

3.3 Chemical signals

3.3.1 ABA signaling and stomata closure

In *Vitis vinifera*, ABA increases in leaves under drought or salinity stress conditions. *In vitro* ABA supply to leaves causes a rapid stomata closure. It is not known whether ABA accumulation in plant leaves is indeed a trigger for stomatal closure, or merely an additional signal involved in its long-term maintenance during the initial stages of prolonged drought and/or post-drought recovery (Tombesi et al., 2015). As evidence, studies on other species reported that enhanced ABA level after drought is likely influenced by hydraulic factor (McAdam & Brodribb, 2014). The study in grape concluded that passive hydraulic control of stomatal closure appears to be dominant over ABA signaling at the early phases of drought stress. The increase of ABA concentration in leaves after stomatal closure was addressed at inhibiting stomatal opening under transient rehydration (Tombesi et al., 2015). Both the xylem ABA and ABA concentrations in leaves and root are possibly associated with stomatal regulation, recent studies have demonstrated that leaf ABA conductance responds correlatively with xylem ABA concentration. Therefore, xylem ABA concentration reflects true signal strength (Hartung, Sauter, & Hose, 2002).

3.3.1.1 Action sites of ABA

We suspect that the ABA concentration at the action sites should be considered as the true ABA signal. ABA biosynthesis related genes were highly expressed in leaf vascular system. Grafting experiments on ABA-deficient mutants suggest that ABA synthesis in leaves, but not roots, plays an important role in regulating stomatal closure under dehydration stress (McAdam, Brodribb, & Ross, 2016). This indicated that ABA is required in response to soil-borne water stress, but not as the long-distance signal. In the shoot of *Arabidopsis*, a 10-fold up to more than 20-foldinduction of ABA levels was observed during water stress, paralleled by a strong activation of ABA signaling (Christmann et al., 2007). Using *Arabidopsis* and a non-invasive imaging technique for ABA action, researchers (Christmann et al., 2007) examine the model that ABA plays as a long-distance signal that communicates water shortage from the root to the shoot. Thus, changes in stomatal conductance related to root ABA export and shoot accumulation, as well as stomatal sensitivity to these hormones, present a good example of the common response to availability of water and nutrients regulating water balance and nutrient flow into the plants.

ABA is thought to function as a cell-to-cell and medium-distance signaling molecule from the leaf vasculature to stomata and other leaf tissues to mediate stress response and tolerance (Takahashi & Shinozaki, 2019). Although ABA has been widely accepted to be a predominant signal for this chemical signaling, it should be noted that root-to-shoot signaling is not just a simple matter of the transmission of ABA from the root to the shoot (Li et al., 2011). It has been increasingly suggested that the intensity of the ABA signal can be modified by many factors, and the same modifications may have important effects on stomatal sensitivity to the root-sourced ABA signal (Jiang & Hartung, 2008). One of the important factors that alter the ABA signal is vascular bundle-sheath cells (BSC). Water and dissolved minerals flow from roots to shoots within the dead xylem trachea, then from the xylem into the mesophyll (radical flow) across a pipe-like structure consisting of dense parenchymatous tissue of the leaf, except for the leaf edges (Shatil-Cohen, Attia, & Moshelion, 2011). Thus, the anatomically intermediate location of the BS layer should expose it to long-distance (chemical or hydraulic) signals coming from the root to leaf. When plants are exposed to drought or salt stress, root borne signals are transported to the leaves via the xylem, resulting in stomatal closure (Schachtman & Goodger, 2008; Shatil-Cohen et al., 2011).

In a study with *Arabidopsis*, Shatil-Cohen et al. (2011) established a role for BSCs as a stress signal-sensing 'control center' for the water flow from root to shoot, which converts the xylem chemical signal ABA. Drought-induced ABA concentration in xylem diminishing BSC osmotic water permeability via down regulated activity of their aquaporins (AQPs) and led to stomatal closure. Another study has shown that tissue-specific ABA biosynthetic enzymes are mainly located in vascular bundles in leaves (Kuromori, Seo, & Shinozaki, 2018). Evidence for root-derived ABA as a long-distance signal has been obtained from split-root experiments with whole plants in which only one part of the root system experienced water deficit (Holbrook, Shashidhar, James, & Munns, 2002).

In order to elucidate the relationship between hydraulic signal and ABA, we must recognize that generation of the hydraulic signal neither depended on ABA biosynthesis nor on ABA signaling. In response to water stress, the *Arabidopsis* mutants *aba2* and *abi1* revealed a hydraulic response comparable to that of wild-type; however, the stomata did not close in the mutants. This finding revealed a signal pathway in which ABA acts downstream of the hydraulic signal in communicating water stress between root and shoot.

3.3.1.2 pH and ABA

Alkalization of xylem SAP under the nitrate deficit condition may also result in stomatal closure, independent of increased xylem ABA concentration, although the mechanism of pH change in this case is unclear (Dodd, Tan, & He, 2003). Gradients of pH along the stem may also modify ABA signal intensity, making stomata at the top of the canopy more sensitive to soil drying. Using the pH image technique, Davies, Wilkinson, and Loveys (2002) found that water stress can change the apoplastic pH in leaf and the change was significantly modify stomatal sensitivity to the xylem ABA signal. It should be noted that the response of leaf conductance to ABA signals varies greatly among plant species and under drought conditions (Jiang & Hartung, 2008). This difference may be caused by drought stress unable to induce enough ABA or leaf conductance is not sensitive to ABA signal (Li et al., 2011).

The mechanism of ABA signaling has long been debated. Christmann et al. (2007) conclude that the associated stomatal control involves both ABA and hydraulics in *Arabidopsis thaliana*. In tomato, the root–shoot signal involves a dialogue between ABA and other compounds, such as strigolactones (SLs) (Visentin et al., 2016) and ethylene precursor ACC (Sobeih, Dodd, Bacon, Donald, & Davies, 2004). In addition, should we specifically exclude ABA as a root-shoot chemical messenger? This may well be the case in tomato and *Arabidopsis thaliana*, but maybe not in maize, sunflower and *Nicotiana*. The results of reciprocal grafting of the latter were opposite to the results in tomato, in which ABA deficient shoots grafted on WT closed stomata during water deficiency.

3.3.1.3 Dual effects of ABA

The dual effects of ABA were demonstrated in *Arabidopsis* by using an *ost2* mutant which guard cells were insensitive to ABA (Pantin et al., 2013). However, this mutant could close stomata in response to ABA by hydraulic regulation in leaf vein cells. Thus, under water stress, ABA inhibits and enhances water transport (aquaporin activity) in fascicles and guard cells, respectively. These opposite effects all contribute to stomatal closure (Maurel et al., 2016).

3.3.2 Ca^{2+} and ROS wave

In plants, the presence of the cell wall between cells constitutes a physical and chemical barrier that keeps neigh-boring cells at a larger physical distance from each other. The cell wall is likely to limit or prevents extracellular vesicle transport between cells (Choi et al., 2017). Cell wall contains high levels of Ca^{2+} and other ions, provides a unique chemical environment for signal transmission.

3.3.2.1 RBOH (respiratory burst oxidase homolog) as a source of ROS signal

Many plant cells are connected by plasmodesmata, which provided a direct way for cell-to-cell communications. Various signal elements such as ions, hormones, proteins, and miRNAs can move through plasmodesmata. Even some large proteins (including transcription factors) can be mobile through plasmodesmata to control developmental programs (Vaten et al., 2011).

A pivotal study by Miller (Fichman & Mittler, 2020) identified a self-propagating mechanism via RBOHD-derived apoplastic ROS, which is a central response to multiple internal ROS signaling mechanisms.

Individual cell-to-cell ROS-relaying stations in the apoplast could also more rapidly diffuse signals than symplastic movement via plasmodesmata (Vestergaard, Flyvbjerg, & Moller, 2012). Several key questions still need to be answered: (a) how ROS signals in the apoplast reach and influence cytoplasmic signaling components? And (b) how these initial ROS signals then relate to downstream hormone signals?

It has already been proved that MAMP (Microbe/pathogen-associated molecular patterns) and ROS-induced phosphorylation of RBOHD via the Ca^{2+} sensor calcium-dependent protein kinases (CPK5) (Dubiella et al., 2013); phosphorylation of RBOHD and RBOHF by CIPK26, CBL1, and CBL9 (Steinhorst & Kudla, 2014); and the discovery of the ROS-activated tonoplastic channel two-pore channel 1 (TPC1), which releases Ca^{2+} into the cytosol during stress (Choi, Toyota, Kim, Hilleary, & Gilroy, 2014). This suggests that Ca^{2+} and ROS signaling coordination occurs through Ca^{2+} dependent NADPH oxidase phosphorylation; In addition to orchestrating rapid acclimation or defense responses at the local stress exposed level, rapid responses could be also important for triggering rapid systemic signaling of plants (Kollist et al., 2019). This response, the ABA and was orchestrated by the ROS wave, indicated that plants can rapidly coordinate their rapid responses in a 'whole-plant' strategy. Many of these responses were dependent on the function of the RBOHD protein. This RBOHD sourced wave could be detected not only at the local position, but also along the systemic signal path in tissues (Devireddy, Zandalinas, Gomez-Cadenas, Blumwald, & Mittler, 2018).

3.3.2.2 ROS wave

The ROS wave is an auto-propagating process of systemic signaling. Once triggered in a single cell, it causes the enhanced production of ROS by this cell (Miller et al., 2009). The result of ROS accumulation at the apoplast, which is sensed by neighboring cells, causes them to enhance their own production of ROS via their RBOHD proteins (Fichman & Mittler, 2020). Recent advances in imaging of ROS, at the whole plant level, made it possible to detect the ROS wave in live whole plants grown in soil (Fichman, Miller, & Mittler, 2019). These studies, as well as the initial characterization of the ROS wave (Miller et al., 2009), indicated that the systemic signal can be triggered by various abiotic stresses. ROS waves are considered important signals that alert cells and tissues to impending stress, along with other signals that may convey specificity. One of the novel functions of the ROS wave is to coordinate the whole-plant systemic stomatal response of plants to different stresses (Devireddy, Arbogast, & Mittler, 2019). Systematic control of stomatal aperture not only affects the water balance and CO_2 exchange rate required for photosynthesis, but also affects water wave and its propagation. The systemic stomatal responses of plants to different treatments highlight a key role for the ROS wave in triggering and coordinating plant acclimation. Systemic adaptation to light stress can be prevented by blocking ROS wave propagation and suppressing systemic stomatal response (Devireddy et al., 2018). Different systemic signals propagate through different plant tissues. For example, it was recently shown that during a systemic wound response (SWR), the ROS wave propagated through the vascular system (Fichman et al., 2019). In contrast, response to light stress showed the wave propagated primarily through non-vascular tissues.

ROS waves increase itself production in almost all cells of plants (Fichman & Mittler, 2020) and this pattern agrees with the tissue-specific expression of RBOHD (Morales, Kadota, Zipfel, Molina, & Torres, 2016), suggesting that RBOHD in all plant tissues is involved in ROS production during similar process.

The root RBOH-dependentH_2O_2 production operates as an early signal to regulate salt tolerance of grafted cucumber through Na^+ exclusion and stomatal closure has been clarified (Niu et al., 2018). Elevated relative expression of *RbohD* and *RbohF* and a higher NADPH oxidase activity have been found in the salt tolerant pumpkin grafted cucumber than in salt sensitive self-grafted cucumber. Early stomatal closure was also observed in the pumpkin-grafted cucumber plants, reducing water loss, and maintaining the plant's hydration status. Inhibition of the NADPH oxidase-mediated H_2O_2 signaling in the root also abolished a rapid stomatal closure in the pumpkin-grafted plants. These results indicated that the root-sourced respiratory burst oxidase homologue-dependent H_2O_2 production enhanced Na^+ exclusion from the root and promotes an early stomatal closure (Niu et al., 2018).

3.3.3 mRNA and miRNA

Root-to-shoot signals that move rapidly via the plant vascular system include RNA species. An increasing number of long mRNA transcripts have been identified in phloem (David-Schwartz, Runo, Townsley, Machuka, & Sinha, 2008). Phloem companion cells (CCs) are highly metabolically active and very well connected to the sieve elements (SEs) and are the most likely site of translation of root-derived mRNAs in the shoot (Shabala et al., 2016). At present, the clearest examples of long-distance mRNA are shoot-to-root regulation of tuberization in potato by movement of the *BEL5* transcript (Hannapel, 2010). Endogenous signaling is most clearly evidenced in the small RNAs induced by abiotic stresses. The long-distance mobility of small RNAs has been demonstrated in numerous grafting experiments (Canales, Henriquez-Valencia, & Brauchi, 2018; Zhang et al., 2014). Although phloem flow is dominantly from shoot-to-root, many miRNAs also move from root to shoot, presumably in the phloem (Zhang et al., 2014). MiRNAs arise from many tissues in the plant body; it is often difficult to discern which root-specific signals are then transported to the shoot. Recent high-throughput biological and genomic techniques have enabled the identification of many miRNAs, as long-distance mediators in plant signaling.

3.3.4 Peptides

Mobile peptides are another type of mediators in long distance signaling in response to abiotic stress. A recent study has shown that the peptide CLE25 is a mobile signaling molecule regulating ABA biosynthesis and stomatal closure in root-to-shoot signaling (Christmann & Grill, 2018). Root-derived CLE25 moves from roots to leaf and induces expression of NCED3 to enhance ABA accumulation in *Arabidopsis* leaves in response to dehydration stress. CLE25 plays an important role in transmitting dehydration signal through the vasculature, mainly from roots to leaves. As a functional feature of these peptides, their perception regulated by the presence of specific receptors in each tissue, enabling peptide–receptor modules to precisely transmits environmental stimuli from sensing tissues to specific target tissues. Such precise systems control plant growth and stress response. Hydraulic change is thought to function rapidly, whereas the movement of peptides and phytohormones constitutes rather slow signals to amplify the expression of stress-inducible genes.

The characteristics of different long-distance signals are summarized in the following Table 1.

Table 1 Different long-distance signals in plants.

Signal type	Route of transmission	Velocity	Species	References
Hydraulic signals	Pressure sensors in apoplast	The speed of sound	*Arabidopsis thaliana*	Wildon et al. (1992)
	Vascular bundles	$<0.03\,cm\,s^{-1}$	*Arabidopsis thaliana*	Christmann et al. (2007)
Electrical signal	Apoplast	$2.1\,cm\,s^{-1}$	*Vitis vinifera*	Galle et al. (2015)
	Blade ends(SP initiated by wounding)	$0.08–0.2\,cm\,s^{-1}$	*Vicia faba*	Zimmermann et al. (2009)
	Along the stems from root to shoot (SWP caused by inflammation)	$0.1–0.2\,cm\,s^{-1}$	*Populus trichocarpa*	Lautner, Grams, Matyssek, and Fromm (2005)
ABA	Vascular bundles	Phloem (several $mm\,s^{-1}$) or xylem ($cm\,s^{-1}$)	*Ricinus communis* *Vitis riparia* *Vitis labrusca*	Peuke (2016)
Ca^{2+}	Vascular tissues	$0.18–0.4\,mm\,s^{-1}$	*Arabidopsis thaliana*	Xiong et al. (2014)
	Through the cortex and endodermal cell layers	Up to $0.4\,mm\,s^{-1}$	*Arabidopsis thaliana*	Choi et al. (2014)
ROS wave	Apoplastic and plasmodesmata	$0.14\,cm\,s^{-1}$	*Arabidopsis thaliana*	Miller et al. (2009)
Peptides	Vascular tissues	Unknown	*Arabidopsis thaliana*	Takahashi et al. (2018)

4. Local control of stomatal opening and closing under salinity

4.1 Guard cell and aquaporins

Stomatal aperture itself is determined by the volume and mechanics of the guard cells. Increased guard cell osmotic pressure leads to water intake, guard cell flushing, and stomata opening (Maurel et al., 2016).

Many studies have shown that the regulation of aquaporin expression in guard cells is linked to components of stomatal regulation. The transcript abundance of several PIPs was enhanced after 4 h in response to spray of ABA in *Arabidopsis* guard cells (Leonhardt et al., 2004). In tobacco, expression of two TIP genes was reduced by drought after withholding water for 4 days (Smart, Moskal, Cameron, & Bennett, 2001).

4.2 Functions of H_2O_2

H_2O_2 is produced in the apoplasm by ABA-dependent activation of the plasma membrane NADPH oxidases and plays a key role in ABA signaling (Murai-Hatano et al., 2015). The sites of H_2O_2 action and its modes of diffusion into the cells are still undetermined. Some studies thought the aquaporins, which were shown to transport H_2O_2 via functional expression in yeast (Dynowski, Schaaf, Loque, Moran, & Ludewig, 2008), represent a candidate for this function.

Guard cells have evolved various mechanisms to sense and integrate various environments to optimize the balance between CO_2 exchange for and water loss via transpiration (Kollist et al., 2019).

Both RBOHD and RBOHF are necessary for executing stomatal closure in response to abscisic acid (ABA) and high levels of CO2 (Chater et al., 2015). Recent data indicate that the entry of apoplastic ROS into the cytoplasm of guard cells is facilitated by aquaporin PIP2;1 as *pip2;1* guard cell failed to accumulate H2O2 in response to ABA and flflg22 (Rodrigues et al., 2017). Accumulation of ROS in the apoplast activates the still unidentified H_2O_2-dependent Ca^{2+} influx channels (Pei et al., 2000), and the increase in cytoplasmic Ca^{2+} concentration triggers a secondary Ca^{2+} transients in chloroplasts as well as activation of multiple Ca^{2+}-dependent protein kinases (Sierla, Waszczak, Vahisalu, & Kangasjarvi, 2016).

Signaling pathways leading to stomatal closure always involve the function of Open Stomata1 (OST1), which serves as information center

(Sierla et al., 2016). It was hypothesized that OST1 might act upstream of apoplastic ROS production, and it has been found that in *ost1* mutants could be related to impaired activation of the ROS-regulated Ca^{2+} channels and consequential compromised activation of CPKs (Sierla et al., 2016). In addition, OST1 activity is negatively regulated by protein phosphatase, and some of protein phosphatase is inhibited by H^2O_2, suggesting that a self-amplifying loop between OST1 and H_2O_2 exists during stomatal closure (Meinhard, Rodriguez, & Grill, 2002).

4.3 ABA and guard cell

Several studies have focused on the prominent role played by ABA accumulation in stomatal closure regulation. ABA is thought to be synthesized under water deficit conditions, and it would lead to the depolarization of membranes in guard cell which triggers osmotic ion efflux (Kudla et al., 2018).

In recent years, the role of ABA in the regulation of stomatal active closure induced by low air humidity has been continuously debated (McAdam & Brodribb, 2016). Although ABA plays an important role in controlling total water conductance through stomata, OST1 is required for rapid stomatal closure under low air humidity conditions (Zhao et al., 2018), suggesting that additional mechanisms of ABA independent activation of OST1 exist.

Several reports have demonstrated the interaction of Ca^{2+} and ROS in guard cell regulation and their potential co-amplification effects with each other (Li, de Ollas, & Dodd, 2018; Merilo et al., 2018; Negi et al., 2008; Tombesi et al., 2015). With the further research on the function of Ca^{2+}, ROS, pH related genes; it provides a good tool to study the spatio-temporal role of these messengers in defining the rapid stomatal response.

4.4 Sensitivity of stomata

However, arguments related to ABA transport from the roots to the shoots and the regulation of stomatal movement still exists (Choi et al., 2017; Li et al., 2018). Apart from ABA concentration, ABA sensitivity in response to water deficit or salinity in the leaves affects stomatal movement. Niu et al. (2019) found that stomatal sensitivity to a root-sourced ABA signals gradually increased from the base to the apex along the vine of grapevine. Interestingly, the sensitivity of stomatal movement to ABA is affected by the type of roots. Cucumber grafted onto luffa is more sensitive to ABA than cucumber grafted onto cucumber (Liu et al., 2015). Thus, in addition to ABA biosynthesis, the increased sensitivity of stomata to ABA might play an important role in rootstock-induced salt tolerance. Our previous study indicated that the pumpkin rootstock promoted stomatal closure in cucumber scion leaf depends on an increased sensitivity to ABA under salinity. This hypersensitivity to ABA is related to enhanced tolerance to salt through the regulation of stomatal closure and expression of stress-related genes. Plants exposed to osmotic stress utilize the ABA signaling transduction pathway to initiate the expression of defense genes. Over expression of some stress-related genes, such as OsZIP72 and OsABI5, results in abiotic-stress tolerance and causes transgenic plants to be hypersensitive to exogenous ABA. ABA sensing and signaling are mediated by several key proteins, such as PP2C and SnRK2. According to a previous study, the improved ABA sensitivity of stomata in luffa grafted cucumber is associated with the differential expression of these two genes. SnRK2 interacts with several proteins, including kinases, phosphatases, transcription factors, and metabolic enzymes, and stomatal opening mediated by KAT1 might be negatively regulated by phosphorylation by SnRK2. Alternatively, stomatal closure in response to nutrient starvation was correlated with increased stomatal sensitivity to ABA which may occur in stressed plants due to apoplastic alkalization (Wilkinson & Davies, 2008).

5. Conclusions and prospects

5.1 How plants integrate different signals participate in stomatal regulation?

Vascular plants have evolved complex physiological systems with multiple organ types and complex signaling processes that optimize their growth and stress resistance in a variety of stressful environments. In recent years, high-throughput sequencing and omics methods have made it possible to identify many long-distance signaling molecules, including small peptides, proteins and RNAs. Under stress, these molecules are transported by different routes through different tissues, including the apoplast, xylem, phloem, endodermis, and cortex. ROS/Ca^{2+} waves are also thought to be long-distance signal mediators in stress responses, suggesting rapid systemic signal transduction throughout the plant. Thus, different signaling systems play a role in cell-to-cell, organ-to-organ communication, and local cellular responses. How plants integrate these signals participate in stomatal regulation needs to be further studied.

A major question in the field of root-to-shoot signaling is how all the signals mediating long distance communication are integrated with each other. Although some details of this process are starting to emerge, a lot of information is still to be gained. The growing availability of mutants defective in long distance signal synthesis, translocation and perception, combined with increasing sensitivity of mass spectrometry to detect minute quantities of signal molecules will help in drawing connections between physical and chemical signals in the plant.

5.2 The plant vascular system: Plant signaling highways

Plants are unable to migrate under adverse environmental conditions, relying instead on complex signaling pathways that work at a local and systemic level to resist stress. At the local level, stress-related signals are transmitted from cell to cell via plasmodesmata or diffused into extracellular space. The communication processes are efficient for signal transmission between adjacent cells or tissues, but not sufficient for systematic communication between organs (Lucas et al., 2013). In order to travel long distances between distant tissues, large molecules often enter the vascular system of plants; A system of specialized transport pipes connecting all the organs of a plant. The xylem ducts and phloem screens of the vascular system connect aboveground and underground organs, enabling vascular plants to extract resources from two distinct resource pools. In addition to playing a role in resource allocation, vascular system of plants are also important conduits for information exchange between organs. Xylem sap primarily containing mineral nutrients, peptides, proteins, and hormones. Many have been implicated as signals or signal chaperones in several developmental and stress related long-distance signaling pathway, such as systemic acquired resistance (SAR), induced systemic resistance (ISR), wound responses, RNA silencing. Xylem-mediated long-distance signaling has been implicated in certain abiotic stress responses. Besides that, the elevated ABA or other chemical signals and hydraulic signals all transmitted through this way. These responses illustrate how the plant vasculature acts a conduit for information sharing between distant tissues under salinity (Fig. 1).

5.3 Grafting as a tool to study root-sourced signal

Grafting is an ancient technology that has undergone a modern renaissance which is widely used in agriculture to increase yield, modify branch architecture, and enhance biotic and abiotic stress tolerance. Independent of its application in agriculture, grafting has been used as a powerful tool to

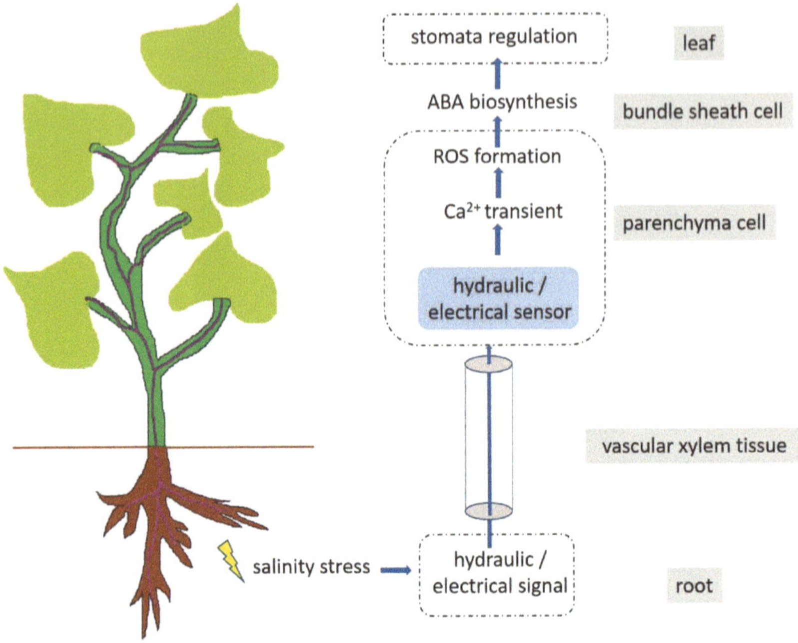

Fig. 1 Hydraulic signaling under salinity stress: conversion of a long-distance signal from root to shoot. A hydraulic signal is transferred throughout the plant in the xylem and causes water gradient, which is out of the parenchyma cells. Salinity induced water loss leads to turgor pressure change in parenchyma cells. Initial signal cascade comprises calcium transients and formation of reactive oxygen species (ROS) and targets ABA biosynthesis. ABA synthesized in parenchyma cells is exported to bundle sheath cells and trigger stomata regulation.

discover long distance signals. Studies using grafting have identified signaling molecules that coordinate root–shoot responses to drought stress (Takahashi et al., 2018) and salinity (Niu et al., 2018). Grafting experiments demonstrated that wild-type roots are capable of complementing drought-induced responses in *cle25* scions by inducing the expression of NCED3, an essential enzyme in the ABA biosynthetic pathway in shoots, thus triggering stomatal closure.

5.4 How root itself perceive salt stress or osmotic stress?

Chemical signals from the roots travel through the xylem to the ground and are involved in stomatal closure. This has been demonstrated using split-root systems, by pressurization of the root system that brings leaf water potential to zero even if plants are grown in dry soil, by exchange of sap between adequate water supply and water deficit plants, and by reciprocal grafting. However, the mechanism of how roots in soil environment perceive salt stress or osmotic stress is still not clear. Would the root signal itself be the downstream signal of a signal in shoot? Based on this, it is necessary to find a specific receptor in root and study its function in detail. This would be an evidence for the existence of the root signal. It is possible to speculate that the root message does not necessarily participate directly in stomatal closure, but it does affect its mechanisms. If we look at it from an evolutionary view, the basic principle is simple: when soil water absorption cannot meet evaporation needs under salinity, stomata must be closed.

Acknowledgments

This research was funded by the National Natural Science Foundation of China (32002112, 31772357), National Key Research and Development Program of China (2018YFD1000800), China Agriculture Research System of MOF and MORA (CARS-25), and Hubei Provincial Natural Science Foundation of China (2019CFA017).

References

Barberon, M., Vermeer, J. E. M., De Bellis, D., Wang, P., Naseer, S., Andersen, T. G., et al. (2016). Adaptation of root function by nutrient-induced plasticity of endodermal differentiation. *Cell*, *164*, 447–459.

Basu, D., & Haswell, E. S. (2017). Plant mechanosensitive ion channels: An ocean of possibilities. *Current Opinion in Plant Biology*, *40*, 43–48.

Bauer, H., Ache, P., Lautner, S., Fromm, J., Hartung, W., Al-Rasheid, K. A. S., et al. (2013). The stomatal response to reduced relative humidity requires guard cell-autonomous ABA synthesis. *Current Biology*, *23*, 53–57.

Canales, J., Henriquez-Valencia, C., & Brauchi, S. (2018). The integration of electrical signals originating in the root of vascular plants. *Frontiers in Plant Science*, *8*, 15.

Chater, C., Peng, K., Movahedi, M., Dunn, J. A., Walker, H. J., Liang, Y. K., et al. (2015). Elevated CO_2-induced responses in stomata require ABA and ABA signaling. *Current Biology*, *25*, 2709–2716.

Choi, W. G., Miller, G., Wallace, I., Harper, J., Mittler, R., & Gilroy, S. (2017). Orchestrating rapid long-distance signaling in plants with Ca^{2+}, ROS and electrical signals. *Plant Journal*, *90*, 698–707.

Choi, W. G., Toyota, M., Kim, S. H., Hilleary, R., & Gilroy, S. (2014). Salt stress-induced Ca^{2+} waves are associated with rapid, long-distance root-to-shoot signaling in plants. *Proceedings of the National Academy of Sciences of the United States of America*, *111*, 6497–6502.

Christmann, A., & Grill, E. (2018). Peptide signal alerts plants to drought. *Nature*, *556*, 178–179.

Christmann, A., Grill, E., & Huang, J. (2013). Hydraulic signals in long-distance signaling. *Current Opinion in Plant Biology*, *16*, 293–300.

Christmann, A., Weiler, E. W., Steudle, E., & Grill, E. (2007). A hydraulic signal in root-to-shoot signalling of water shortage. *Plant Journal*, *52*, 167–174.

Cuin, T. A., Dreyer, I., & Michard, E. (2018). The role of potassium channels in *Arabidopsis thaliana* long distance electrical signalling: AKT2 modulates tissue excitability while GORK shapes action potentials. *International Journal of Molecular Sciences*, *19*, 926.

David-Schwartz, R., Runo, S., Townsley, B., Machuka, J., & Sinha, N. (2008). Long-distance transport of mRNA via parenchyma cells and phloem across the host-parasite junction in Cuscuta. *New Phytologist, 179*, 1133–1141.
Davies, W. J., Wilkinson, S., & Loveys, B. J. N. P. (2002). Stomatal control by chemical signalling and the exploitation of this mechanism to increase water use efficiency in agriculture. *New Phytologist, 153*, 449–460.
Demidchik, V., Shabala, S., Isayenkov, S., Cuin, T. A., & Pottosin, I. (2018). Calcium transport across plant membranes: mechanisms and functions. *New Phytologist, 220*, 49–69.
Devireddy, A. R., Arbogast, J., & Mittler, R. (2019). Coordinated and rapid whole-plant systemic stomatal responses. *New Phytologist, 225*, 21–25.
Devireddy, A. R., Zandalinas, S. I., Gomez-Cadenas, A., Blumwald, E., & Mittler, R. (2018). Coordinating the overall stomatal response of plants: rapid leaf-to-leaf communication during light stress. *Science Signaling, 11*, eaam9514.
Dodd, I. C., Tan, L. P., & He, J. (2003). Do increases in xylem sap pH and/or ABA concentration mediate stomatal closure following nitrate deprivation? *Journal of Experimental Botany, 54*, 1281–1288.
Duan, B., Ma, Y., Jiang, M., Yang, F., Ni, L., & Lu, W. (2015). Improvement of photosynthesis in rice (*Oryza sativa* L.) as a result of an increase in stomatal aperture and density by exogenous hydrogen sulfide treatment. *Plant Growth Regulation, 75*, 33–44.
Dubiella, U., Seybold, H., Durian, G., Komander, E., Lassig, R., Witte, C. P., et al. (2013). Calcium-dependent protein kinase/NADPH oxidase activation circuit is required for rapid defense signal propagation. *Proceedings of the National Academy of Sciences of the United States of America, 110*, 8744–8749.
Dynowski, M., Schaaf, G., Loque, D., Moran, O., & Ludewig, U. (2008). Plant plasma membrane water channels conduct the signalling molecule H_2O_2. *Biochemical Journal, 414*, 53–61.
Fichman, Y., Miller, G., & Mittler, R. (2019). Whole-plantlive imaging of reactive oxygen species. *Molecular Plant, 12*, 1203–1210.
Fichman, Y., & Mittler, R. (2020). Rapid systemic signaling during abiotic and biotic stresses: is the ROS wave master of all trades? *Plant Journal, 102*, 887–896.
Fromm, J., & Lautner, S. (2007). Electrical signals and their physiological significance in plants. *Plant Cell and Environment, 30*, 249–257.
Galle, A., Lautner, S., Flexas, J., & Fromm, J. (2015). Environmental stimuli and physiological responses: The current view on electrical signalling. *Environmental and Experimental Botany, 114*, 15–21.
Gil, P. M., Saavedra, J., Schaffer, B., Navarro, R., Fuentealba, C., & Minoletti, F. (2014). Quantifying effects of irrigation and soil water content on electrical potentials in grapevines (*Vitis vinifera*) using multivariate statistical methods. *Scientia Horticulturae, 173*, 71–78.
Hannapel, D. J. (2010). A model system of development regulated by the long-distance transport of mRNA. *Journal of Integrative Plant Biology, 52*, 40–52.
Hare, P. D., Cress, W. A., & van Staden, J. (1999). Proline synthesis and degradation: a model system for elucidating stress-related signal transduction. *Journal of Experimental Botany, 50*, 413–434.
Hartung, W., Sauter, A., & Hose, E. (2002). Abscisic acid in the xylem: where does it come from, where does it go to? *Journal of Experimental Botany, 53*, 27–32.
Hilleary, R., & Gilroy, S. (2018). Systemic signaling in response to wounding and pathogens. *Current Opinion in Plant Biology, 43*, 57–62.
Holbrook, N. M., Shashidhar, V. R., James, R. A., & Munns, R. (2002). Stomatal control in tomato with ABA-deficient roots: response of grafted plants to soil drying. *Journal of Experimental Botany, 53*, 1503–1514.

Huber, A. E., & Bauerle, T. L. (2016). Long-distance plant signaling pathways in response to multiple stressors: the gap in knowledge. *Journal of Experimental Botany, 67*, 2063–2079.
Jiang, F., & Hartung, W. (2008). Long-distance signalling of abscisic acid (ABA): the factors regulating the intensity of the ABA signal. *Journal of Experimental Botany, 59*, 37–43.
Jiang, Z., Zhou, X., Tao, M., Yuan, F., Liu, L., Wu, F., et al. (2019). Plant cell-surface GIPC sphingolipids sense salt to trigger Ca^{2+} influx. *Nature*, 572.
Kinnersley, A. M., & Turano, F. J. (2000). Gamma aminobutyric acid (GABA) and plant responses to stress. *Critical Reviews in Plant Sciences, 19*, 479–509.
Kollist, H., Zandalinas, S. I., Sengupta, S., Nuhkat, M., Kangasjarvi, J., & Mittler, R. (2019). Rapid responses to abiotic stress: priming the landscape for the signal transduction network. *Trends in Plant Science, 24*, 25–37.
Kudla, J., Becker, D., Grill, E., Hedrich, R., Hippler, M., Kummer, U., et al. (2018). Advances and current challenges in calcium signaling. *New Phytologist, 218*, 414–431.
Kudoyarova, G. R., Kholodova, V. P., & Veselov, D. S. (2013). Current state of the problem of water relations in plants under water deficit. *Russian Journal of Plant Physiology, 60*, 165–175.
Kuromori, T., Seo, M., & Shinozaki, K. (2018). ABA transport and plant water stress responses. *Trends in Plant Science, 23*, 513–522.
Kurusu, T., Kuchitsu, K., Nakano, M., Nakayama, Y., & Iida, H. (2013). Plant mechanosensing and Ca^{2+} transport. *Trends in Plant Science, 18*, 227–233.
Lautner, S., Grams, T. E. E., Matyssek, R., & Fromm, J. (2005). Characteristics of electrical signals in poplar and responses in photosynthesis. *Plant Physiology, 138*, 2200–2209.
Leonhardt, N., Kwak, J. M., Robert, N., Waner, D., Leonhardt, G., & Schroeder, J. I. (2004). Microarray expression analyses of *Arabidopsis* guard cells and isolation of a recessive abscisic acid hypersensitive protein phosphatase 2C mutant. *Plant Cell, 16*, 596–615.
Li, W., de Ollas, C., & Dodd, I. C. (2018). Long-distance ABA transport can mediate distal tissue responses by affecting local ABA concentrations. *Journal of Integrative Plant Biology, 60*, 16–33.
Li, B., Feng, Z., Xie, M., Sun, M., Zhao, Y., Liang, L., et al. (2011). Modulation of the root-sourced ABA signal along its way to the shoot in *Vitis riparia* × *Vitis labrusca* under water deficit. *Journal of Experimental Botany, 62*, 1731–1741.
Liu, Z. X., Bie, Z. L., Huang, Y., Zhen, A., Niu, M. L., & Lei, B. (2013). Rootstocks improve cucumber photosynthesis through nitrogen metabolism regulation under salt stress. *Acta Physiologiae Plantarum, 35*, 2259–2267.
Liu, S., Li, H., Lv, X., Ahammed, G. J., Xia, X., Zhou, J., et al. (2015). Grafting cucumber onto luffa improves drought tolerance by increasing ABA biosynthesis and sensitivity. *Scientific Reports, 6*, 20212.
Longman, K. A., Zimmermann, M. H., Brown, C. L., & Tyree, M. T. (1972). Trees: structure and function. *Journal of Ecology, 60*, 942.
Lucas, W. J., Groover, A., Lichtenberger, R., Furuta, K., Yadav, S. R., Helariutta, Y., et al. (2013). The plant vascular system: Evolution, development and functions. *Journal of Integrative Plant Biology, 55*, 294–388.
Maurel, C., Verdoucq, L., & Rodrigues, O. (2016). Aquaporins and plant transpiration. *Plant, Cell & Environment, 39*, 2580–2587.
McAdam, S. A., & Brodribb, T. J. (2014). Separating active and passive influences on stomatal control of transpiration. *Plant Physiology, 164*, 1578–1586.
McAdam, S. A. M., & Brodribb, T. J. (2016). Linking turgor with ABA biosynthesis: implications for stomatal responses to vapor pressure deficit across land plants. *Plant Physiology, 171*, 2008–2016.
McAdam, S. A. M., Brodribb, T. J., & Ross, J. J. (2016). Shoot-derived abscisic acid promotes root growth. *Plant, Cell & Environment, 39*, 652–659.

Medina, V., & Gilbert, M. E. (2016). Physiological trade-offs of stomatal closure under high evaporative gradients in field grown soybean. *Functional Plant Biology*, *43*, 40–51.

Meinhard, M., Rodriguez, P. L., & Grill, E. (2002). The sensitivity of ABI2 to hydrogen peroxide links the abscisic acid-response regulator to redox signalling. *Planta*, *214*, 775–782.

Merilo, E., Yarmolinsky, D., Jalakas, P., Parik, H., Tulva, I., Rasulov, B., et al. (2018). Stomatal VPD response: There is more to the story than ABA. *Plant Physiology*, *176*, 851–864.

Miller, G., Schlauch, K., Tam, R., Cortes, D., Torres, M. A., Shulaev, V., et al. (2009). The plant NADPH oxidase RBOHD mediates rapid systemic signaling in response to diverse stimuli. *Science Signaling*, *2*, 45–48.

Monshausen, G. B., & Haswell, E. S. (2013). A force of nature: Molecular mechanisms of mechanoperception in plants. *Journal of Experimental Botany*, *64*, 4663–4680.

Morales, J., Kadota, Y., Zipfel, C., Molina, A., & Torres, M. A. (2016). The *Arabidopsis* NADPH oxidases RbohD and RbohF display differential expression patterns and contributions during plant immunity. *Journal of Experimental Botany*, *67*, 1663–1676.

Munns, R., James, R. A., Xu, B., Athman, A., Conn, S. J., Jordans, C., et al. (2012). Wheat grain yield on saline soils is improved by an ancestral Na^+ transporter gene. *Nature Biotechnology*, *30*, 360–364.

Murai-Hatano, M., Kuwagata, T., Hayashi, H., Ishikawa-Sakurai, J., Moriyama, M., & Okada, M. (2015). Rice plants sense daily weather and regulate aquaporin gene expressions in the roots -close correlation with potential evaporation. *Journal of Agricultural Meteorology*, *71*, 124–135.

Negi, J., Matsuda, O., Nagasawa, T., Oba, Y., Takahashi, H., Kawai-Yamada, M., et al. (2008). CO_2 regulator SLAC1 and its homologues are essential for anion homeostasis in plant cells. *Nature*, *452*, 483–486.

Neto, A. D. D., Prisco, J. T., Eneas, J., de Abreu, C. E. B., & Gomes, E. (2006). Effect of salt stress on antioxidative enzymes and lipid peroxidation in leaves and roots of salt-tolerant and salt-sensitive maize genotypes. *Environmental and Experimental Botany*, *56*, 87–94.

Niu, M. L., Huang, Y., Sun, S. T., Sun, J. Y., Cao, H. S., Shabala, S., et al. (2018). Root respiratory burst oxidase homologue-dependent H_2O_2 production confers salt tolerance on a grafted cucumber by controlling Na^+ exclusion and stomatal closure. *Journal of Experimental Botany*, *69*, 3465–3476.

Niu, M. L., Sun, S. T., Nawaz, M. A., Sun, J. Y., Cao, H. S., Lu, J. Y., et al. (2019). Grafting cucumber onto pumpkin induced early stomatal closure by increasing ABA sensitivity under salinity conditions. *Frontiers in Plant Science*, *10*, 1290.

Pantin, F., Monnet, F., Jannaud, D., Costa, J. M., Renaud, J., Muller, B., et al. (2013). The dual effect of abscisic acid on stomata. *New Phytologist*, *197*, 65–72.

Pei, Z. M., Murata, Y., Benning, G., Thomine, S., Klusener, B., Allen, G. J., et al. (2000). Calcium channels activated by hydrogen peroxide mediate abscisic acid signalling in guard cells. *Nature*, *406*, 731–734.

Peuke, A. D. (2016). ABA flow modelling in Ricinus communis exposed to salt stress and variable nutrition. *Journal of Experimental Botany*, *67*, 5301–5311.

Peyronnet, R., Tran, D., Girault, T., & Frachisse, J. M. (2014). Mechanosensitive channels: feeling tension in a world under pressure. *Frontiers in Plant Science*, *5*, 558.

Rodrigues, O., Reshetnyak, G., Grondin, A., Saijo, Y., Leonhardt, N., Maurel, C., et al. (2017). Aquaporins facilitate hydrogen peroxide entry into guard cells to mediate ABA- and pathogen-triggered stomatal closure. *Proceedings of the National Academy of Sciences of the United States of America*, *114*, 9200–9205.

Sade, N., Shatil-Cohen, A., Attia, Z., Maurel, C., Boursiac, Y., Kelly, G., et al. (2014). The role of plasma membrane aquaporins in regulating the bundle sheath-mesophyll continuum and leaf hydraulics. *Plant Physiology*, *166*, 1609–1620.

Salvador-Recatala, V. (2016). The AKT2 potassium channel mediates NaCl induced depolarization in the root of *Arabidopsis thaliana*. *Plant Signaling & Behavior*, *11*, e1165381.
Schachtman, D. P., & Goodger, J. Q. (2008). Chemical root to shoot signaling under drought. *Trends Plant Science*, *13*, 281–287.
Secchi, F., & Zwieniecki, M. A. (2014). Down-regulation of plasma intrinsic protein1 aquaporin in poplar trees is detrimental to recovery from embolism. *Plant Physiology*, *164*, 1789–1799.
Shabala, S., Pang, J. Y., Zhou, M. X., Shabala, L., Cuin, T. A., Nick, P., et al. (2009). Electrical signalling and cytokinins mediate effects of light and root cutting on ion uptake in intact plants. *Plant, Cell & Environment*, *32*, 194–207.
Shabala, S., White, R. G., Djordjevic, M. A., Ruan, Y. L., & Mathesius, U. (2016). Root-to-shoot signalling: integration of diverse molecules, pathways and functions. *Functional Plant Biology*, *43*, 87–104.
Shabala, S., Wu, H., & Bose, J. (2015). Salt stress sensing and early signalling events in plant roots: Current knowledge and hypothesis. *Plant Science*, *241*, 109–119.
Shatil-Cohen, A., Attia, Z., & Moshelion, M. (2011). Bundle-sheath cell regulation of xylem-mesophyll water transport via aquaporins under drought stress: a target of xylem-borne ABA? *Plant Journal*, *67*, 72–80.
Sierla, M., Waszczak, C., Vahisalu, T., & Kangasjarvi, J. (2016). Reactive oxygen species in the regulation of stomatal movements. *Plant Physiology*, *171*, 1569–1580.
Smart, L. B., Moskal, W. A., Cameron, K. D., & Bennett, A. B. (2001). MIP genes are down-regulated under drought stress in Nicotiana glauca. *Plant & Cell Physiology*, *42*, 686–693.
Sobeih, W. Y., Dodd, I. C., Bacon, M. A., Donald, G., & Davies, W. J. (2004). Long-distance signals regulating stomatal conductance and leaf growth in tomato (*Lycopersicon esculentum*) plants subjected to partial root-zone drying. *Journal of Experimental Botany*, *55*, 2353–2363.
Spanswick, R. M., & Costerton, J. W. F. (1967). Plasmodesmata in nitella translucens: structure and electrical resistance. *Journal of Cell Science*, *2*, 451–464.
Steinhorst, L., & Kudla, J. (2014). Signaling in cells and organisms—Calcium holds the line. *Current Opinion in Plant Biology*, *22*, 14–21.
Steinhorst, L., & Kudla, J. (2019). How plants perceive salt. *Nature*, *572*, 318–320.
Stolarz, M., & Dziubinska, H. (2017). Osmotic and salt stresses modulate spontaneous and glutamate-induced action potentials and distinguish between growth and circumnutation in *Helianthus annuus* seedlings. *Frontiers in Plant Science*, *8*, 1766.
Sun, J. Y., Cao, H. S., Cheng, J. T., He, X. M., Sohail, H., Niu, M. L., et al. (2018). Pumpkin CmHKT1;1 Controls Shoot Na^+ accumulation via limiting Na^+ transport from rootstock to scion in grafted cucumber. *International Journal of Molecular Sciences*, *19*, 20.
Takahashi, F., & Shinozaki, K. (2019). Long-distance signaling in plant stress response. *Current Opinion in Plant Biology*, *47*, 106–111.
Takahashi, F., Suzuki, T., Osakabe, Y., Betsuyaku, S., Kondo, Y., Dohmae, N., et al. (2018). A small peptide modulates stomatal control via abscisic acid in long-distance signalling. *Nature*, *556*, 235–238.
Tombesi, S., Nardini, A., Frioni, T., Soccolini, M., Zadra, C., Farinelli, D., et al. (2015). Stomatal closure is induced by hydraulic signals and maintained by ABA in drought-stressed grapevine. *Scientific Reports*, *5*, 12449.
Vaten, A., Dettmer, J., Wu, S., Stierhof, Y. D., Miyashima, S., Yadav, S. R., et al. (2011). Callose biosynthesis regulates symplastic trafficking during root development. *Developmental Cell*, *21*, 1144–1155.
Vestergaard, C. L., Flyvbjerg, H., & Moller, I. M. (2012). Intracellular signaling by diffusion: Can waves of hydrogen peroxide transmit intracellular information in plant cells? *Frontiers in Plant Science*, *3*, 295.

Visentin, I., Vitali, M., Ferrero, M., Zhang, Y., Ruyter-Spira, C., Novák, O., et al. (2016). Low levels of strigolactones in roots as a component of the systemic signal of drought stress in tomato. *New Phytologist*, *212*, 954–963.

Vodeneev, V., Akinchits, E., & Sukhov, V. (2015). Variation potential in higher plants: Mechanisms of generation and propagation. *Plant Signaling & Behavior*, *10*, e1057365.

Wegner, L. H., Sattelmacher, B., Läuchli, A., & Zimmermann, U. (1999). Trans-root potential, xylem pressure, and root cortical membrane potential of 'low-salt' maize plants as influenced by nitrate and ammonium. *Plant, Cell & Environment*, *22*, 1549–1558.

Wegner, L. H., Stefano, G., Shabala, L., Rossi, M., Mancuso, S., & Shabala, S. (2011). Sequential depolarization of root cortical and stelar cells induced by an acute salt shock—Implications for Na^+ and K^+ transport into xylem vessels. *Plant, Cell & Environment*, *34*, 859–869.

Wildon, D. C., Thain, J. F., Minchin, P. E. H., Gubb, I. R., Reilly, A. J., Skipper, Y. D., et al. (1992). Electrical signalling and systemic proteinase inhibitor induction in the wounded plant. *Nature*, *360*, 62–65.

Wilkinson, S., & Davies, W. J. (2008). Manipulation of the apoplastic pH of intact plants mimics stomatal and growth responses to water availability and microclimatic variation. *Journal of Experimental Botany*, *59*, 619–631.

Xiong, T. C., Ronzier, E., Sanchez, F., Corratge-Faillie, C., Mazars, C., & Thibaud, J. B. (2014). Imaging long distance propagating calcium signals in intact plant leaves with the BRET-based GFP-aequorin reporter. *Frontiers in Plant Science*, *5*, 43.

Zhang, W. N., Kollwig, G., Stecyk, E., Apelt, F., Dirks, R., & Kragler, F. (2014). Graft-transmissible movement of inverted-repeat-induced siRNA signals into flowers. *Plant Journal*, *80*, 106–121.

Zhao, Y., Zhang, Z. J., Gao, J. H., Wang, P. C., Hu, T., Wang, Z. G., et al. (2018). *Arabidopsis* duodecuple mutant of PYL ABA receptors reveals PYL repression of ABA-independent SnRK2 activity. *Cell Reports*, *23*, 3340–3351.

Zhu, J. K. (2003). Salt and drought stress signal transduction in plants. *Annual Review of Plant Biology*, *53*, 247–273.

Zhu, J. K. (2016). Abiotic stress signaling and responses in plants. *Cell*, *167*, 313–324.

Zimmermann, M. R., Maischak, H., Mithofer, A., Boland, W., & Felle, H. H. (2009). System potentials, a novel electrical long-distance apoplastic signal in plants, induced by wounding. *Plant Physiology*, *149*, 1593–1600.

CHAPTER NINE

Salinity and night-time transpiration under current climate scenarios

Wieland Fricke*
School of Biology and Environmental Sciences, Science Centre West, University College Dublin (UCD), Belfield, Dublin, Republic of Ireland
*Corresponding author: e-mail address: wieland.fricke@ucd.ie

Contents

1. Introduction 222
2. Transport processes: Getting things from A to B 224
3. Aquaporins 228
4. Night-time transpiration and growth 233
5. Night-time transpiration and energy 238
6. Outlook: Night-time transpiration in a changing climate 240
Acknowledgments 243
References 243

Abstract

Plant growth occurs during both day and night and is associated with water loss via stomata. This chapter looks at the processes which are associated with water uptake and transpirational water loss during day and night, and how these processes relate to leaf growth and may be affected differently by the salt stress. Potential advantages and disadvantages of a night-time water loss and growth are discussed, particularly in the context of salinity and the uptake of excess Na^+ and Cl^- by the root system. The potential importance of aquaporins to enable a more selective uptake of water, compared with Na^+ and Cl^-, by the root under salinity is addressed. Both stomata and root aquaporins control plant water flow. Climate change and associated increases in temperature during the day and night period have the potential to impact differently on day- and night-time processes and on the adaptation of plants to salt stress.

Abbreviations

ABA abscisic acid
AQP aquaporin
CAM crassulacean acid metabolism
g stomatal conductance
L hydraulic conductance

Advances in Botanical Research, Volume 103
ISSN 0065-2296
https://doi.org/10.1016/bs.abr.2022.02.015

Lp hydraulic conductivity
WAE water acquisition efficiency
WUE water use efficiency

1. Introduction

Water use efficiency (WUE) provides a measure of how much CO_2 is taken up through stomata and assimilated through photosynthesis at the expense of water being lost into the environment. As CO_2 is taken up through stomata and water is lost through stomata, and since stomata are located in leaves, the focus is on leaves. However, one could also argue that for many plants growing in a saline environment, the tougher task is to optimize the amount of CO_2 being assimilated by the shoot and water taken up by the root per unit of Na^+ and Cl^- ions taken up in parallel. We refer to this feature here as "Water Acquisition Efficiency (WAE)." This WAE could be extended to K^+ and NO_3^-, as the uptake of these mineral nutrients is impeded by high salt, yet both nutrients are central to the metabolism of cells and to obtain the full benefit of assimilated CO_2. The more K^+ and NO_3^-, and the less Na^+ and Cl^-, are taken up per unit water taken up and C assimilated, the better. The uptake of water and ions occurs through roots. The substance which is common to WUE and WAE is water. That means that to understand WUE, we need to consider WUE also in the context of WAE and relevant root properties.

Plants have a limited amount of energy available at any one time, and the above-mentioned processes cost energy (for review, see Munns et al., 2020; Tyerman et al., 2020). The best balance needs to be struck between the use of energy for growth and the "status quo" (base metabolism) and the use of energy for stress tolerance mechanisms. Some processes are thermodynamically spontaneous, such as the evaporation of water from a high (leaf-internal air space) to a low (ambient air) water potential and water vapor pressure environment. A spontaneous process as such does not require energy, yet the regulation of process requires energy too.

A key question in relation to optimizing the assimilation of CO_2 and uptake of K^+ and NO_3^-, while minimizing the uptake of Na^+ and Cl^- and loss of water is: to which extent can these processes be uncoupled from each other (Fig. 1)? This question leads to the transport processes of

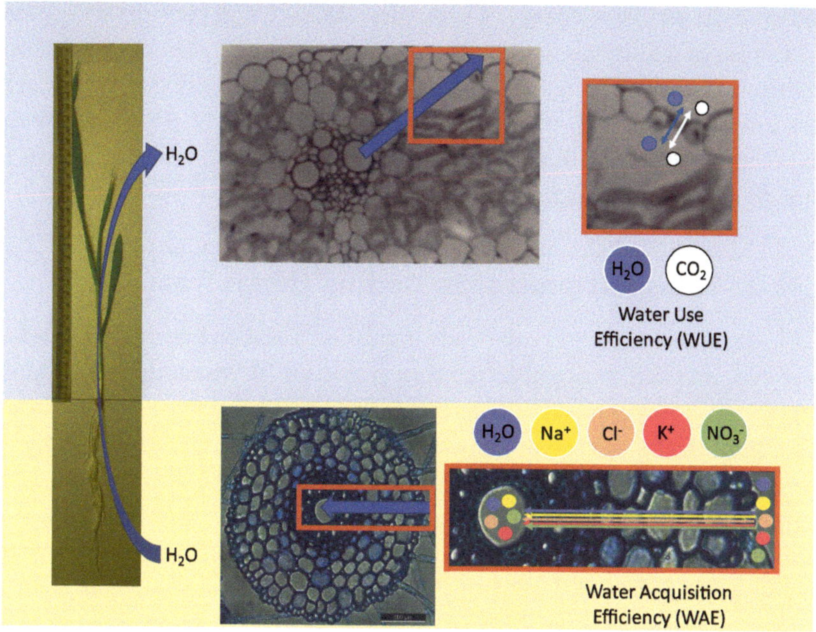

Fig. 1 Control of water flow through an annual plant (barley as an example). The flow of water through the plant can be controlled at shoot and root level. Water is mostly a limited resource in the field. It is therefore an advantage to lose as little as possible water through stomata into the air while taking up as much as possible CO_2 for photosynthesis. This water use efficiency (WUE) is affected by changes in the local driving forces next to the stomatal pore, driving forces, which cause water to exit and CO_2 to enter the plant. In analogy to WUE in the shoot, one can consider a water acquisition efficiency (WAE) in the root. Uptake of water through the root system is associated to various extent with the uptake of mineral nutrients such as K^+ and NO_3^-, and in a saline environment, with the uptake of a significant amount of Na^+ and Cl^-. Particularly concerning the latter, it may desirable under salinity to keep the ratio of water molecules to Na^+ and Cl^- ions taken up high (high WAE). The picture on the left shows a 16d-old barley plant, the ruler is 30cm long. The picture at the bottom shows a cross-section of a barley root (scale bar 100 μm), together with an enlarged portion of it, which indicates the movement of solutes from the external medium across the root cylinder to the large central, late metaxylem vessel. The pictures at the top show a cross-section of a mature barley leaf, together with a portion highlighting the area around a stomatal pore (the arrows, which run between a pair of guard cells, point into the ambient air and substomatal cavity). The large blue arrows shown in the pictures reflect the net movement of water through the system.

substances, and how they can be modified, at the molecule, cell, tissue, and organ (root/shoot) level. We will look in the following first at some basic aspects of transport processes of interest, then move on to aquaporins and water flow through plants, have a closer look at night-time related processes, and raise the aspect of energy. Throughout this we will consider the potential impact of salinity on these processes. We will finally consider everything in the context of a changing climate. The aim is to identify gaps in our current knowledge of how plants can optimize WUE and WAE, and how the ability to optimize WUE and WAE may be affected by predicted changes in climate and weather patterns.

2. Transport processes: Getting things from A to B

Transport processes can be distinguished based on various criteria. For example, one can distinguish between processes (i) which directly require (active) or do not require (passive) energy; (ii) which involve transport across membranes facilitated by specialized transport/channel proteins or involving membrane vesicles (e.g., endo/exocytosis); or (iii) which occur by diffusion or bulk flow, two different modes of transport. The latter distinction is important when considering the movement of substances over shorter and longer distances. While diffusion is fast enough to get solutes from A to B at the dimension of organelles and smaller cells, bulk, or mass flow is required to move substances over longer distances—and cytoplasmic streaming does compensate already at cell level for the comparatively slow movement of solutes by diffusion.

Transport by mass/bulk flow has the advantage that it is fast. The solvent water moves typically along a rather hydrophilic apoplast and does not encounter hydrophobic barriers such as the lipid bilayer of plasma membrane. Also, in the xylem and to some extent phloem, bulk flow occurs along voluminous tubes. The downside of mass flow is that it is not selective, as substances are moved with the solvent. In contrast, transport across membranes is generally more costly and slowed down by the presence of a hydrophobic lipid bilayer, yet it is more or less selective and enables the distinction between the solvent water and dissolved substance(s).

The flow, or transport rate of a substance from location A to B is in analogy to Ohm's law the numeric product of the conductance of the path and the driving force for the movement (Fig. 2). The driving force can be related directly to the chemical nature of substance, e.g., a difference in concentration of substance between locations A and B; or it relates to the movement of solvent, e.g., a difference in hydrostatic pressure between locations A and B.

Gradients in hydrostatic pressure which drive bulk flow between A and B are typically provided through either xylem tension (day), large positive pressures in phloem or to some extent also positive xylem (root)

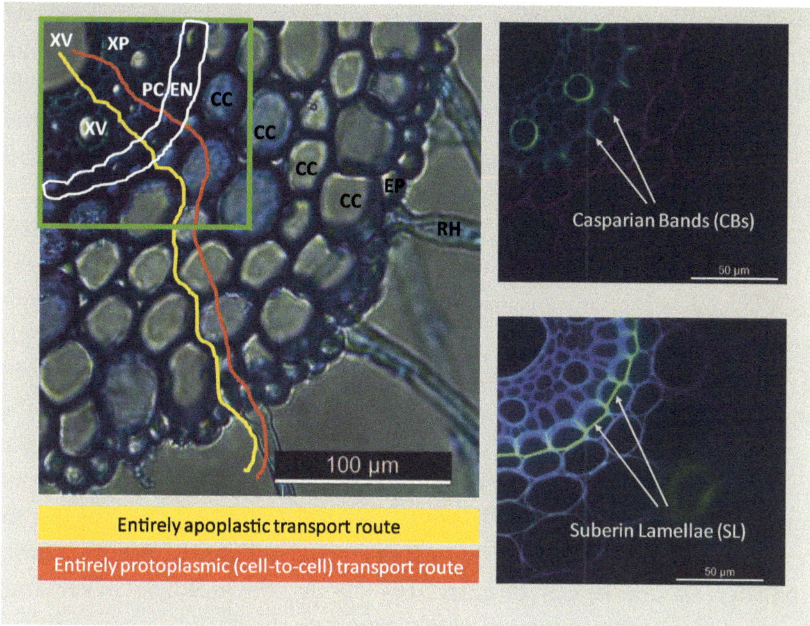

Fig. 2 Transport paths of water and solutes across the root cylinder. Water and solute can move across the root cylinder along either, or both, of two types of transport paths. One transport path runs entirely along the continuous apoplast between root epidermal cells (with root hairs) and central xylem vessel. In this case, water and solutes will never cross a (hydrophobic) lipid bilayer. That speeds up the movement of substances yet also means that any selectivity is reduced to ion absorbing and size-filtering properties of the apoplast. The alternative pathway, which runs entirely along the protoplasm of cells across the root cylinder, forces substances to cross membranes and is therefore highly selective, yet the presence of hydrophobic lipid bilayers slows down the speed of movement. This cell-to-cell path can involve the crossing of lipid bilayer as such, it can also involve the transport of molecules through specialized transporters and channels, such as aquaporins when considering water. Furthermore, the cell-to-cell path can also involve transport of substances through plasmodesmata, from cell to cell, in which case one refers to "symplasmic" transport. The picture on the left shows a quarter of the cross-section of a barley root. The two pictures on the right also show parts of a cross-section of barley root, yet they highlight two structures which can obstruct the transport of substances along the apoplast of endodermis. These structures are Casparian bands (CBs) and suberin lamellae (SL). Casparian bands (seen here as "simple lines") are made of hydrophobic material and form a band around the radial walls of the endodermis. This slows down or may even stop movement of substances as they approach the endodermis from the periphery in the apoplast. Suberin lamellae are also hydrophobic, and they are located between the plasma membrane and wall of the outer-facing (toward cortex) tangential walls of the endodermis. Note that being located like this, suberin lamellae impact not only on the movement of solutes along an entirely apoplastic path, but they also impact on the protoplasmic path: any substances which exit the innermost cortex cells through the plasma membrane must pass the suberin lamellae first before crossing the plasma membrane of endodermal cells. *Abbreviations*: CC, cortex cell: EP, epidermis; RH, root hair; XV, xylem vessel; XP, xylem parenchyma; PC, pericycle; EN, endodermis. Scale bars: large picture on left, 100 μm; two smaller pictures on right, 50 μm.

pressures in the night, which can lead to guttation. Bulk flow makes it difficult to enable a solute-specific type of transport, unless the matrix, such as a cell wall, has different ion absorption or filtering properties for solutes of different charge density or size. In contrast, transport processes where the driving force is solute specific, make it far easier to enable a solute-specific transport. A good example is not just a membrane-associated transport process, but also the diffusion of CO_2 and water vapor through stomata. Both chemical species follow their own difference in energy, and this makes it possible to alter the ratio of the speed of movement of the two, for example, by altering the water vapor concentration of boundary layer or the CO_2 concentration within the substomatal cavity—means which enable the adjustment of WUE.

In contrast, in roots, water and nutrients, or salt, which enter the apoplast of rhizodermal cells must at some point be separated from each other to enable a substance-specific transport. This is achieved not so much by adjusting the driving force but by adjusting the conductance in a substance-specific manner. In practice, this means that the entry of solutes into the protoplast of cells must be made "easy" at the entry point of solutes into the root cylinder (root epidermal cells with root hairs); or that the apoplast must be "sealed" through impregnation with hydrophobic substances somewhere along the path across the root cylinder. The consequence is that water and dissolved solutes are forced across membranes, where substance-specific transport proteins take over. The best studied apoplastic barrier which occurs in roots is at the cortex/stele boundary, where a ring of cells forms the endodermis and where radial and tangential walls form Casparian bands and suberin lamellae, respectively (see Fig. 3) (Barberon, 2017; Barberon et al., 2016; Enstone, Peterson, & Ma, 2003; Geldner, 2013). The former can also be found at the boundary between the rhizodermis and outermost cortex cell layer, leading to the formation of an exodermis (Freundl, Steudle, & Hartung, 2000; Hose, Clarkson, Steudle, Schreiber, & Hartung, 2001; Schreiber & Franke, 2011).

Roots are not only the organ where plants can regulate the uptake of water, they are also the organ where the overall load of plant with soil-borne substances is determined. There can be some reshuffling of substances between xylem and phloem, and therefore root and shoot (Wolf, Munns, Tonnet, & Jeschke, 1990). However, most of what enters the root xylem ends up in the shoot and accumulates with time. This applies also to Na^+ and Cl^- taken up by the majority of crop plants under saline conditions (Munns, 1985; Wolf et al., 1990). The relationship between the concentration of solutes in

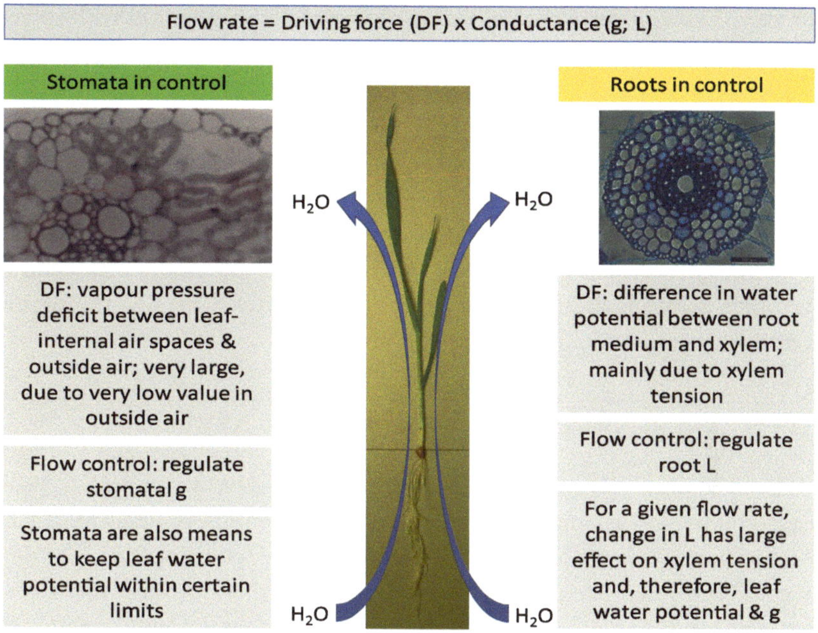

Fig. 3 Stomata and roots control the flow of water through the plant. Stomata provide the largest hydraulic resistance along the soil/root medium—plant—atmosphere continuum. The flow rate of water through a plant is the numeric product of the hydraulic conductance and the biophysical force which drives the movement of water. Using data from Saini and Fricke (2020) for barley, it can be calculated that the hydraulic resistance, which is the reverse of conductance, of stomata accounts for >99% of the total plant resistance to water flow; roots make up less than 0.5% of that resistance. From that point, there is no question that stomata are in control of plant water flow. However, small as the hydraulic resistance in roots may be, it does have a potentially large effect on the water potential in xylem. This is because the driving force for water movement across the root cylinder is the difference in water potential, or hydrostatic pressure (depending on whether water moves mainly along a cell-to-cell or apoplast path), between root xylem and root medium. Root xylem water potential is dominated by the tension component in a transpiring plant during the day. If the conductance of the radial flow path across roots is reduced by say 80%, the resistance increases fivefold. To maintain the same flow rate, requires now a xylem tension which is 5 times more negative. Xylem tension in the root and shoot may reach such a negative value that stomata start to close, in which case, changes in root hydraulics, such as through closure of aquaporins (AQPs), impose a control on plant water flow.

the xylem and the flow rate of liquid in the xylem, which approximates the transpiration rate, has been shown to be inverse (Munns & Passioura, 1984a, 1984b). In other words, the total solute load, which arrives in the shoot per unit time varies little, even though the amount of water which is transpired can vary. This implies that the loading of xylem with water and with solutes can be uncoupled from each other. Xylem parenchyma cells which are located close to xylem vessels within the stele enable a solute-specific loading or unloading of xylem. Upstream, the endodermis with Casparian bands and suberin lamellae makes it possible to load the stele in a solute-specific manner and to alter the ratio between solutes and water passing into the stele. To achieve the latter, transport proteins are required which allow specifically water but not solutes to cross the plasma membrane of endodermal cells, or any other cells which are located upstream the flow path. Such transport, or channel proteins are aquaporins (AQPs).

3. Aquaporins

Aquaporins are membrane-intrinsic proteins which form typically water-conducting channels. Aquaporins can also transport small-molecular weight solutes such as boron, urea, ammonia, silicon, glycerol and CO_2; they have also been suggested to transport ions. Aquaporins are gated, that means they exist in either a closed or open conformation (Maurel, Verdoucq, Luu, & Santoni, 2008; Törnroth-Horsefield et al., 2006). Aquaporins occur in all living plant cells and in all organismal kingdoms. There exist different isoforms of AQPs, and those AQPs which are located within the plasma membrane of cells (plasma membrane intrinsic proteins, PIPs) are of most interest to us here. By forcing liquid to cross membranes, such as at the endodermis, but also by having water flowing from cell-to-cell across the entire root cortex and epidermis, the rate of water flow into the central stele can be regulated through AQPs—and the ratio of water and specific solutes entering the stele can be regulated as well (Fig. 3).

The activity of AQPs can be adjusted particularly through the abundance of AQP proteins in the membrane and through gating of AQP proteins. For example, AQP activity can be increased through phosphorylation and decreased through dephosphorylation (Guenther et al., 2003; Johansson, Larsson, Ek, & Kjellbom, 1996; Wei et al., 2007); it can also be regulated through intracellular pH and acidification (Gerbeau et al., 2002; Maurel et al., 2015; Tournaire-Roux et al., 2003). In addition, H_2O_2 can affect AQP activity through oxidative gating which involves the Fenton reaction (Henzler, Ye, & Steudle, 2004; Ye & Steudle, 2006; Fenton reaction: $Fe^{2+} + H_2O_2 = Fe^{3+} + OH + {}^{*}OH$); H_2O_2 can also cause the internalization of AQP from the plasma membrane (Boursiac et al., 2008). Externally applied H_2O_2 causes root and cell Lp to decrease

(Boursiac et al., 2008; del Martínez-Ballesta, Silva, López-Berenguer, Cabañero, & Carvajal, 2006; Ye & Steudle, 2006). In contrast, externally applied abscisic acid (ABA) increases AQP activity and root and cell Lp (Freundl et al., 2000; Hose, Steudle, & Hartung, 2000).

Many studies have supported the significance of AQPs in the regulation of root water uptake, and its adjustment to stress (e.g., Boursiac et al., 2005; Coffey et al., 2018; Gitto & Fricke, 2018; Knipfer, Danjou, Vionne, & Fricke, 2020; Li, Pan, Melzer, & Fricke, 2020; Meng & Fricke, 2017; Saini & Fricke, 2020; Vandeleur et al., 2009; Zhu, Schraut, Hartung, & Schäffner, 2005; for reviews, see Aroca, Porcel, & Ruiz-Lozano, 2012; Chaumont & Tyerman, 2014; Gambetta, Knipfer, Fricke, & McElrone, 2017; Maurel et al., 2015; Tyerman, Wignes, & Kaiser, 2017; Wang et al., 2016). How much AQPs contribute to water uptake across the root cylinder and into the stele varies between plant species (e.g., Bramley, Turner, Turner, & Tyerman, 2009; Vandeleur et al., 2009). Values can be close to 100% and are generally highest in herbaceous species (Javot & Maurel, 2002). One could argue that the more water and solutes are forced to take a membranous path due to the presence of apoplastic barriers, the more AQPs are likely to contribute to root water uptake. This can be visualized by the so-called composite model of water transport (Fig. 3). This model distinguishes between a purely, or predominantly apoplastic path across the root cylinder and a cell-to-cell path, where water moves from cell to cell by traversing the protoplasm of each cell (Frensch & Steudle, 1989; Knipfer & Fricke, 2010; Steudle, 2000; Steudle & Jeschke, 1983). The cell-to-cell path can involve three different means to get water from one to the next cell. Water can cross the membrane lipid bilayer as such, water can move through AQPs, or water can move through plasmodesmata. How much water moves through AQPs depends on the conductance of the membrane lipid bilayer and plasmodesmata. We do not know how much water flows through plasmodesmata, as we cannot experimentally measure it directly. Studies which use the AQP inhibitor Hg (which also has cytotoxicity) point to AQPs as being the major means through which water moves from cell-to-cell in roots. We should also point out here that although AQPs are important for water transport in roots they have received increasing attention in the study of water transport in leaves too, between leaf xylem and substomatal cavity (Maurel et al., 2015). As some AQP isoforms appear to transport CO_2 (Maurel et al., 2015), AQPs have added an additional means through which the net rate of CO_2 assimilation and WUE can potentially be adjusted.

The best size to characterize the intrinsic water transport (hydraulic) property of a flow path is hydraulic conductivity, Lp (Suku, Knipfer, & Fricke, 2014). The Lp describes how much water is transported per unit time and biophysical driving force across a unit area which is perpendicular to the direction of flow path. In roots of annual grasses/cereals such as barley and wheat, Lp describes the hydraulic property of the radial flow path between peripheral epidermis and central xylem; the axial transport path along the main axis of root, which involves xylem, is generally considered to be not rate-limiting for water uptake (Frensch & Steudle, 1989; for review, see Gambetta et al., 2017). Applied to roots, the surface area equals the root surface area, and the driving force is the difference in water potential and hydrostatic pressure between epidermis and xylem. For a flow path which runs entirely from cell-to-cell, the driving force is the water potential gradient, whereas for a flow path which has a significant apoplastic component, merely the difference in hydrostatic pressure matters, as the apoplast, being non-selective for solutes, does not transmit osmotic forces.

Application of AQP inhibitors generally reduce root Lp (for reviews, see Javot & Maurel, 2002; Knipfer & Fricke, 2014; Maurel et al., 2015). Similarly, experimental treatments such as drought, cold or salinity, which cause a reduction in either gene expression or protein level of AQPs also reduce root Lp (Aroca et al., 2012). These studies are a strong indicator that much of root Lp is regulated through AQP activity and that changes in root Lp in response to experimental treatments involve changes in AQP activity. Maybe an even more interesting observation is that treatments which are applied to plants grown under "optimum" control conditions, hardly ever increase root Lp. It appears, that root Lp and AQP activity are run by default at a maximum of capacity (see also, Ding et al., 2020). We do not know why this is the case. One possible explanation is as follows. In a plant, where most water flows along the cell-to-cell path and through AQPs, any significant increase in AQP activity leads to an increase in root Lp. This in turn allows more water to be taken up per unit existing root surface area, more water can be transpired through open stomata, and more CO_2 can be assimilated. The result is more growth. The alternative to having more functional AQPs would be to either increase the driving force for radial root water uptake or to increase the absorbing root surface area. The former consists mainly of xylem tension in the day. Increasing the driving force and tension could render the tensions so low (negative pressures) that they become limiting for stomatal opening. Increasing the root surface area instead requires the investment of dry matter and energy; it is also an irreversible process. Thus, by having AQPs run by default at a maximum of their capacity, xylem tensions are kept at a minimum (not so negative), as is the resource allocation to roots. There must be a finite amount of AQP protein which can be inserted per unit plasma membrane without impacting negatively on transport processes of other substances, particularly mineral nutrients. Also, AQP protein drains resources. This probably explains why there exist an upper "default" threshold of AQP activity. It also explains why during plant development, root water uptake increases mainly through an increase in root surface area. Changes in Lp (potentially through regulation of AQPs) are only observed early in development (bean, Fiscus & Markhart III, 1979; barley, Suku, Knipfer, & Fricke, 2014).

If the majority of root water uptake occurs in most plant species through AQPs (Gambetta et al., 2017; Vandeleur et al., 2009), then one can predict that AQP activity and root Lp should decrease under conditions where plants transpire less water and take up less water. This is the case during the diurnal cycle, where AQP activity and root Lp are lower during the night compared with day period; it can also be observed under conditions where changes in the soil environment, such as cold, drought and salinity, make it harder for plants to absorb water and where plants transpire less (for reviews, see Aroca et al., 2012; Maurel et al., 2015; Gambetta et al., 2017). Under these conditions, stomata tend to be less open too. The question is: are stomata less open because a reduced AQP activity in the root causes less water to arrive per unit time in the shoot; or do AQPs in the root require less activity because reduced stomatal opening in the shoot results in a reduced rate of water loss and requires less water uptake by the root? It is a bit a chicken-and-egg question, but what is important here is that both, shoot and root have the setup which is required to govern the response of each other. Shoots can regulate the rate of water flow through plants through microscopic pores (stomata), roots can achieve the same through molecular pores (AQPs). People who work on root aquaporins may say in this context "beat that!." Stomatal control and therefore also to some extent WUE can be affected by AQPs in roots.

As was mentioned above, AQP activity and Lp in roots can be expected to impact on the ratio of water molecules and Na^+ and Cl^- ions arriving in the shoot under saline conditions. Close to 100% of the water molecules which arrive in the shoot are lost through transpiration, while CO_2 is taken up through stomatal pores. That means that AQP activity in the root ultimately affects the ratio of CO_2 which has been assimilated by the shoot and of Na^+ and Cl^- which has accumulated in the shoot. Most of the Na^+ and Cl^- requires compartmentalization in the vacuole (Fricke, Leigh, & Tomos, 1996; Leigh & Tomos, 1993). This process drains energy—energy which would otherwise be available for growth and is provided through newly-assimilated carbon. The activity of root AQPs may not impact directly on WUE, but it does have the potential to impact on the efficacy with which the newly assimilated carbon can be used for growth (Fig. 4).

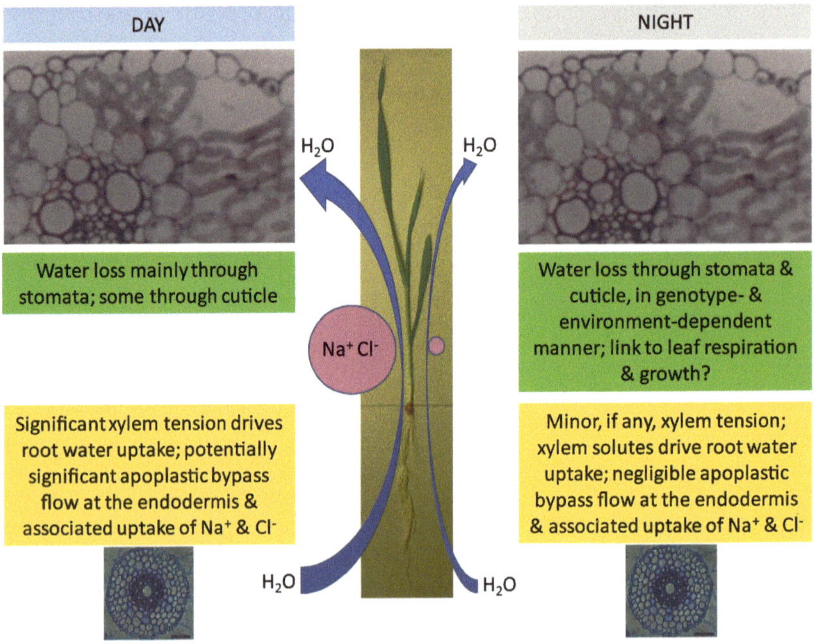

Fig. 4 Transpiration and associated xylem tensions during the day and night. Day-time transpirational water loss occurs mainly (>90%) through stomata and is associated with the build-up of significant tensions in the xylem. These tensions can cause significant bypass flow of solutes along the apoplast of endodermis. Under saline conditions, this could result in significant uptake of Na^+ and Cl^-, due to a poor uncoupling of water and ion flows. Transpirational water loss during the night in an annual cereal such as barley (shown), grown on hydroponics and in a controlled environment, accounts for about 10–15% the rate of day-time water loss. Some of this water loss occurs through the cuticle, some of it through partially open stomata. The relation between the rate of night- and day-time water loss and the contribution of each flow path (cuticle, stomata) to this water loss depends on species, genotype and environment. Water loss in the night is associated with significantly smaller, if any, xylem tensions compared with the day. In plants, such as rice, which have a high intrinsic bypass flow, this could mean that the uncoupling of water from ion uptake is better during the night, and that the ratio of water molecules to Na^+ and Cl^- ions delivered to the shoot is more favorable compared with the day period. While night-time transpirational water loss through partially open stomata does not provide any carbon gain in exchange, it does potentially provide a means through which the loading of shoot with Na^+ and Cl^- parallel to water uptake can be minimized. This will save energy, as less Na^+ and Cl^- need to be compartmentalized.

4. Night-time transpiration and growth

Most of the water which is taken up by plants is lost through transpiration, at least in most plant species and under most conditions (Caird, Richards, & Donovan, 2007). For example, in barley plants grown on hydroponics, more than 98% of the taken-up water by the root system and transported to the shoot is lost through transpiration. The remaining <2% of this water are "stored" in the shoot through leaf cell expansive growth. Salt stress changes the percentage of water stored little during the day (Fricke, 2020). Transpirational water loss aids the movement of mineral nutrients within the transpiration stream from root to shoot; it also aids the cooling of aerial surfaces. Yet, the main reason why plants loose water through transpiration is the inevitable consequence of having to open stomata to assimilate CO_2. One would expect that plants do not transpire significant quantities of water in the night; however, they can do (Caird et al., 2007). The rates of this "night-time" transpirational water loss account typically for 5–15% of day rates and can be as high as 50% and more in some wheat and grapevine genotypes, yet also account for less than 1% of day rates (for reviews, see Caird et al., 2007; Fricke, 2019). Cuticular water loss cannot be "turned off," as is possible through closing of stomatal pores, and some, maybe most of the night-time water loss may occur through the cuticle. However, there are also indications that night-time water loss occurs through partially open stomata (for review, see Fricke, 2019) (Fig. 5).

Plants grow in size during the night too, and depending on the environmental conditions and species, plants may even grow more in the night compared with day (for review, see Walter & Schurr, 2005). Night-time growth must be supported through respiration, which in turn requires O_2 and produces CO_2. It can be concluded from the night-time respiration rates and cuticle conductance to CO_2, that the latter is insufficient to allow the produced CO_2 to escape into the air during the night at sufficiently high rates (Even et al., 2018). Rather, CO_2 would accumulate to very high levels in intercellular air spaces and likely lead to cellular acidosis and death. Partially open stomata offer a solution to this dilemma.

If a night-time water loss is the result of having to facilitate the exit of respiratory CO_2 at sufficiently high rates, and the latter is required to support night-time growth, then night-time transpiration can be viewed not as a detrimental trait of plants which results in the "waste" of water (no carbon gain), but instead in a trait which enables growth in size at a time (night)

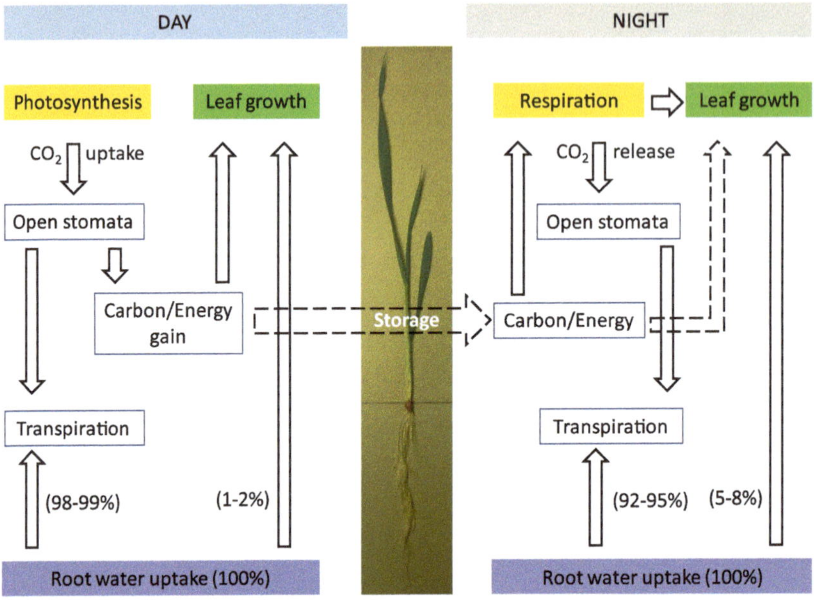

Fig. 5 Water flow through the plant and shoot gas exchange are linked with each other during the day and night and support leaf growth. The scheme and figures are given for an annual plant such as barley, grown in a laboratory environment, on hydroponics. Photosynthesis provides carbon and energy during the day, part of which is invested into growth and part of which is used for the synthesis of storage compounds. More than 98% of the water taken up through the root system and transported to the shoot is lost through transpiration, the remaining 1–2% are "stored" in the plant through leaf cell expansive growth (as the growth ratio of root and shoot of the plant shown is close to 1, another 1–2% of water taken up will be stored through cell expansion in the root). Leaf expansive growth during the night accounts for 50% and more of the growth during the day. This night-time expansive growth is supported through the carbon and energy contained in storage compounds. Some of these compounds need to be respired. This creates CO_2, which needs to escape from the shoot to avoid acidosis of cells. It can be calculated based on respiration rates and cuticle conductance to CO_2, that stomata need to be partially open to enable the respiratory CO_2 to escape at sufficiently high rates during the night (Even et al., 2018). If so, night-time transpiration through stomata is the direct consequence of night-time leaf growth. Contrary to the day, a much higher portion, ca. 5–8%, of the water taken up and transported to the shoot is retained through leaf expansive growth. The taken-up water is used more "efficiently" to increase the water content of plant.

when soil water potential is less negative and renders water uptake easier. This could apply the more, the more challenging the soil environment is during the day for the uptake of water, such as during drought and salinity.

Night-time transpiration is certainly inferior to day-time transpiration in terms of the carbon balance, as CO_2 is net lost not gained parallel to water loss. However, one can also look at the amount of water taken up by a plant and ask: how many percent of the taken-up water is retained in the plant during day and night through leaf cell expansive growth and organ growth in size (Fricke, 2020)?

Leaves show a diurnal pattern in expansive growth (Poiré et al., 2010; Rozema, Arp, van Diggelen, Kok, & Letschert, 1987; Waldron, Terry, & Nemson, 1985). Expansion rates in the night can not only match but even succeed rates during the day. This applies particularly to dicotyledonous plants (Poiré et al., 2010; Rozema et al., 1987; Ruts, Matsubara, Wiese-Klinkenberg, & Walter, 2012). The ratio of leaf expansion rate during day and night responds to changes in the environment, such as air temperature and soil water availability, yet also changes with leaf developmental state. Leaf expansion during the night consumes carbon storage reserves, it also produces leaf surface area and the capacity to assimilate more carbon through photosynthesis during the subsequent day. Night-time leaf expansion provides a more efficient use of taken-up water. A plant which transpires water at night at say ca. 10% the day-rate (Caird et al., 2007; Even et al., 2018), and where leaves expand at 50–100% the day-rate, stores between 5 times (0.5/0.10) and 10 times (1/0.10) as much of the taken-up water through leaf expansive growth during the night compared with day (see also Fig. 6). it should also apply to expanding leaves of dicotyledonous plants, as these show a shift from a metabolic to hydraulic limitation of growth during development, which is accompanied by a several fold increase in the night-to-daytime ratio of leaf expansive growth (Tardieu, Parent, & Simonneau, 2010).

The night-time transpiration could provide a stress acclimation process, so that leaf expansive growth can occur increasingly during that portion (night) of 24 h-diurnal cycle, where plants encounter less shortage in water. Other mechanisms operate as well. Several studies suggest that a reduction in night-time transpirational water loss aids the restoration of plant water status during the night, which in turn enables more growth during the following day under water-deficit conditions (Caldeira, Jeanguenin, Chaumont, & Tardieu, 2014; Coupel-Ledru et al., 2016). Studies on trees have shown that sap flow during the night can be used almost entirely for the restoration of plant water status and the avoidance of xylem cavitation in some species, whereas in other species sap flow during the night supports significant (>30% of taken up water) rates of night-time transpiration (Coupel-Ledru

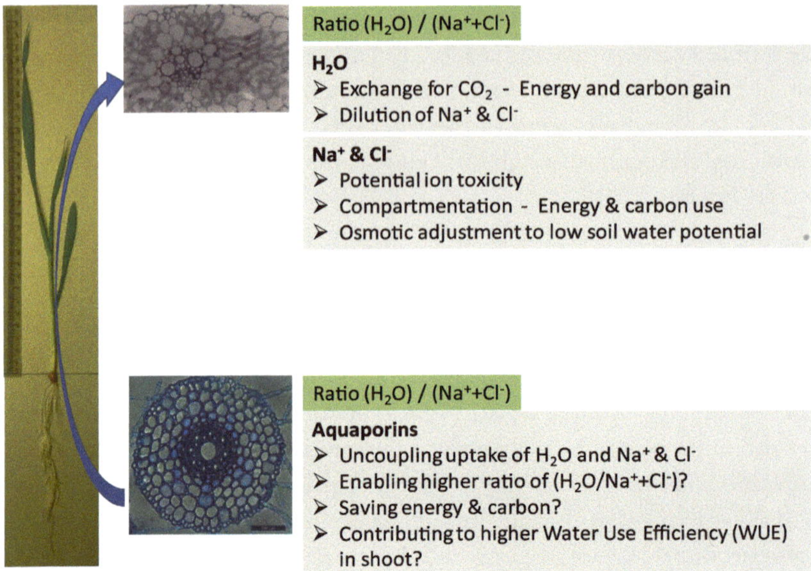

Fig. 6 Root aquaporins can save energy under salinity. Water use efficiency (WUE), the amount of carbon fixed per unit water molecules transpired, has received much attention when studying stress tolerance mechanisms in plants, particular to stresses which impact directly on the availability of water in the root environment (drought, salinity). The efficiency through which water is taken up by the root system in terms of discriminating against potentially harmful substances such as high Na^+ and Cl^- during salinity, needs to be considered too. Let's take two extreme scenarios. In scenario 1, all substances move across the root cylinder by mass flow along a low-resistance and non-selective apoplast. This allows fast uptake of water, yet it also results in a fixed ratio between water and ions (here Na^+, Cl^-) being delivered to the shoot. The water which is delivered allows the opening of stomata in exchange for CO_2 assimilation, so effectively allows the fixation of energy and carbon. The Na^+ and Cl^- which accumulate in the shoot need to be compartmentalized, this costs energy and carbon. At the same time, these ions aid the osmotic adjustment and maintenance of turgor in leaf cells in a low soil water potential environment. In scenario 2, all substances have to cross membranes and pass substance-specific transporter/channels molecules, such as aquaporins (AQPs) in the case of water. The provision of the transporter/channel molecules and their regulation costs energy and carbon, yet it does enable a more selective delivery of water toward the shoot. The ratio of the molecule (water) which indirectly enables the fixation of carbon and energy to the molecules (Na^+, Cl^-) which consume carbon and energy in the shoot due to their compartmentation increases. Having water flowing mainly through AQPs in roots and sealing off the apoplast to minimize bypass flow of Na^+ and Cl^- at the endodermis, should save energy, particularly under salinity.

et al., 2016; Daley & Phillips, 2006; Wang et al., 2012). The extent to which night-time transpiration can provide a stress acclimation process depends on the susceptibility of plants to xylem cavitation, the capacity to store water and have leaves expanding during the night. Grapevine (*Vitis vinifera* L.) genotypes showed a negative correlation between growth and the rate of night-time transpiration under well-watered and water-deficit conditions. This was possibly because the decrease in night-time transpiration favored a replenishment of plant water content and reduced the risk of embolism in xylem, both aiding growth during the subsequent day period (Coupel-Ledru et al., 2016). In contrast, paper birch (*Betula papyrifera* March.), which showed the largest night-time transpiration rate of three tree species studied, also showed the highest growth rate. In this case, growth may have been prioritized over water use, in support of paper birch being an early successional species and shade intolerant (Daley & Phillips, 2006).

Research on *Arabidopsis* points to a tight link between the amount of storage carbohydrate which is available to support growth and the actual growth rates during the night. This link even applies under conditions of sudden changes in period length, which are imposed experimentally (Graf & Smith, 2011; Smith & Stitt, 2007; Sulpice et al., 2014). One could argue that the "natural" equivalent would be the random occurrence and frequency of cloudy and sunny periods during a day. The ability of plants to somehow "measure" how much storage carbohydrates are available to support night-time growth assures that carbon and energy reserves are not drained prematurely during the night, as the latter could lead to the starvation and damage of cells toward the end of dark period (Graf, Schlereth, Stitt, & Smith, 2010; Graf & Smith, 2011; Mora-García, de Leone, & Yanovsky, 2017; Scialdone & Howard, 2015; Webb & Satake, 2015). Similarly, experimentally induced changes in the rate of carbon assimilation and sugar production during the day feed forward on the growth rate during the subsequent night period (Sulpice et al., 2014; Wiese, Christ, Virnich, Schurr, & Walter, 2007).

There exist interspecific differences between plants in the relation between the supply and use of storage carbon and diurnal growth rates (Graf & Smith, 2011; Habte, Müller, Shtaya, Davis, & von Korff, 2014; Müller, von Korff, & Davis, 2014), and the developmental status also affects the day-to-night ratio of leaf expansion. Leaves of *Arabidopsis* at their very early developmental stage are metabolically limited in area expansion, as they rely on the import of sugars from older leaves (Pantin, Simonneau, Rolland, Dauzat, & Muller, 2011). As the leaves grow larger and increase their own photosynthetic output, and transpiring leaf area, leaf expansion is increasingly limited through hydraulic means and reduction in cell turgor. At the same time, the ratio between night-time and day-time leaf expansive growth increases several-fold (Pantin et al., 2011). The availability of water and the developmental stage of leaves affect the amplitude of growth; the internal clock affects the periodicity of growth.

5. Night-time transpiration and energy

It is tempting to look at night-time transpiration as a "waste" of water, as plants loose water into the air without increasing their carbon budget through photosynthesis. However, as mentioned above, experiments on barley show that night-time leaf expansion provides a more efficient use of water which is taken up as part of the night-time transpiration stream. Night-time transpiration is associated with smaller xylem tensions, and this will reduce any apoplastic bypass leakage of Na^+ and Cl^- into the stele of roots. This should render roots more selective in their transport properties and enable roots to be more discriminate against Na^+ and Cl^- under saline conditions. It would be great to have data on any diurnal changes of the ratio of Na^+, Cl^- and water molecules arriving in the shoot. Also more data are needed on the diurnal usage of taken-up water in support of leaf growth in a range of species which differ in their ecophysiology and mode of photosynthesis (C_3, C_4, CAM). Furthermore, we have considered here merely water use for leaf growth. What about water use for root growth? This question is more difficult to address, simply because root compared with leaf growth is far more difficult to quantify within single day- and night-periods (Walter & Schurr, 2005).

The relation between the rate of day- and night-time transpirational water loss changes little in barley plants when they are exposed to high external salt concentration or to an osmotic up- and downshock through sudden addition and removal of salt (Even et al., 2018). Neither does the relation change between the net rate of day-time photosynthesis and night-time respiration. These data could be explained through a tight link between the day-time synthesis and the night-time degradation of storage carbohydrates (Graf & Smith, 2011; Smith & Stitt, 2007). At any given photosynthetically active radiation and vapor pressure difference, a higher rate of day-time transpiration reflects a higher rate of photosynthesis, and vice versa. More photosynthesis means that more sugars being produced and more sugars being available for the synthesis of storage carbohydrates. A higher carbon storage during the day facilitates a higher carbon respiration in the night. Considering that night-time transpiration may serve the release of respiratory CO_2, a higher net rate of photosynthesis and carbon assimilation during the day will ultimately lead to a higher night-time transpirational water loss. This could explain why the ratio between day- and night-time transpirational water loss for a given species and genotype is rather invariant, and while a stress such as salinity has little effect on it either. As salinity impacts on the net rate of carbon assimilation during the day, it indirectly also impacts on the respiration of some of this carbohydrate and transpirational water loss during the night. An implicit assumption is that the allocation of newly assimilated carbon in the day toward the synthesis of storage carbohydrates for use in the night is tightly regulated, also under conditions of, e.g., salt stress.

The existence of a tight link between the production of carbohydrate during the day and allocation for consumption during the night is supported through work on the diurnal coordination of the breakdown of carbon storage reserves and growth (for reviews, see Graf & Smith, 2011; Graf et al., 2010; Webb & Satake, 2015; Scialdone & Howard, 2015; Mora-García et al., 2017). *Arabidopsis*, which stores mainly starch, rapidly adjusts the consumption of stored carbon during the night to any changes in the production of storage-carbon during the preceding day. Dark respiration and leaf growth are adjusted to the availability of resources. Pasture grasses such as *Lolium perenne* and *Festuca rubra*, but also cereals such as barley, accumulate a large portion of stored carbohydrate as fructans. They too show a tight regulation between the day-time synthesis and night-time breakdown of carbon (Roth, Lüscher, Sprenger, Boller, & Wiemken, 1997; Schnyder, Ostler, & Lehmeier, 2017; Sicher, Kremer, & Harris, 1984). An explanation alternative to that that carbon sources regulate the diurnal growth of leaves is that control could be due to hormonal signals. Munns, Guo, Passioura, and Cramer (2000) observed for barley plants that a reduction in leaf elongation growth during the night was not caused by a depletion of assimilates.

A previous evaluation of the energetic and solute transport processes associated with night-time transpiration in barley under conditions of salt stress concluded that forces in addition to osmotic ones must contribute to night-time water uptake by roots. The most likely alternative force is tension in the xylem, lowering xylem water potential. If so, night-time transpirational water flow, is similar to day-time transpiration, driven by a vapor pressure difference between leaf internal air spaces and ambient air. It is also possible that a water-ion cotransport mechanism, similar to the one described originally for epithelia in animals (Zeuthen, 1994, 2010), contribute to water uptake into the xylem during the night in salt-stressed plants. Alternative mechanisms which have been proposed to explain water uptake by roots against an osmotic gradient and in the absence of any significant xylem tensions may also operate (for reviews, see Wegner, 2014, 2015; Fricke, 2015, 2017).

The acclimation of glycophytic plants to salt stress is associated with the ability to control the delivery of Na^+ and Cl^- to shoot tissues and also the compartmentation of these ions between and within shoot tissues. This applies irrespective whether plants adapt mainly through a salt-exclusion mechanism at the level of root and leaf sheath (e.g., wheat; Takahashi et al., 2015) or whether the main means to cope with salt involves the compartmentation of Na^+ and Cl^- between leaf tissues (e.g., barley, Leigh & Storey, 1993; Fricke et al., 1996). A bypass flow along the apoplast, which results as consequence of an incomplete "seal" at the endodermis, causes "leakage" of Na^+ and Cl^- into the root stele. Having less bypass flow should make it easier to keep the load of root xylem and leaf with Na^+ and Cl^- low. Bypass flow is driven by xylem tension. These tensions result from differences in hydrostatic pressure between root medium (higher) and xylem (lower hydrostatic pressure) (Faiyue, Al-Azzawi, & Flowers, 2010; Steudle, 2000; Steudle & Peterson, 1998). Smaller xylem tensions, as they will occur during the night compared with day should therefore favor less leakage of Na^+ and Cl^- into the stele, and enable a more selective uptake of water through membrane-associated processes (e.g., AQPs). If so, night-time transpiration and associated leaf growth under salt stress render the root more selective. This should reduce the metabolic costs to deal with any excess Na^+ and Cl^- in leaf tissues, and the benefit should be the larger, the higher bypass flow is in a species under day-time transpiring conditions. For example, barley has an inherently low bypass flow (Knipfer, Besse, Verdeil, & Fricke, 2011; Knipfer & Fricke, 2010; Steudle & Jeschke, 1983), whereas species such as maize and rice show a much higher bypass flow (e.g., Meng & Fricke, 2017; for reviews, see Gambetta et al., 2017; Steudle, 2000).

6. Outlook: Night-time transpiration in a changing climate

The climate is changing on planet Earth. The changes are globally and locally and occur on a palaeontological rapid scale. The clear trend is toward a warmer average climate, more frequent extreme weather events and more long-lasting weather constellations. Also, the seasonality of extreme weather events such as storms seems to change. So, how would these changes in climate impact on plant transpiration during day & night (Fig. 7)?

Predictions are that the average night-time temperature increases more than the average day-time temperature increases in most areas (Cox, Maclean, Gardner, & Gaston, 2020). This will have a profound effect on the availability of water, as the prediction for these areas is that an increased cloud cover during the day leads to a higher relative humidity and water availability. Areas, which experience a preferential warming during the day compared with night, will have contrary changes, leading to a reduced supply of water (Cox et al., 2020). Therefore, depending on the relative increases in temperature during the day and night period, one can predict different effects on the relation between day- and night-time transpirational water loss and growth. If we consider an area where temperatures are predicted to increase more during the night compared with day, we should obtain a higher percentage of growth achieved during a 24 h cycle in the night—provided that the provision of storage carbohydrates during the day period does not become the rate-limiting factor. If we consider an area where temperatures are predicted to increase more during the day compared with night, then we can predict a lower availability of water during the day, which may offset the stimulating effect of temperature on leaf growth and result in minor changes in leaf growth, or may even decrease leaf growth due to a hydraulic limitation. This could result in a partial shift to more leaf growth during the night, as growth is then not any more limited hydraulically. In either of the two cases, it is possible that leaf growth during the night will increase more compared with leaf growth during the day (Fig. 7).

The key greenhouse gas is CO_2. Atmospheric CO_2 levels have increased from about 350 ppm a few decades ago to a current around 430 ppm. As concerns ambient CO_2 and night-time transpiration, one can hypothesize that larger ambient CO_2 concentration in air should make it more difficult for respiratory CO_2 to escape through stomata during the night. This either leads to the production of respiratory CO_2 being downregulated, and leaf growth, being fuelled by respiration, to decrease in the night; or stomatal conductance increases, leading to an increase in night-time transpirational water loss. During the day, increases in ambient CO_2 levels potentially lead to increased leaf and plant growth, provided other factors which limit growth, particularly mineral nutrients and water, are in ample supply. If these factors are limiting growth, particularly water, increased leaf growth in size during the (more humid) night, may be an adaptive mechanism to

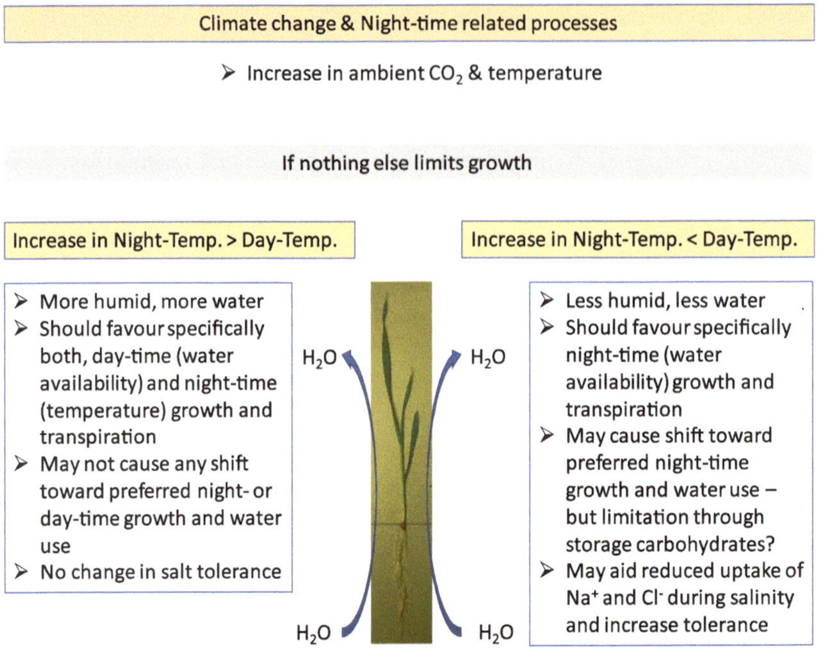

Fig. 7 Night-time processes in a changing climate. Ambient CO_2 and temperature increase as result of climate change. How these changes affect the relation between day- and night-time related processes is difficult to predict. For simplicity, we consider here a situation where the overall water availability and mineral nutrient supply do not limit plant growth. We still consider possible differences in the water supply between the day and night period. Also, the latest prediction is that most areas of our planet will experience a larger increase in night- compared with day-time temperature, which is associated with a larger availability of water, while the remaining areas will experience the opposite trend (see left- and right-hand column of figure). In principle, an increase in temperature should increase growth, as should an increase in ambient CO_2. If night-time temperature increases more compared with day-time temperature, we may see an overall increase in plant growth and no change in the tolerance of plants to salt stress. Under the contrary scenario, a lower availability of water may specifically favor night-time growth (which is hydraulically less limited) and could lead, through reduced apoplastic tension associated with root water uptake during the night, to a reduced uptake of Na^+ and Cl^- and an increased tolerance to salt stress. However, there are other factors which come into play, particularly how any change in day-time growth affects the provision of storage carbohydrates for night-time growth, and how the altered cloud cover which is associated with the two scenarios affects productivity. Finally, a higher ambient CO_2 may make it more difficult for respiratory CO_2 to escape from leaves during the night and lead to an increased stomatal aperture and night-time water loss, irrespective of relative changes in temperature during day and night.

make up for any reduced growth during the day. Again, this will depend on the supply of carbohydrates provided through day-time photosynthesis (Fig. 7).

Also, plants can adopt a rather isohydric or anisohydric regulation of their shoot water relations (Tardieu & Simonneau, 1998), that is trying to keep leaf water potential within a narrow range at the expense of stomatal conductance or running a more "risky" approach by allowing leaf water potential to reach more negative values to the benefit of stomatal conductance. This behavior is likely to impact on how climate change affects the relation between day- and night-time growth and transpiration.

We do not know how the above is impacted on by salinity. If plants can be more selective for water compared with Na^+ and Cl^- during uptake by roots in the night compared with day, one would predict that a higher portion of transpiration and leaf growth in the night supports better adaptation to a saline environment. We need data about such a selective uptake during night compared with day and in dependence of relative changes in night- and day-time temperatures.

Acknowledgments

The author would like to thank all former and present team members for their contribution to the research which was cited in this chapter.

References

Aroca, R., Porcel, R., & Ruiz-Lozano, J. M. (2012). Regulation of root water uptake under abiotic stress conditions. *Journal of Experimental Botany*, *63*, 43–57.

Barberon, M. (2017). The endodermis as a checkpoint for nutrients. *New Phytologist*, *213*, 1604–1610.

Barberon, M., Vermeer, J. E. M., De Bellis, D., Wang, P., Naseer, S., Andersen, T. G., et al. (2016). Adaptation of root function by nutrient-induced plasticity of endodermal differentiation. *Cell*, *164*, 447–459.

Boursiac, Y., Boudet, J., Postaire, O., Luu, D. T., Tournaire-Roux, C., & Maurel, C. (2008). Stimulus-induced downregulation of root water transport involves reactive oxygen species-activated cell signalling and plasma membrane intrinsic protein internalization. *Plant Journal*, *56*, 207–218.

Boursiac, Y., Chen, S., Luu, D. T., Sorieul, M., van den Dries, N., & Maurel, C. (2005). Early effects of salinity on water transport in Arabidopsis roots. Molecular and cellular features of aquaporin expression. *Plant Physiology*, *139*, 790–805.

Bramley, H., Turner, N. C., Turner, D. W., & Tyerman, S. D. (2009). Roles of morphology, anatomy, and aquaporins in determining contrasting hydraulic behavior of roots. *Plant Physiology*, *150*, 348–364.

Caird, M. A., Richards, J. H., & Donovan, L. A. (2007). Nighttime stomatal conductance and transpiration in C_3 and C_4 plants. *Plant Physiology*, *143*, 4–10.

Caldeira, C., Jeanguenin, L., Chaumont, F., & Tardieu, F. (2014). Circadian rhythms of hydraulic conductance and growth are enhanced by drought and improve plant performance. *Nature Communications*, *5*, 5365.
Chaumont, F., & Tyerman, S. D. (2014). Aquaporins: Highly regulated channels controlling plant water relations. *Plant Physiology*, *164*, 1600–1618.
Coffey, O., Bonfield, R., Corre, F., Sirigiri, J. A., Meng, D., & Fricke, W. (2018). Root and cell hydraulic conductivity, apoplastic barriers and aquaporin gene expression in barley (*Hordeum vulgare* L.) grown at low supply of potassium. *Annals of Botany*, *122*, 1131–1141.
Coupel-Ledru, A., Lebon, E., Christophe, A., Gallo, A., Gago, P., Pantin, F., et al. (2016). Reduced nighttime transpiration is a relevant breeding target for high water-use efficiency in grapevine. *Proceedings of the National Academy of Sciences of the United States of America*, *113*, 8963–8968.
Cox, D. T. C., Maclean, I. M. D., Gardner, A. S., & Gaston, K. J. (2020). Global variation in diurnal asymmetry in temperature, cloud cover, specific humidity and precipitation and its association with leaf area index. *Global Change Biology*, *26*, 7099–7111.
Daley, M. J., & Phillips, N. G. (2006). Interspecific variation in nighttime transpiration and stomatal conductance in a mixed New England deciduous forest. *Tree Physiology*, *26*, 411–419.
del Martínez-Ballesta, M. C., Silva, C., López-Berenguer, C., Cabañero, F. J., & Carvajal, M. (2006). Plant aquaporins: New perspectives on water and nutrient uptake in saline environment. *Plant Biology*, *8*, 535–546.
Ding, L., Milhiet, T., Couvreur, V., Nelissen, H., Meziane, A., Parent, B., et al. (2020). Modification of the expression of the aquaporin ZmPIP2;5 affects water relations and plant growth. *Plant Physiology*, *182*, 2154–2165.
Enstone, D. E., Peterson, C. A., & Ma, F. (2003). Root endodermis and exodermis: Structure, function, and responses to the environment. *Journal of Plant Growth Regulation*, *21*, 335–351.
Even, M., Sabo, M., Meng, D., Kreszies, T., Schreiber, L., & Fricke, W. (2018). Nighttime transpiration in barley (*Hordeum vulgare* L.) facilitates respiratory carbon dioxide release and is regulated during salt stress. *Annals of Botany*, *122*, 569–582.
Faiyue, B., Al-Azzawi, M. J., & Flowers, T. J. (2010). The role of lateral roots in bypass flow in rice (*Oryza sativa* L.). *Plant, Cell and Environment*, *33*, 702–716.
Fiscus, E. L., & Markhart, A. H., III. (1979). Relationships between root system water transport properties and plant size in *Phaseolus*. *Plant Physiology*, *64*, 770–773.
Frensch, J., & Steudle, E. (1989). Axial and radial hydraulic resistance to roots of maize (*Zea mays* L.). *Plant Physiology*, *91*, 719–726.
Freundl, E., Steudle, E., & Hartung, W. (2000). Apoplastic transport of abscisic acid through roots of maize: Effect of the exodermis. *Planta*, *210*, 222–231.
Fricke, W. (2015). The significance of water cotransport for sustaining transpirational water flow in plants – A quantitative approach. *Journal of Experimental Botany*, *66*, 731–739.
Fricke, W. (2017). Water transport and energy. *Plant, Cell & Environment*, *40*, 977–994.
Fricke, W. (2019). Night-time transpiration – Favouring growth? *Trends in Plant Science*, *24*, 311–317.
Fricke, W. (2020). Energy costs of salinity tolerance in crop plants: Night-time transpiration and growth. *New Phytologist*, *225*, 1152–1165.
Fricke, W., Leigh, R. A., & Tomos, A. D. (1996). The intercellular distribution of vacuolar solutes in the epidermis and mesophyll of barley leaves changes in response to NaCl. *Journal of Experimental Botany*, *47*, 1413–1426.
Gambetta, G., Knipfer, T., Fricke, W., & McElrone, A. J. (2017). In F. Plant Aquaporins, & S. T. Chaumont (Eds.), *Aquaporins and root water uptake* (pp. 133–153). Berlin Heidelberg: Springer Verlag. https://doi.org/10.1007/978-3-319-49395-4_6.

Geldner, N. (2013). The endodermis. *Annual Review of Plant Biology, 64*, 531–558.
Gerbeau, P., Amodeo, G., Henzler, T., Santoni, V., Ripoche, P., & Maurel, C. (2002). The water permeability of Arabidopsis plasma membrane is regulated by divalent cations and pH. *Plant Journal, 30*, 71–81.
Gitto, A., & Fricke, W. (2018). Zinc treatment of hydroponically-grown barley (*H. vulgare*) plants causes a reduction in root and cell hydraulic conductivity and isoform-dependent decrease in aquaporin gene expression. *Physiologia Plantarum, 164*, 176–190.
Graf, A., Schlereth, A., Stitt, M., & Smith, A. M. (2010). Circadian control of carbohydrate availability for growth in *Arabidopsis* plants at night. *Proceedings of the National Academy of Sciences of the United States of America, 107*, 9458–9463.
Graf, A., & Smith, A. M. (2011). Starch and the clock: The dark side of plant productivity. *Trends in Plant Science, 16*, 169–175.
Guenther, J. F., Chanmanivone, N., Galetovic, M. P., Wallace, I. S., Cobb, J. A., & Roberts, D. M. (2003). Phosphorylation of soybean nodulin 26 on serine 262 enhances water permeability and is regulated developmentally and by osmotic signals. *Plant Cell, 15*, 981–991.
Habte, E., Müller, L. M., Shtaya, M., Davis, S. J., & von Korff, M. (2014). Osmotic stress at the barley root affects expression of circadian clock genes in the shoot. *Plant, Cell & Environment, 37*, 1321–1337.
Henzler, T., Ye, Q., & Steudle, E. (2004). Oxidative gating of water channels (aquaporins) in Chara by hydroxyl radicals. *Plant, Cell & Environment, 27*, 1184–1195.
Hose, E., Clarkson, D. T., Steudle, E., Schreiber, L., & Hartung, W. (2001). The exodermis: A variable apoplastic barrier. *Journal of Experimental Botany, 52*, 2245–2264.
Hose, E., Steudle, E., & Hartung, W. (2000). Abscisic acid and hydraulic conductivity of maize roots: A study using cell- and root-pressure probes. *Planta, 211*, 874–882.
Javot, H., & Maurel, C. (2002). The role of aquaporins in root water uptake. *Annals of Botany, 90*, 301–313.
Johansson, I., Larsson, C., Ek, B., & Kjellbom, P. (1996). The major integral proteins of spinach leaf plasma membranes are putative aquaporins and are phosphorylated in response to Ca2+ and apoplastic water potential. *Plant Cell, 8*, 1181–1191.
Knipfer, T., Besse, M., Verdeil, J.-L., & Fricke, W. (2011). Aquaporin-facilitated water uptake in barley (*Hordeum vulgare* L.) roots. *Journal of Experimental Botany, 62*, 4115–4126.
Knipfer, T., Danjou, M., Vionne, C., & Fricke, W. (2020). Salt stress reduces root water uptake in barley (*Hordeum vulgare* L.) through modification of the transcellular transport path. *Plant, Cell & Environment, 44*, 458–475.
Knipfer, T., & Fricke, W. (2010). Root pressure and a solute reflection coefficient close to unity exclude a purely apoplastic pathway of radial water transport in barley (*Hordeum vulgare* L.). *New Phytologist, 187*, 159–170.
Knipfer, T., & Fricke, W. (2014). Root aquaporins. Chapter 13. In A. Morte, & A. Varma (Eds.), *Root engineering, soil biology 40*. Berlin Heidelberg: Springer Verlag.
Leigh, R. A., & Storey, R. (1993). Intercellular compartmentation of ions in barley leaves in relation to potassium nutrition and salinity. *Journal of Experimental Botany, 44*, 755–762.
Leigh, R. A., & Tomos, A. D. (1993). Ion distribution in cereal leaves: Pathways and mechanisms. *Philosophical Transactions of the Royal Society of London B, 341*, 75–86.
Li, L., Pan, S., Melzer, R., & Fricke, W. (2020). Apoplastic barriers, aquaporin gene expression and root and cell hydraulic conductivity in phosphate-limited sheepgrass plants. *Physiologia Plantarum, 168*, 118–132.
Maurel, C., Boursiac, Y., Luu, D. T., Santoni, V., Shahzad, Z., & Verdoucq, L. (2015). Aquaporins in plants. *Physiological Reviews, 95*, 1321–1358.
Maurel, C., Verdoucq, L., Luu, D. T., & Santoni, V. (2008). Plant aquaporins: Membrane channels with multiple integrated functions. *Annual Review of Plant Biology, 59*, 595–624.

Meng, D., & Fricke, W. (2017). Changes in root hydraulic conductivity facilitate the overall hydraulic response of rice (*Oryza sativa* L.) cultivars to salt and osmotic stress. *Plant Physiology and Biochemistry, 113*, 64–77.
Mora-García, S., de Leone, M. J., & Yanovsky, M. (2017). Time to grow: Circadian regulation of growth and metabolism in photosynthetic organisms. *Current Opinion in Plant Biology, 35*, 84–90.
Müller, L. M., von Korff, M., & Davis, S. J. (2014). Connections between circadian clocks and carbon metabolism reveal species-specific effects on growth control. *Journal of Experimental Botany, 65*, 2915–2923.
Munns, R. (1985). Na^+, K^+ and cl^- in xylem sap flowing to shoots of NaCl-treated barley. *Journal of Experimental Botany, 36*, 1032–1042.
Munns, R., Day, D. A., Watt, M., Arsova, B., Barkla, B. J., Bose, J., et al. (2020). Tansley review: Energy costs of salt tolerance in crop plants. *New Phytologist, 225*, 1072–1090.
Munns, R., Guo, J., Passioura, J. B., & Cramer, G. R. (2000). Leaf water status controls day-time but not daily rates of leaf expansion in salt-treated barley. *Australian Journal of Plant Physiology, 27*, 949–957.
Munns, R., & Passioura, J. B. (1984a). Effect of prolonged exposure to NaCl on the osmotic pressure of leaf xylem sap from intact, transpiring barley plants. *Australian Journal of Plant Physiology, 11*, 497–507.
Munns, R., & Passioura, J. B. (1984b). Hydraulic resistance of plants. III. Effects of NaCl in barley and lupin. *Australian Journal of Plant Physiology, 11*, 351–359.
Pantin, F., Simonneau, T., Rolland, G., Dauzat, M., & Muller, B. (2011). Control of leaf expansion: A developmental switch from metabolics to hydraulics. *Plant Physiology, 156*, 803–815.
Poiré, R., Wiese-Klinkenberg, A., Parent, B., Mielewczik, M., Schurr, U., Tardieu, F., et al. (2010). Diel time-courses of leaf growth in monocot and dicot species: Endogenous rhythms and temperature effects. *Journal of Experimental Botany, 61*, 1751–1759.
Roth, A., Lüscher, M., Sprenger, N., Boller, T., & Wiemken, A. (1997). Fructan and fructan-metabolizing enzymes in the growth zone of barley leaves. *New Phytologist, 136*, 73–79.
Rozema, J., Arp, W., van Diggelen, J., Kok, E., & Letschert, J. (1987). An ecophysiological comparison of measurements of the diurnal rhythm of the leaf elongation and changes of the leaf thickness of salt-resistant dicotyledonae and monocotyledonae. *Journal of Experimental Botany, 38*, 442–453.
Ruts, T., Matsubara, S., Wiese-Klinkenberg, A., & Walter, A. (2012). Diel patterns of leaf and root growth: Endogenous rhythmicity or environmental response? *Journal of Experimental Botany, 63*, 3339–3351.
Saini, G., & Fricke, W. (2020). Photosynthetically active radiation impacts significantly on root and cell hydraulics in barley (*Hordeum vulgare* L.). *Physiologia Plantarum, 170*, 357–372.
Schnyder, H., Ostler, U., & Lehmeier, C. A. (2017). Respiratory turn-over and metabolic compartments: From the design of tracer experiments to the characterization of respiratory substrate-supply systems. In G. Tcherkez, & J. Ghashghaie (Eds.), *Vol. 43. Plant respiration: Metabolic fluxes and carbon balance. Advances in photosynthesis and respiration (including bioenergy and related processes)*. Cham: Springer.
Schreiber, L., & Franke, R. B. (2011). *Endodermis and exodermis in roots*. Chichester: John Wiley & Sons.
Scialdone, A., & Howard, M. (2015). How plants manage food reserves at night: Quantitative models and open questions. *Frontiers in Plant Science, 6*, 204. https://doi.org/10.3389/fpls.2015.00204.
Sicher, R. C., Kremer, D. F., & Harris, W. G. (1984). Diurnal carbohydrate metabolism of barley primary leaves. *Plant Physiology, 76*, 165–169.

Smith, A. M., & Stitt, M. (2007). Coordination of carbon supply and plant growth. *Plant, Cell & Environment*, *30*, 1126–1149.
Steudle, E. (2000). Water uptake by plant roots: An integration of views. *Plant and Soil*, *226*, 46–56.
Steudle, E., & Jeschke, W. D. (1983). Water transport in barley roots. *Planta*, *158*, 237–248.
Steudle, E., & Peterson, C. A. (1998). How does water get through roots? *Journal of Experimental Botany*, *49*, 775–788.
Suku, S., Knipfer, T., & Fricke, W. (2014). Do root hydraulic properties change during the early vegetative stage of plant development in barley (*Hordeum vulgare*)? *Annals of Botany*, *113*, 385–402.
Sulpice, R., Flis, A., Ivakov, A. A., Apelt, F., Krohn, N., Encke, B., et al. (2014). *Arabidopsis* coordinates the diurnal regulation of carbon allocation and growth across a wide range of photoperiods. *Molecular Plant*, *7*, 137–155.
Takahashi, F., Tilbrook, J., Trittermann, C., Berger, B., Roy, S. J., Seki, M., et al. (2015). Comparison of leaf sheath transcriptome profiles with physiological traits of bread wheat cultivars under salinity stress. *PLoS One*, *10*, e0133322. https://doi.org/10.1371/journal.pone.0133322.
Tardieu, F., Parent, B., & Simonneau, T. (2010). Control of leaf growth by abscisic acid: Hydraulic or non-hydraulic processes? *Plant, Cell & Environment*, *33*, 636–647.
Tardieu, S., & Simonneau, T. (1998). Variability among species of stomatal control under fluctuating soil water status and evaporative demand: Modelling isohydric and anisohydric behaviours. *Journal of Experimental Botany*, *49*, 419–432.
Törnroth-Horsefield, S., Wang, Y., Hedfalk, K., Johanson, U., Karlsson, M., Tajkhorshid, E., et al. (2006). Structural mechanism of plant aquaporin gating. *Nature*, *439*, 688–694.
Tournaire-Roux, C., Sutka, M., Javot, H., Gout, E., Gerbeau, P., Luu, D. T., et al. (2003). Cytosolic pH regulates root water transport during anoxic stress through gating of aquaporins. *Nature*, *425*, 393–397.
Tyerman, S. D., Munns, R., Fricke, W., Arsova, B., Barkla, B. J., Bose, J., et al. (2020). Energy costs of salinity tolerance in crop plants. *New Phytologist*, *221*, 25–29.
Tyerman, S. D., Wignes, J. A., & Kaiser, B. N. (2017). Root hydraulic and aquaporin responses to N availability. In F. Chaumont, & S. Tyerman (Eds.), *Plant Aquaporins*. Springer, Cham: Signaling and Communication in Plants.
Vandeleur, R. K., Mayo, G., Shelden, M. C., Gilliham, M., Kaiser, B. N., & Tyerman, S. D. (2009). The role of plasma membrane intrinsic protein aquaporins in water transport through roots: Diurnal and drought stress responses reveal different strategies between isohydric and anisohydric cultivars of grapevine. *Plant Physiology*, *149*, 445–460.
Waldron, L. J., Terry, N., & Nemson, J. A. (1985). Diurnal cycles of leaf extension in unsalinized and salinized *Beta vulgaris*. *Plant, Cell & Environment*, *8*, 207–211.
Walter, A., & Schurr, U. (2005). Dynamics of leaf and root growth: Endogenous control versus environmental impact. *Annals of Botany*, *95*, 891–900.
Wang, M., Ding, L., Gao, L., Li, Y., Shen, Q., & Guo, S. (2016). The interactions of aquaporins and mineral nutrients in higher plants. *International Journal of Molecular Sciences*, *17*, 1229–1245.
Wang, H., Zhao, P., Hölscher, D., Wang, Q., Lu, P., Cai, X. A., et al. (2012). Nighttime sap flow of *Acacia mangium* and its implications for nighttime transpiration and stem water storage. *Journal of Plant Ecology*, *5*, 294–304.
Webb, A. A. R., & Satake, A. (2015). Understanding circadian regulation of carbohydrate metabolism in *Arabidopsis* using mathematical models. *Plant Cell Physiology*, *56*, 586–593.
Wegner, L. H. (2014). Root pressure and beyond: Energetically uphill water transport into xylem vessels? *Journal of Experimental Botany*, *65*, 381–392.

Wegner, L. H. (2015). A thermodynamic analysis of the feasibility of water secretion into xylem vessels against a water potential gradient. *Functional Plant Biology*, *42*, 828–835.
Wei, W., Alexandersson, E., Golldack, D., Miller, A. J., Kjellbom, P. O., & Fricke, W. (2007). HvPIP1;6, a barley (*Hordeum vulgare* L.) plasma membrane water channel particularly expressed in growing compared with non-growing leaf tissues. *Plant Cell Physiology*, *48*, 1132–1147.
Wiese, A., Christ, M. M., Virnich, O., Schurr, U., & Walter, A. (2007). Spatio-temporal leaf growth patterns of *Arabidopsis thaliana* and evidence for sugar control of the diel leaf growth cycle. *New Phytologist*, *174*, 752–761.
Wolf, O., Munns, R., Tonnet, M. L., & Jeschke, W. D. (1990). Concentrations and transport of solutes in xylem and phloem along the leaf Axis of NaCl-treated *Hordeum vulgare*. *Journal of Experimental Botany*, *41*, 1133–1141.
Ye, Q., & Steudle, E. (2006). Oxidative gating of water channels (aquaporins) in corn roots. *Plant, Cell & Environment*, *29*, 459–470.
Zeuthen, T. (1994). Cotransport of K+, cl– And H2O by membrane proteins from chloroid plexus epithelium of *Necturus maculosus*. *Journal of Physiology*, *478*, 203–219.
Zeuthen, T. (2010). Water-transporting proteins. *Journal of Membrane Biology*, *234*, 57–73.
Zhu, C., Schraut, D., Hartung, W., & Schäffner, A. R. (2005). Differential responses of maize MIP genes to salt stress and ABA. *Journal of Experimental Botany*, *56*, 2971–2981.

CHAPTER TEN

Application of omics technologies in single-type guard cell studies for understanding the mechanistic basis of plant adaptation to saline conditions

Fatemeh Rasouli[a,b,c,*], Ali Kiani-Pouya[b,c], Yuan Wang[b], Ali Movahedi[d], Min Yu[a], Ayesha Tahir[e], and Sergey Shabala[a,c,*]

[a]International Research Centre for Environmental Membrane Biology, Foshan University, Foshan, Guangdong, China
[b]State Key Laboratory of Molecular Plant Genetics, Shanghai Centre for Plant Stress Biology, Centre for Excellence in Molecular Plant Sciences, Chinese Academy of Sciences, Shanghai, China
[c]Tasmanian Institute of Agriculture, College of Science and Engineering, University of Tasmania, Hobart, TAS, Australia
[d]Co-Innovation Centre for Sustainable Forestry in Southern China, Key Laboratory of Forest Genetics & Biotechnology, Ministry of Education, Nanjing Forestry University, Nanjing, China
[e]Department of Biosciences, COMSATS University Islamabad, Islamabad, Pakistan
*Corresponding authors: e-mail address: fatemeh.rasouli@utas.edu.au; sergey.shabala@utas.edu.au

Contents

1. Stomata and salinity 250
2. Application of omics approaches in single-cell-type studies in plants 252
3. Guard cell proteomics 254
4. Guard cell transcriptome 256
5. Guard cell metabolomics 257
6. Single cell isolation methods 258
7. Optimization of GC-enriched epidermal fragments for omics studies under saline conditions 259
8. Case studies of application of GC-enriched epidermal fragments method for proteomics analysis under saline conditions 261
9. Future prospects 264
References 266

Abstract

Stomatal pores, representing around 1% of total leaf surface, are the main gates for plant gas exchange, thus playing an important role in photosynthesis and water relations. Under saline conditions, operation of stomata is severely affected by adverse effects of salt on biochemical and physiological processes in guard cells (GCs), reducing CO_2 assimilation and causing yield loss. Next generation sequencing (NGS) technologies

Advances in Botanical Research, Volume 103
ISSN 0065-2296
https://doi.org/10.1016/bs.abr.2022.02.016

such as transcriptome and proteome analyses may be powerful tools in understanding the mechanistic basis of GC operation under saline conditions. As such studies require a large number of cells, developing an appropriate isolation method to provide adequate supply of GC with a sufficient purity is a critical step in this process. The most widely used method is isolation of GC protoplasts although it comes with some certain limitations. A viable and better alternative may be using GC-enriched epidermal. The modified protocol showed more efficacy and was more enriched in GC-specific genes suggesting the suitability of this approach for exploring responses of GCs to environmental and chemical stimuli at the molecular level. Application of GC-enriched epidermal fragments method for proteomics analysis proved to be efficient in understanding the mechanistic basis of stomata operation under saline conditions. Omics studies could be applied in conjunction with other research tools such as genome editing at GC level to identify new genes or proteins conferring stomata function which can then be targeted by breeding program to produce plants with better stomatal performance under adverse environmental conditions.

1. Stomata and salinity

Biomass production in plants is ultimately dependent on their ability for CO_2 assimilation, which is regulated by the fine-tuning of the stomatal aperture (Kiani-Pouya et al., 2020). Although stomatal pores represent a small amount (around 1%) of total leaf surface, they are the main gates for leaf gas exchange and responsible for about 95% of water loss through transpiration (Hedrich & Shabala, 2018). Thus, efficient stomatal operation plays a critical role in photosynthesis and plant water relations.

There is a diverse range of anatomical and morphological characteristics in stomata where they have a significant variation in shape, size, and number across plant species. In addition to these stomatal factors, there are many other non-stomatal factors such as biochemical processes which are performed in the cytosol, chloroplast and mitochondria determine the capacity of a plant for the net rate of photosynthesis (Sharkey, Bernacchi, Farquhar, & Singsaas, 2007). As another source of complexity, there are many environmental factors particularly environmental stresses that affect stomatal performance. These broad variations have the potential to influence stomatal movement and, consequently, plant photosynthesis and water use efficiency (WUE) (Bertolino, Caine, & Gray, 2019).

Salinity stress severely affects stomatal operation by imposing osmotic, oxidative and ionic stresses (Liu et al., 2017). Because of these adverse effects, CO_2 acquisition capacity in plants is reduced and thus salt stress becomes a limiting factor in photosynthesis (Chaves, Flexas, & Pinheiro, 2009). Effect of salinity on stomatal conductance is complex, as both stomata size and density are affected by salinity stress (Liu et al., 2014; Peng et al., 1998). Under normal conditions, a positive correlation between transpiration rate and stomatal density exist (e.g., barley, Miskin, Rasmusson, & Moss, 1972; wheat, Wang & Clarke, 1993). Also, larger GCs and high stomatal aperture have a positive association with plant grain yield (Liu et al., 2017). Under saline stress, however, plants are facing a dilemma of keeping their stomata open, to maintain normal growth, or reduce in stomatal conductance, to conserve water under hyperosmotic saline conditions. Accordingly, two strategies may be employed. The strategy employed by halophytes is to have a fewer (Shabala, 2013) but faster (Rasouli, Kiani-Pouya, Shabala, et al., 2021; Rasouli, Kiani-Pouya, Tahir, et al., 2021) stomata. Under moderate salinity levels, salt tolerant cultivars and species that are able to osmotically adjust to saline environment show lesser reduction if stomatal conductance, compared with their sensitive counterparts. This is further illustrated by comparing changes in stomatal conductance in cultivated (sensitive) and wild (tolerant) barley (a reduction by 72% and 59%, respectively; Kiani-Pouya et al., 2020).

Stomatal movement is regulated by numerous environmental and internal signals. Under stressful conditions, the production of primary signals such as ABA and H_2O_2 are altered, affecting stomata function. Salinity stress increases the ABA concentration that triggers stomatal closure (Park et al., 2009; Pei, Ward, Harper, & Schroeder, 1997). In this context, although progress has been made regarding the role of ABA in stomatal closure in the model plant Arabidopsis, both basal ABA levels and their modulation by salinity appear to be strikingly different between halophytes and glycophyte species (Hedrich & Shabala, 2018). The reason for this difference is unknown, given the paucity of available data on single GC metabolism, especially in halophytic species. This gap in the knowledge should be filled, as understanding the mechanistic basis of guard cell responses to salt stress in halophytes which have unique tolerance to high salt concentrations, represents an underutilized resource for improving adaptive responses to salinity in major staple crops.

In this context, application of the next-generation sequencing (NGS) technologies which provide a comprehensive picture of the biological molecules at cellular level could be considered as steppingstone in this direction. As such, transcriptomic analysis of stress-induced changes in GC under saline conditions (by RNA-seq technology) may facilitate identification of genes related to salt stress at the single-cell level (De Vos et al., 2019). Furthermore, post-translational regulations of these genes could be investigated in complex biological processes such as under salt stress through proteomics approaches (Salekdeh, Siopongco, Wade, Ghareyazie, & Bennett, 2002). In this regard, mass spectrophotometry-based proteomic methods can determine specific protein isoforms where they might have an important key role in GC operation. In the following sections, the current knowledge in the field is comprehensively summarized and analyzed.

2. Application of omics approaches in single-cell-type studies in plants

Single-cell-type samples defined as a population of the identical cells developing from common origin and having the same physiological role (Trapnell, 2015). Application of omics research in single cell-type studies are effective methods for detecting key genes and proteins and related pathways in biological systems (Dai & Chen, 2012; Specht & Slavov, 2018). While tissue-level studies can be informative, single cell type information is lost in tissues composed of multiple cell types. On the contrary to multiple cell types such as whole leaf, isolated single cell types such as GCs which is a population of specialized differentiated cells, have the potential to decipher novel biological processes underlying specific gene or protein functions (Fig. 1).

Investigators have recently focused on analysis of this type of cells from different field studies, related to both mammalian (Navin & Hicks, 2011) and plant systems (Geilfus, Lan, & Carpentier, 2018; Geng, Yu, Zhu, Dufresne, & Chen, 2017). Plants are made up of various highly specialized cell types that have different physiological and biochemical functions. Single cell-type studies have been used to improve our understanding of many biological processes (Lafzi, Moutinho, Picelli, & Heyn, 2018). In plants, only spores and pollen grains are readily available single-cells and other cell types such as EBCs, root hair and GCs are not easily available single cells. As both transcriptomics (e.g., RNAseq) and proteomics methods require substantial amounts of high-quality material, this poses a significant challenge for single-cell studies (Dai & Chen, 2012). Yet, application of omics approaches for investigating of various single-cell types such as epidermal bladder cells (EBC) (Barkla & Vera-Estrella, 2015), root hair (Li & Lan, 2015) and stomata (Geilfus et al., 2018; Misra, Assmann, & Chen, 2014; Misra, de Armas, Tong, & Chen, 2015; Misra, Reichman, & Chen, 2018) have been conducted.

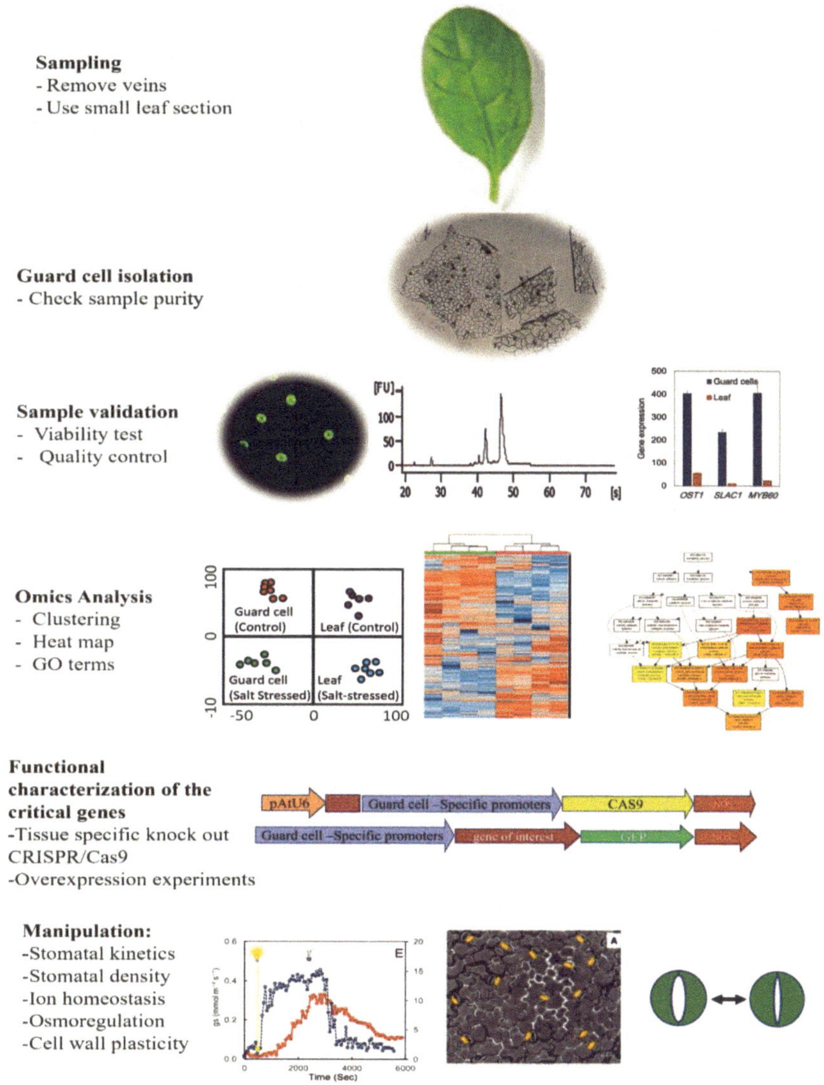

Fig. 1 The flowchart showing key steps in OMICS study on guard cells.

Good examples of applications of transcriptomics and proteomics are studies conducted by Barkla, Vera-Estrella, and Pantoja (2012) and Oh et al. (2015) demonstrating differential abundance of proteins and genes involved in EBC in *Mesembryanthemum crystallinum* under saline conditions. EBCs are specialized epidermal cells responsible for defense against environmental stresses such as salt stress and are an ideal model for single cell studies

(Oh et al., 2015; Van Cutsem et al., 2011). Aforementioned proteomic study on EBC under saline condition showed a significant difference in protein profile between plants grown under saline and non-saline conditions (Barkla et al., 2012). This study revealed a diverse biological function and cellular locations of identified proteins where a significant proportion of differently expressed proteins were involved in biological processes such as ion and water homeostasis, carbohydrate metabolism, and photosynthesis (Barkla et al., 2012). A transcriptome analysis of EBCs isolated from intact plants of *M. crystallinum,* also revealed the key role of EBCs in salinity stress where 7% of transcript-contigs differently expressed between salt-treated and control plants (Oh et al., 2015). This study also demonstrated that some biological processes such as ion transport, metabolism related to energy generation and osmolyte accumulation, and stress signaling in response to salinity stress had significantly altered between saline and non-saline conditions (Oh et al., 2015). These studies exemplify the potential of transcriptomics and proteomics investigations in single cell-type to understand salt-adaptive mechanisms in plants.

3. Guard cell proteomics

Although omics technologies have been advanced in recent years and low and its affordability has made it possible to apply this technology in many areas of plant science, but it has not been extensively applied to GC studies, mostly due to difficulties of generating substantial amounts of purified material (Rasouli, Kiani-Pouya, Zhang, & Shabala, 2020). In applying the proteomics approach in guard cell studies, few studies have used proteomics to investigate the protein profile of GCs (Table 1).

In one of the earliest proteomics investigations on GCs, protoplasts of GCs were isolated from Arabidopsis epidermis through enzymatic digestion of epidermal fragments (Zhao, Zhang, Stanley, & Assmann, 2008). In another proteomics analysis in *Brassica napus*, 1458 proteins related to biological processes such as cell structure, signaling energy, transport, transcription were identified (Zhu, Dai, McClung, Yan, & Chen, 2009). As this study was comparing protein profile of GC and mesophyll, the results of this investigation highlighted the differences between GC and mesophyll cells where in contrast to GC, many proteins related to defense, disease, starch synthesis and photosynthesis were significantly enriched in mesophyll cells (Zhu et al., 2009).

Table 1 Different omics studies on the guard cells of plants carried out using various isolation methods.

Analysis	Plant species	Guard cell isolation method	Treatment	References
Proteomics	Arabidopsis	Protoplasts	None	Zhao et al. (2008)
	Brassica napus	Protoplasts	None	Zhu et al. (2009)
	Arabidopsis	Protoplasts	ABA	Zhao et al. (2010)
	Arabidopsis	Protoplasts	None	Zhang et al. (2016)
	Brassica napus	Protoplasts	Reduced CO_2	Geng et al. (2017)
	Arabidopsis	Guard-cell-enriched epidermal fragment	Light	Geilfus et al. (2018)
	Sugar beet	Guard-cell-enriched epidermal fragment	Salt	Rasouli, Kiani-Pouya, Li, et al. (2020)
	Quinoa	Guard-cell-enriched epidermal fragment	Salt	Rasouli, Kiani-Pouya, Shabala, et al. (2021) and Rasouli, Kiani-Pouya, Tahir, et al. (2021)
Microarray or RNA-seq	Arabidopsis	Protoplasts	ABA	Leonhardt et al. (2004)
	Arabidopsis		ABA	Pandey et al. (2010)
	Arabidopsis		ABA	Wang et al. (2011)
	Arabidopsis	Dissected guard cells from lyophilized leaves	Sucrose	Bates et al. (2012)
	Arabidopsis	Guard-cell-enriched epidermal fragment	Reduced humidity	Bauer et al. (2013)
	B. napus	Protoplasts	ABA	Yoo et al. (2016)
	Arabidopsis	Fluorescence-activated cell sorting (FACS)	Developmental stage	Adrian et al. (2015)
Metabolomics	*Brassica napus*	Protoplasts	Reduced CO_2	Geng et al. (2017)
	Brassica napus	Protoplasts	Increased CO_2	Misra et al. (2014)
	Arabidopsis	Epidermis	None	Zhang et al. (2014)
	Arabidopsis	Protoplasts	None	Jin et al. (2013)

Some proteomics studies have also investigated the effects of different environmental cues on the GCs. The responses of GC to environmental factors such as CO_2, light and ABA have been studied and the obtained results have been shown that various proteins and pathways are involved in response to these various signals (Geilfus et al., 2018; Geng et al., 2017; Zhao, Stanley, Zhang, & Assmann, 2010). For example, in a proteomic and metabolomic investigation on GC of *B. napus* in response to low CO_2, 1397 proteins and 411 metabolites were identified and the majority of them were involved in redox regulation, starch, and sucrose metabolism, and fatty acid metabolism (Geng et al., 2017).

4. Guard cell transcriptome

Similar to proteomics studies, few transcriptome studies have also been conducted on GCs (Table 1). The following case studies showing how transcriptome investigation could improve our understanding of GC functions.

Wang et al. (2011) performed a transcriptome investigation to compare gene expression levels in leaf and GC in Arabidopsis. The findings of this investigation revealed that GO terms such as response to abiotic stimulus and response to endogenous stimulus were enriched in both GC and leaf. On the other hand, GO terms such as nucleobase metabolism and protein metabolism were relatively enriched higher in leaf than GCs. In regard with the GC specific transcription factors, the results of this study showed that while some important promoter motifs which regulate the genes in response to ABA such as ABREs and LTREs were upregulated in both leaf and GCs, GC promotors were enriched in DRE/CRT and MYB elements (Wang et al., 2011). The results also showed that the transcription factor activity was enriched only in the GCs, suggesting that GC may utilize specific transcription factors differently.

Trascriptome analysis has also used as a supporting evidence for this hypothesis that sugars derived from photosynthesis may play a role in stomata opening. It was belied that much smaller number of chloroplasts in GCs compared with mesophyll is an indicator of GCs reliance on imported energy for stomatal movement (Lu, Outlaw, Smith, & Freed, 1997). Through a trascriptome analysis (Bates et al., 2012) it was showed that the metabolism of sugar in GC is different from that in mesophyll cells. The results of this study revealed that in sucrose treated Arabidopsis GCs, 440 genes were differentially expressed in response to sucrose

(Bates et al., 2012). They argued that sugars synthesized in mesophyll cells are then exported to apoplast and transported to GC via phloem. As GCs have low potential for carbon reduction, they take up the sugars from the apoplast. This study further argued that sugars can be stored as starch or metabolized to generate required energy fro GC metabolism and movement (Bates et al., 2012).

5. Guard cell metabolomics

In combination with proteomics and transcriptomics studies, metabolomics analysis provides information to better understand the underlying mechanism of an organism under different conditions. When applied to plants, this knowledge can be associated with crop plants performance under stressful conditions to then release crop varieties that can efficiently cope with these adverse environmental conditions (Barkla & Vera-Estrella, 2015; Benjamin, Lucini, Jothiramshekar, & Parida, 2019).

Metabolomics is a fast-developing technology and a broad-spectrum of metabolites have been determined in plants under various growing conditions including salt stress (Barkla & Vera-Estrella, 2015; Kumari, Das, Parida, & Agarwal, 2015). For instance, some metabolites such as mono-, di-, oligo-, and polysaccharides such as glucose, fructose, sucrose, trehalose, raffinose and polyol compounds such as sorbitol, mannitol, glycerol, inositol have yet been determined in plants. Also, metabolites such as proline and in methylated proline-related group, and compounds such as methyl-proline, proline betaine, and hydroxyproline betaine pipecolic acid are among amino acid group metabolites that have been identified (Kumari et al., 2015).

Although metabolomics studies on GCs are relatively rare, these have improved our understanding and shed a light on metabolic regulation of stomatal movement in response to different environmental stimuli and factors such as ABA, CO_2, sucrose, organic acids, and plant cell-microbe interaction (Table 1; reviewed by Medeiros et al., 2018).

Jin et al. (2013) applied metabolomics approach to investigate the association among different metabolic components under drought stress in Arabidopsis wild type and *gpa1* mutant (with ABA-hyposensitive stomata due to mutation in G-protein subunit). The results of this study revealed that abundance of cyclic adenosine diphosphate ribose and Ca^{2+}-mobilizing agent, sphingosine-1-phosphate were increased in response to ABA with weaker ABA-stimulated increases in *gpa1* mutant. ABA also regulated other hormones specially IAA in GCs where IAA was reduced in wild type but not in *gpa1* mutant. The same pattern was not manifested in mesophyll cells. A lower level of IAA could be a component mechanism for promotion of stomatal closure and ABA inhibition of stomatal opening. The fact that ABA hyposensitivity was more evident in the GC metabolome of *gpa1* than in its transcriptome (Pandey et al., 2010) suggests that identification of metabolome is a necessary component of integrated omics analyses.

In another metabolomics study on GCs and mesophyll cells (Misra et al., 2015), contrasting metabolic responses were achieved between GCs and mesophyll cells in response to elevated bicarbonate. The GCs demonstrated differential increases of different metabolites such as defense-responsive pathways metabolites (flavonoids, alkaloids, and phenolics), primary C-metabolites, N-metabolites (amino acids and purines), as compared to the mesophyll cells, suggesting differential carbon to nitrogen (C/N) homeostasis in GC and mesophyll cells. In this study, the GCs also showed decreased levels of auxins, cytokinins, phenylpropanoids, and redox metabolites where all of which were increased in mesophyll cells when leaves exposed to bicarbonate.

Despite application of metabolomics methods in aforementioned studies, this approach has not yet been applied to investigate the metabolome profile of GC in responses to salinity stress.

6. Single cell isolation methods

As mentioned above, one of the major limitations for applications of the omics approach at the single-cell level was a difficulty of generating sufficient amounts of purified material. Several single cell isolation methods have been developed during the past years. Fluorescent activated cell sorting (FACS) is one of the methods that has been used in both medical and plant sciences though this approach has not been sufficiently developed for plant science field (Gautam & Sarkar, 2015). This method uses a reporter gene such as green fluorescent protein (GFP) to tag a specific cell type and collect the marked cells separately.

Laser assisted microdissection (LAM) is another method that extensively used in RNA-sequencing and gene expression profiling. This microscope-based method selects the cells of interest from a cell population mixture using a pulsed infrared laser (Gautam & Sarkar, 2015). Although this method has also been used in other research areas such as exploring cell or tissue specific transcriptomes in plants (Brooks et al., 2009) but it has not developed enough thus there is no general protocol to be appropriate for all cell types and plant species (Gautam & Sarkar, 2015). Another limitation of application of this method is that there is no possibility for amplification of proteins and metabolites extracted by this method and hence could not be used for proteomics and metabolomics approaches (Wilson & Nairn, 2018).

Concerning GC isolation in plants, it dates back more than four decades when a protocol was developed by Zeiger and Hepler (1976) to isolate protoplast from onion and tobacco leaves. Since then, GC isolated protoplasts have been used for proteomics (Zhao et al., 2008) and RNA-sequencing analyses but only in a very few studies (Pandey et al., 2010; Wang et al., 2011).

Preparation of GC protoplast for downstream application, e.g., proteomics analysis is a multistep process that requires prolonged enzymatic digestion for dissolving cell walls. It has been shown that this method may impose kinds of chemical and osmotic stresses to GC that consequently alters gene expression. Bates et al. (2012) showed that GC protoplast samples prepared through enzymatic digestion method had higher levels of stress-related genes than protoplast samples isolated with non-enzymatic method (Bates et al., 2012). Although some improvement such as reduced digestion time by using of appropriate cellulose enzyme (Yao et al., 2018) and cell wall digestion in the presence of transcription inhibitors (Obulareddy, Panchal, & Melotto, 2013) has been made to enhance the efficiency of this method but it still remains questionable approach. In this regard, it has been argued that the enzymatic solutions used in this method are derived from fungal cell-wall preparations and may result in GC samples contamination with fungal elicitors (Bauer et al., 2013; Fujikawa et al., 2012).

Manually isolating GC from lyophilized leaf strips is another technique used for GC preparation for omics analysis. Compared to other GC isolation methods, this technique is less frequently used, and it may also result in low RNA yield (Bates et al., 2012). The other disadvantage of this method that it is not appropriate for other omics analysis such as proteomics.

7. Optimization of GC-enriched epidermal fragments for omics studies under saline conditions

Compared to aforementioned methods, using GC-enriched epidermal fragments offers several competitive advantages and could be used as an alternative approach for isolation of GCs for omics studies (Rasouli, Kiani-Pouya, Zhang, & Shabala, 2020). This method is amazingly fast and also retains the GC wall (and associated proteins) which plays a significant role in stomatal operation (Hunt et al., 2017).

Our previous study of applying the adopted protocol of this method (Jalakas, Yarmolinsky, Kollist, & Brosché, 2017) for different crops showed that the obtained GC samples contained a high percentage of contamination with non-guard cell materials that was especially high in the salt-grown samples (Rasouli, Kiani-Pouya, Zhang, & Shabala, 2020). It has been argued that growing conditions such as differences in the cultural media (Yao et al., 2018) and the type of studied stress can affect the efficiency of the isolation method (Rasouli, Kiani-Pouya, Zhang, & Shabala, 2020). Regarding salt stress, as this stress increases leaf cuticle, parenchyma thickness and vascular bundle sheath thickness (Akcin, Akcin, & Yalcin, 2014), it can affect the GC isolation method (Rasouli, Kiani-Pouya, Zhang, & Shabala, 2020). Thus, modifications are required to develop efficient protocols for GC samples isolation from salt-grown plants for various omics studies. These modifications included using smaller leaf sections, removing all the visible veins in a leaf, increasing blending time and number of runs, and using a basic solution instead of distilled water (Rasouli, Kiani-Pouya, Zhang, & Shabala, 2020). These alterations resulted in production of GC-enriched epidermal fragments that were not contaminated by any kind of pavement, mesophyll or vascular cells. GC-specific genes such as OST1 and SLAC1, were highly expressed in the GCs, confirming that our GC preparation was highly enriched in GC gene transcripts (Fig. 1).

One of the advantages of GC-enriched epidermal fragments over the commonly used protoplasting protocol is that in the latter the GC sample is subjected to substantial manipulation before tissue fixation mainly because of prolonged enzymatic digestion. This preparation method consequently results in some stress-related genes being remarkably upregulated in the protoplast method, which is not the case with non-enzymatic approaches such as GC-enriched epidermal fragments.

Another advantage of GC-enriched epidermal fragments method is that this technique includes cell wall of GC samples and thus GC wall is included in the omics analysis. Given the importance of GC walls and its extracellular matrix in the stomatal movements (Jones, Milne, Ashford, & McQueen-Mason, 2003), including GC walls into the omics studies related to GC would be very advantageous. Furthermore, contrary to the other cell wall types, GC walls are exceptionally strong and tensile, and they have a key role in stomatal opening and closure (Rui et al., 2018). The fact that stomatal dynamic depends on the structural and functional proteins embedded in the GC wall (Hunt et al., 2017) existing proteins in the GC wall become more important in stomatal kinetics. Therefore, using the GC-enriched epidermal fragments method for isolating of GCs which incorporates the cell wall for downstream applications such as omics analysis provides this opportunity to include the function of GC wall proteins in the analysis.

Comparisons of some GC-specific genes enrichment in the GC between GC-enriched epidermal fragments (Rasouli, Kiani-Pouya, Zhang, & Shabala, 2020) with the original protocol in Arabidopsis (Bauer et al., 2013) revealed that the former preparation approach was more enriched in GC-specific genes, indicating that the modified protocol resulted in a more purified GC sample (Rasouli, Kiani-Pouya, Zhang, & Shabala, 2020). The tissue prepared in this manner could be the source of protein and RNA extractions for omics analysis (Fig. 1), facilitating stomatal function research and improving experimental efficiency. Altogether, the results showed that this modified protocol was more efficient and more enriched in GC-specific genes suggesting the suitability of this isolation approach for exploring responses of GCs to stimuli and various types of stress at the molecular level.

8. Case studies of application of GC-enriched epidermal fragments method for proteomics analysis under saline conditions

After successfully optimizing GC isolation method from the leaf epidermis, we applied this approach for a proteomics analysis of sugar beet GC under saline and control conditions with high quality and purity samples. The obtained results showed of the 2088 proteins determined in sugar beet GC, 82 were differentially expressed by salinity stress (Rasouli et al., 2020). Gene ontology enrichment analysis and protein classification demonstrated that these proteins were involved in different biological processes such as lipid metabolism, cell wall modification, ATP biosynthesis, and signaling. From these significant differentially expressed proteins, several proteins classified as "stress proteins" included non-specific lipid transfer protein, chaperone proteins, heat shock proteins, inorganic pyrophosphatase 2, which is responsible for energized vacuole membrane for ion transportation. Furthermore, several antioxidant enzymes such as peroxide, superoxidase dismutase were highly upregulated. As mentioned in the preceding section, one of the excellence advantages of GC-enriched epidermal fragments over frequently used protoplast digestion is that the former keeps the GC wall.

This advantage was well represented in this study through overexpression of cell wall proteins demonstrating the importance role of cell wall proteins in stomata function. The cell wall proteins detected in GCs provided some evidence that GC walls were more flexible in response to salt stress (Rasouli, Kiani-Pouya, Li, et al., 2020). This study also revealed that some proteins such as L-ascorbate oxidase that was constitutively high under both saline and non-saline conditions might have a role in the ability of sugar beet GCs to adapt to salt stress conditions through mitigating salinity-induced oxidative stress (Rasouli, Kiani-Pouya, Li, et al., 2020).

The majority of studies that applied proteomics approach have been used to study the response of GCs to, e.g., environmental factors have employed the model plant Arabidopsis. Halophytes on the other hand, as highly salt-tolerant plant species have evolved mechanisms to efficiently benefit from adverse saline conditions (Flowers & Colmer, 2015). It is noteworthy that most research investigating the response of GC to salinity stress in halophytic species have focused on physiological or anatomical aspects. Thus, studies of the salt-induced proteins in GCs of halophytic plants will direct us toward identification of salt responsive proteins in GC involved in salt tolerance and adaptation mechanisms of halophytes at the GC level. In this context, quinoa as a cultivated halophytic plant which has a high tolerance to salinity and drought stresses (Kiani-Pouya et al., 2019; Shabala, 2013) may be ideal candidate to apply a proteomics analysis through epidermal fragments enriched in isolated GCs under saline conditions.

In our study, in total 2471 proteins were identified, of which 36% were differentially expressed in response to 300 mM NaCl in GCs (Rasouli, Kiani-Pouya, Shabala, et al., 2021). Bioinformatics analysis revealed that these differently expressed proteins belong to various biological functions such as signaling molecules, enzyme modulators, transcription factors and oxidoreductases. Furthermore, several most abundant proteins involved in responses were found induced by salinity treatment, included desiccation-responsive protein 29B (50-fold), osmotin-like protein OSML13 (13-fold), polycystin-1, lipoxygenase, alpha-toxin, and triacylglycerol lipase (PLAT) domain-containing protein 3-like (8-fold), and dehydrin early responsive to de-hydration (ERD14) (8-fold). These findings suggest that quinoa as a halophytic plant can employ specific modes of responses to salinity stress at the GC level (Rasouli, Kiani-Pouya, Shabala, et al., 2021). The results also showed that 10 proteins related to the GO term "response to ABA" were upregulated in quinoa GC. These proteins included aspartic protease, phospholipase D, plastid-lipid-associated protein (Rasouli, Kiani-Pouya,

Shabala, et al., 2021). In addition, seven proteins in the sucrose-starch pathway were upregulated in the GC in response to salt stress. Moreover, amino acids tryptophan and L-methionine synthase were induced and accumulated by salinity stress in GC. Exogenous application of tryptophan, L-methionine and sucrose resulted in reduction in stomatal aperture and conductance, suggesting a possible role of aspartic proteinase in mitigating oxidative stress induced by salinity stress (Rasouli, Kiani-Pouya, Shabala, et al., 2021).

To provide additional analysis to the underlying salinity tolerance mechanisms at GC level between sugar beet and quinoa we compared the obtained results of these two proteomics investigations (Rasouli, Kiani-Pouya, Li, et al., 2020; Rasouli, Kiani-Pouya, Shabala, et al., 2021; Rasouli, Kiani-Pouya, Tahir, et al., 2021). The results revealed 20 common differentially abundant protein between these two plant species where 10 proteins were upregulated, and 10 proteins were downregulated. Aspartic protease in guard cell 1, and non-specific lipid-transfer proteins had enhanced abundance while catalase and cationic peroxidase proteins had lower abundances in GCs of both sugar beet and quinoa under saline conditions. Given that catalase is an antioxidant that detoxifies reaction oxygen species such as H_2O_2, lower content of this scavenger may be required in GC under salinity stress as H_2O_2 acts as a signaling molecule in the GC which induces stomatal closure in response to high salt concentration. Comparing proteomics results of quinoa and sugar beet also demonstrated some proteins such as L-ascorbate oxidase that involved in mitigation of oxidative stress were presented at higher levels under non-saline conditions in both plants. This finding suggests that constitutively higher content of such proteins in the GCs of halophytic plant such as quinoa and salt-tolerant sugar beet can confer augmented tolerance to GCs of these species under saline conditions.

Quinoa as a halophytic species showed the larger number of differentially abundant proteins than sugar beet under salinity stress suggesting that quinoa needed modulation of more proteins to deal with salinity stress. We also compared the obtained proteomics results of differentially expressed proteins in GC in response to salt stress in terms of biological function between quinoa and sugar beet based. This analysis showed that both plants had consistent results where the largest single group of proteins included those involved in protein biogenesis and disposal (proteins such as ribosomal subunits, molecular chaperones, and proteasomal subunits) and the second most abundant protein group were identified as "stress" proteins. On the other hand, this analysis revealed that according to overall abundance protein in

GCs, sugar beet and quinoa had different top proteins. The top protein in the GCs of these two plants were also different from what is previously reported in the other species such as Arabidopsis (Zhao et al., 2008) suggesting that salt-responsive mechanisms may be different at single cell levels among various plant species. Thus, further research is required to obtain insight into role of those top proteins in GCs of different plants particularly in GCs of salt-tolerant crop species such as quinoa and sugar beet.

Proteomics analysis of quinoa has also revealed that in addition to seven proteins in the sucrose-starch pathway, sucrose transporter (SUC3) and sugar transporters were additionally overexpressed in the quinoa GC in response to salt stress. Although information in relation to kinetics, regulation and expression of sugar transporters in GCs is still limited and their role in stomata regulations have been matter of debates (Granot & Kelly, 2019; Lawson & Matthews, 2020) but this obtained result from GC of quinoa suggesting that this plant as a halophyte may have different salt tolerance mechanisms at GC level. In this context, exogenous application of sucrose induced stomata closure in our previous study (Rasouli, Kiani-Pouya, Shabala, et al., 2021; Rasouli, Kiani-Pouya, Tahir, et al., 2021); however, it has shown that sucrose and other sugars may have a role in stomatal opening (Lawson & Matthews, 2020). To achieve insight into the function of different genes expression in the GCs, mutants lacking sucrose transporter or sugar transporters genes function in sucrose/starch pathway may provide critical information in this regard and decipher the role of sucrose and sugars functions in the stomatal opening and closure in response to salinity stress. Source of sugars in the GCs and the mechanisms by which sucrose and other sugars enter GCs are also important and required to be the subject of future research on GC.

9. Future prospects

Application of NGS technologies such as RNA-sequencing and proteomics on GC research could increase our understanding of GC functions. By applying omics approach on GC there would be a possibility to identify new genes or proteins mediating stomata operation under stress conditions. This gene or protein could then be targeted by breeding program with the aim of producing plants with better stomatal performance under adverse environmental stresses such as salinity stress.

One of the promising approaches that may be beneficial to stomatal performance is manipulating the ion channels in GC (Fig. 1). Given the importance of ion channels in stomatal movements, this alteration could be directed toward improving both photosynthesis and stomatal conductance in GCs. Recently, by using optogenetics approach, a synthetic channel (BLINK1) was expressed in the GC in Arabidopsis with the aim of increasing ion flux to drive turgor that could affect photosynthesis and stomatal conductance (Papanatsiou et al., 2019). The results of this study showed that BLINK1 enhanced stomatal kinetics and plants with this synthetic channel had a greater biomass with no decrease in water use efficiency. As a complementarity approach to this kind of studies, application of NGS technologies such as proteomics and transcriptome which has the potential to identify new genes or proteins in relation to stomata function can be further accompanied by optogenetics method. For instance, this approach could be used as a tool for manipulating channels in the GC with the aim of promoting salinity tolerance in different crop plants.

Genome editing techniques have now become a routine procedure in many research projects. Genome editing and loss of function mutant alleles which provide the possibility of altered genetic material at particular locations in the genome has been indispensable for functional genomic studies in plants. In this context, in the past decade, there has been an explosion in the generation of mutated lines through application of clustered regularly interspaced short palindromic repeats (CRISPR) and associated Cas9 nuclease protein approach. CRISPR-Cas9 could be used as a powerful tool for gene editing and generating mutated plant lines for further gene functional studies, e.g., under various environmental stresses. Although this method of genome editing is not currently applicable to all plant species, however, alternative techniques such as T-DNA or transposon insertions, RNA interference, or artificial microRNAs can be applied for gene expression studies, hence giving an opportunity to explore the role of specific gene(s) in the GC function. In addition to this, a tissue-specific knockout method for tissue-specific genome modification via a CRISPR system which is called CRISPR-TSKO has recently been developed (Decaestecker et al., 2019) enabling for a tissue specific generation of somatic DNA mutations. CRISPR-TSKO toolset which is a simple method to use, also has high efficacy and allows for multiplexing and large-scale screening approaches (Decaestecker et al., 2019). Given the potential of CRISPR-TSKO for somatic gene knockouts of several essential genes in a specific tissue, this molecular tool can be optimized to study GC specific genes. Indeed, this approach opens possibilities to explore the functions of fundamentally important genes in GCs of various crop plants.

Given the facts that omics approaches are applicable under both short and long-term stress conditions, they could provide valuable information in relation to acclimation of plants to stress conditions as well as signal transduction mechanisms under adverse environmental conditions. While prolonged exposure of plants to stressful conditions such as salt stress provide information in relation to pathways, genes and proteins that are important in acclimation process, short-term salt stress that involve multiple time points on the other hand provide a comprehensive picture on the modulation in the expression profiles of early and late responsive genes in saline conditions (Liu, Horvath, Li, & Liu, 2019). The outcomes of these kind of investigations enable us to identify important underlying pathways as well as gene expressions networks under, e.g., salt stress conditions. Also, turnover and degradation of specific proteins of interest can be investigated over the timeframe of measurements.

In addition to above discussion, some new area of research may be required for application of omics studies in relation to GC investigations. For example, phosphorylation has a key role in the signal transduction pathways in GCs (Zhang et al., 2014) and could become a target of investigation on GCs in future studies through phospho-proteomics analysis of GCs in response to various stimuli including salt stress. In this context, some modifications need to be considered in application of methods and protocols. For example, phosphorylation in response to a stimulus is a quite fast cellular reaction which occurs in the second-to-minute time regime (Blazek, Santisteban, Zengerle, & Meier, 2015), hence, a time resolution of a few seconds for investigating cell-signaling events needs to be considered.

References

Adrian, J., Chang, J., Ballenger, C. E., Bargmann, B. O., Alassimone, J., Davies, K. A., et al. (2015). Transcriptome dynamics of the stomatal lineage: Birth, amplification, and termination of a self-renewing population. *Developmental Cell*, *33*(1), 107–118. https://doi.org/10.1016/j.devcel.2015.01.025.

Akcin, T. A., Akcin, A., & Yalcin, E. (2014). Anatomical adaptations to salinity in *Spergularia marina* (Caryophyllaceae) from Turkey. *Proceedings of the National Academy of Sciences, India Section B: Biological Sciences*, *85*(2), 625–634. https://doi.org/10.1007/s40011-014-0386-8.

Barkla, B. J., & Vera-Estrella, R. (2015). Single cell-type comparative metabolomics of epidermal bladder cells from the halophyte *Mesembryanthemum crystallinum*. *Frontiers in Plant Science*, *6*, 435. https://doi.org/10.3389/fpls.2015.00435.

Barkla, B. J., Vera-Estrella, R., & Pantoja, O. (2012). Protein profiling of epidermal bladder cells from the halophyte *Mesembryanthemum crystallinum*. *Proteomics*, *12*(18), 2862–2865. https://doi.org/10.1002/pmic.201200152.

Bates, G. W., Rosenthal, D. M., Sun, J., Chattopadhyay, M., Peffer, E., Yang, J., et al. (2012). A comparative study of the *Arabidopsis thaliana* guard-cell transcriptome and its modulation by sucrose. *PLoS One*, 7(11), e49641. https://doi.org/10.1371/journal.pone.0049641.

Bauer, H., Ache, P., Lautner, S., Fromm, J., Hartung, W., Al-Rasheid, K. A. S., et al. (2013). The stomatal response to reduced relative humidity requires guard cell-autonomous ABA synthesis. *Current Biology*, *23*(1), 53–57. https://doi.org/10.1016/j.cub.2012.11.022.

Benjamin, J. J., Lucini, L., Jothiramshekar, S., & Parida, A. (2019). Metabolomic insights into the mechanisms underlying tolerance to salinity in different halophytes. *Plant Physiology and Biochemistry*, *135*, 528–545.

Bertolino, L. T., Caine, R. S., & Gray, J. E. (2019). Impact of stomatal density and morphology on water-use efficiency in a changing world. *Fronties in Plant Science*, *10*, 225. https://doi.org/10.3389/fpls.2019.00225.

Blazek, M., Santisteban, T. S., Zengerle, R., & Meier, M. (2015). Analysis of fast protein phosphorylation kinetics in single cells on a microfluidic chip. *Lab on a Chip*, *15*(3), 726–734. https://doi.org/10.1039/c4lc00797b.

Brooks, L., III, Strable, J., Zhang, X., Ohtsu, K., Zhou, R., Sarkar, A., et al. (2009). Microdissection of shoot meristem functional domains. *PLoS Genetics*, *5*(5), e1000476. https://doi.org/10.1371/journal.pgen.1000476.

Chaves, M. M., Flexas, J., & Pinheiro, C. (2009). Photosynthesis under drought and salt stress: Regulation mechanisms from whole plant to cell. *Annals of Botony*, *103*(4), 551–560. https://doi.org/10.1093/aob/mcn125.

Dai, S., & Chen, S. (2012). Single-cell-type proteomics: Toward a holistic understanding of plant function. *Molecular & Cellular Proteomics*, *11*(12), 1622–1630. https://doi.org/10.1074/mcp.R112.021550.

De Vos, S., Van Stappen, G., Sorgeloos, P., Vuylsteke, M., Rombauts, S., & Bossier, P. (2019). Identification of salt stress response genes using the artemia transcriptome. *Aquaculture*, *500*, 305–314. https://doi.org/10.1016/j.aquaculture.2018.09.067.

Decaestecker, W., Buono, R. A., Pfeiffer, M. L., Vangheluwe, N., Jourquin, J., Karimi, M., et al. (2019). CRISPR-TSKO: A technique for efficient mutagenesis in specific cell types, tissues, or organs in Arabidopsis. *The Plant Cell*, *31*(12), 2868–2887. https://doi.org/10.1105/tpc.19.00454.

Flowers, T. J., & Colmer, T. D. (2015). Plant salt tolerance: Adaptations in halophytes. *Annals of Botany*, *115*(3), 327–331. https://doi.org/10.1093/aob/mcu267.

Fujikawa, T., Sakaguchi, A., Nishizawa, Y., Kouzai, Y., Minami, E., Yano, S., et al. (2012). Surface α-1,3-glucan facilitates fungal stealth infection by interfering with innate immunity in plants. *PLoS Pathogens*, *8*(8), e1002882. https://doi.org/10.1371/journal.ppat.1002882.

Gautam, V., & Sarkar, A. K. (2015). Laser assisted microdissection, an efficient technique to understand tissue specific gene expression patterns and functional genomics in plants. *Molecular Biotechnology*, *57*(4), 299–308. https://doi.org/10.1007/s12033-014-9824-3.

Geilfus, C. M., Lan, J., & Carpentier, S. (2018). Dawn regulates guard cell proteins in *Arabidopsis thaliana* that function in ATP production from fatty acid beta-oxidation. *Plant Molecular Biology*, *98*(6), 525–543. https://doi.org/10.1007/s11103-018-0794-x.

Geng, S., Yu, B., Zhu, N., Dufresne, C., & Chen, S. (2017). Metabolomics and proteomics of *Brassica napus* guard cells in response to low CO_2. *Frontiers in Molecular Biosciences*, *4*, 51. https://doi.org/10.3389/fmolb.2017.00051.

Granot, D., & Kelly, G. (2019). Evolution of guard-cell theories: The story of sugars. *Trends in Plant Science*, *24*(6), 507–518. https://doi.org/10.1016/j.tplants.2019.02.009.

Hedrich, R., & Shabala, S. (2018). Stomata in a saline world. *Current Opinion in Plant Biology*, *46*, 87–95. https://doi.org/10.1016/j.pbi.2018.07.015.

Hunt, L., Amsbury, S., Baillie, A., Movahedi, M., Mitchell, A., Afsharinafar, M., et al. (2017). Formation of the stomatal outer cuticular ledge requires a guard cell wall proline-rich protein. *Plant Physiology*, *174*(2), 689–699. https://doi.org/10.1104/pp.16.01715.

Jalakas, P., Yarmolinsky, D., Kollist, H., & Brosché, M. (2017). Isolation of guard-cell enriched tissue for RNA extraction. *Bio-Protocol*, 7(15), e2447. https://doi.org/10.21769/BioProtoc.2447.

Jin, X., Wang, R. S., Zhu, M., Jeon, B. W., Albert, R., Chen, S., et al. (2013). Abscisic acid-responsive guard cell metabolomes of Arabidopsis wild-type and *gpa1* G-protein mutants. *The Plant Cell*, *25*(12), 4789–4811. https://doi.org/10.1105/tpc.113.119800.

Jones, L., Milne, J. L., Ashford, D., & McQueen-Mason, S. J. (2003). Cell wall arabinan is essential for guard cell function. *Proceedings of the National Academy of Sciences of the United States of America*, *100*(20), 11783–11788. https://doi.org/10.1073/pnas.1832434100.

Kiani-Pouya, A., Rasouli, F., Bazihizina, N., Zhang, H., Hedrich, R., & Shabala, S. (2019). A large-scale screening of quinoa accessions reveals an important role of epidermal bladder cells and stomatal patterning in salinity tolerance. *Environmental and Experimental Botany*, *168*. https://doi.org/10.1016/j.envexpbot.2019.103885.

Kiani-Pouya, A., Rasouli, F., Rabbi, B., Falakboland, Z., Yong, M., Chen, Z.-H., et al. (2020). Stomatal traits as a determinant of superior salinity tolerance in wild barley. *Journal of Plant Physiology*, *245*, 153108. https://doi.org/10.1016/j.jplph.2019.153108.

Kumari, A., Das, P., Parida, A. K., & Agarwal, P. K. (2015). Proteomics, metabolomics, and ionomics perspectives of salinity tolerance in halophytes. *Frontiers in Plant Science*, *6*, 537. https://doi.org/10.3389/fpls.2015.00537.

Lafzi, A., Moutinho, C., Picelli, S., & Heyn, H. (2018). Tutorial: Guidelines for the experimental design of single-cell RNA sequencing studies. *Nature Protocols*, *13*(12), 2742–2757. https://doi.org/10.1038/s41596-018-0073-y.

Lawson, T., & Matthews, J. (2020). Guard cell metabolism and stomatal function. *Annual Review of Plant Biology*, *71*, 273–302. https://doi.org/10.1146/annurev-arplant-050718-100251.

Leonhardt, N., Kwak, J. M., Robert, N., Waner, D., Leonhardt, G., & Schroeder, J. I. (2004). Microarray expression analyses of Arabidopsis guard cells and isolation of a recessive abscisic acid hypersensitive protein phosphatase 2C mutant. *The Plant Cell*, *16*(3), 596–615. https://doi.org/10.1105/tpc.019000.

Li, W., & Lan, P. (2015). Re-analysis of RNA-seq transcriptome data reveals new aspects of gene activity in Arabidopsis root hairs. *Frontiers in Plant Science*, *6*, 421. https://doi.org/10.3389/fpls.2015.00421.

Liu, X., Fan, Y., Mak, M., Babla, M., Holford, P., Wang, F., et al. (2017). QTLs for stomatal and photosynthetic traits related to salinity tolerance in barley. *BMC Genomics*, *18*(1), 9. https://doi.org/10.1186/s12864-016-3380-0.

Liu, D., Horvath, D., Li, P., & Liu, W. (2019). RNA sequencing characterizes transcriptomes differences in cold response between northern and southern alternanthera philoxeroides and highlight adaptations associated with northward expansion. *Frontiers in Plant Science*, *10*, 24. https://doi.org/10.3389/fpls.2019.00024.

Liu, X., Mak, M., Babla, M., Wang, F., Chen, G., Veljanoski, F., et al. (2014). Linking stomatal traits and expression of slow anion channel genes *HvSLAH1* and *HvSLAC1* with grain yield for increasing salinity tolerance in barley. *Frontiers in Plant Science*, *5*(634). https://doi.org/10.3389/fpls.2014.00634.

Lu, P., Outlaw, W. H., Smith, B. G., & Freed, G. A. (1997). A new mechanism for the regulation of stomatal aperture size in intact leaves - accumulation of mesophyll-derived sucrose in the guard-cell wall of *Vicia faba*. *Plant Physiology*, *114*(1), 109–118. https://doi.org/10.1104/pp.114.1.109.

Medeiros, D. B., Souza, L. P., Antunes, W. C., Araujo, W. L., Daloso, D. M., & Fernie, A. R. (2018). Sucrose breakdown within guard cells provides substrates for glycolysis and glutamine biosynthesis during light-induced stomatal opening. *The Plant Journal*, *94*(4), 583–594. https://doi.org/10.1111/tpj.13889.

Miskin, K. E., Rasmusson, D. C., & Moss, D. N. (1972). Inheritance and physiological effects of stomatal frequency in barley. *Crop Science*, *12*, 780–783.

Misra, B. B., Assmann, S. M., & Chen, S. (2014). Plant single-cell and single-cell-type metabolomics. *Trends in Plant Science, 19*(10), 637–646. https://doi.org/10.1016/j.tplants.2014.05.005.

Misra, B. B., de Armas, E., Tong, Z., & Chen, S. (2015). Metabolomic responses of guard cells and mesophyll cells to bicarbonate. *PLoS One, 10*(12), e0144206. https://doi.org/10.1371/journal.pone.0144206.

Misra, B. B., Reichman, S. M., & Chen, S. (2018). The guard cell ionome: Understanding the role of ions in guard cell functions. *Progress in Biophysics & Molecular biology, 146*, 50–62. https://doi.org/10.1016/j.pbiomolbio.2018.11.007.

Navin, N., & Hicks, J. (2011). Future medical applications of single-cell sequencing in cancer. *Genome Medicine, 3*(5), 31. https://doi.org/10.1186/gm247.

Obulareddy, N., Panchal, S., & Melotto, M. (2013). Guard cell purification and RNA isolation suitable for high-throughput transcriptional analysis of cell-type responses to biotic stresses. *Molecular Plant-Microbe Interactions, 26*(8), 844–849. https://doi.org/10.1094/Mpmi-03-13-0081-Ta.

Oh, D. H., Barkla, B. J., Vera-Estrella, R., Pantoja, O., Lee, S. Y., Bohnert, H. J., et al. (2015). Cell type-specific responses to salinity - the epidermal bladder cell transcriptome of *Mesembryanthemum crystallinum. New Phytologist, 207*(3), 627–644. https://doi.org/10.1111/nph.13414.

Pandey, S., Wang, R. S., Wilson, L., Li, S., Zhao, Z., Gookin, T. E., et al. (2010). Boolean modeling of transcriptome data reveals novel modes of heterotrimeric G-protein action. *Molecular Systems Biology, 6*, 372. https://doi.org/10.1038/msb.2010.28.

Papanatsiou, M., Petersen, J., Henderson, L., Wang, Y., Christie, J. M., & Blatt, M. R. (2019). Optogenetic manipulation of stomatal kinetics improves carbon assimilation, water use, and growth. *Science, 363*(6434), 1456–1459. https://doi.org/10.1126/science.aaw0046.

Park, S.-Y., Fung, P., Nishimura, N., Jensen, D. R., Fujii, H., Zhao, Y., et al. (2009). Abscisic acid inhibits type 2C protein phosphatases via the PYR/PYL family of START proteins. *Science, 324*(5930), 1068–1071. https://doi.org/10.1126/science.1173041.

Pei, Z. M., Ward, J. M., Harper, J. F., & Schroeder, J. I. (1997). Chloride and malate uptake channel in *Vicia faba* guard cell vacuoles activated by the serine/threonine kinase, CDPK. *Plant Physiology, 114*(3), 1405. https://doi.org/10.1104/pp.116.2.785.

Peng, S., Laza, R. C., Khush, G. S., Sanico, A. L., Visperas, R. M., & Garcia, F. V. (1998). Transpiration efficiencies of indica and improved tropical japonica rice grown under irrigated conditions. *Euphytica, 103*, 103–108. https://doi.org/10.1023/A:1018383612439.

Rasouli, F., Kiani-Pouya, A., Li, L., Zhang, H., Chen, Z., Hedrich, R., et al. (2020). Sugar beet (Beta vulgaris) guard cells responses to salinity stress: A proteomic analysis. *International Journal of Molecular Sciences, 21*(7). https://doi.org/10.3390/ijms21072331.

Rasouli, F., Kiani-Pouya, A., Shabala, L., Li, L., Tahir, A., Yu, M., et al. (2021). Salinity effects on guard cell proteome in *Chenopodium quinoa. International Journal of Molecular Sciences, 22*(1), 428. https://doi.org/10.3390/ijms22010428.

Rasouli, F., Kiani-Pouya, A., Tahir, A., Shabala, L., Chen, Z., & Shabala, S. (2021). A comparative analysis of stomatal traits and photosynthetic responses in closely related halophytic and glycophytic species under saline conditions. *Environmental and Experimental Botany, 181*, 104300. https://doi.org/10.1016/j.envexpbot.2020.104300.

Rasouli, F., Kiani-Pouya, A., Zhang, H., & Shabala, S. (2020). Developing and validating protocols for mechanical isolation of guard-cell enriched epidermal peels for omics studies. *Functional Plant Biology, 47*(9), 803–814. https://doi.org/10.1071/fp20085.

Rui, Y., Chen, Y., Kandemir, B., Yi, H., Wang, J. Z., Puri, V. M., et al. (2018). Balancing strength and flexibility: How the synthesis, organization, and modification of guard cell walls govern stomatal development and dynamics. *Frontiers in Plant Science, 9*, 1202. https://doi.org/10.3389/fpls.2018.01202.

Salekdeh, G. H., Siopongco, J., Wade, L. J., Ghareyazie, B., & Bennett, J. (2002). Proteomic analysis of rice leaves during drought stress and recovery. *Proteomics*, *2*(9), 1131–1145. https://doi.org/10.1002/1615-9861.

Shabala, S. (2013). Learning from halophytes: Physiological basis and strategies to improve abiotic stress tolerance in crops. *Annals of Botany*, *112*(7), 1209–1221. https://doi.org/10.1093/aob/mct205.

Sharkey, T. D., Bernacchi, C. J., Farquhar, G. D., & Singsaas, E. L. (2007). Fitting photosynthetic carbon dioxide response curves for C(3) leaves. *Plant, Cell & Environment*, *30*(9), 1035–1040. https://doi.org/10.1111/j.1365-3040.2007.01710.x.

Specht, H., & Slavov, N. (2018). Transformative opportunities for single-cell proteomics. *Journal of Proteome Research*, *17*(8), 2565–2571. https://doi.org/10.1021/acs.jproteome.8b00257.

Trapnell, C. (2015). Defining cell types and states with single-cell genomics. *Genome Research*, *25*(10), 1491–1498. https://doi.org/10.1101/gr.190595.115.

Van Cutsem, E., Simonart, G., Degand, H., Faber, A. M., Morsomme, P., & Boutry, M. (2011). Gel-based and gel-free proteomic analysis of *Nicotiana tabacum* trichomes identifies proteins involved in secondary metabolism and in the (a)biotic stress response. *Proteomics*, *11*(3), 440–454. https://doi.org/10.1002/pmic.201000356.

Wang, H., & Clarke, J. M. (1993). Relationship of excised-leaf water-loss and stomatal frequency in wheat. *Canadian Journal of Plant Science*, *73*(1), 93–99. https://doi.org/10.4141/cjps93-012.

Wang, R. S., Pandey, S., Li, S., Gookin, T. E., Zhao, Z., Albert, R., et al. (2011). Common and unique elements of the ABA-regulated transcriptome of Arabidopsis guard cells. *BMC Genomics*, *12*, 216. https://doi.org/10.1186/1471-2164-12-216.

Wilson, R. S., & Nairn, A. C. (2018). Cell-type-specific proteomics: A neuroscience perspective. *PRO*, *6*(4). https://doi.org/10.3390/proteomes6040051.

Yao, X. H., Zhao, W. C., Yang, R., Wang, J. L., Zhao, F. K., & Wang, S. H. (2018). Preparation and applications of guard cell protoplasts from the leaf epidermis of *Solanum lycopersicum*. *Plant Methods*, *14*, 26. https://doi.org/10.1186/s13007-018-0294-7.

Yoo, M. J., Ma, T. Y., Zhu, N., Liu, L. H., Harmon, A. C., Wang, Q. M., et al. (2016). Genome-wide identification and homeolog-specific expression analysis of the SnRK2 genes in *Brassica napus* guard cells. *Plant Molecular Biology*, *91*(1–2), 211–227. https://doi.org/10.1007/s11103-016-0456-9.

Zeiger, E., & Hepler, P. K. (1976). Production of guard cell protoplasts from onion and tobacco. *Plant Physiology*, *58*(4), 492–498. https://doi.org/10.1104/pp.58.4.492.

Zhang, L., Foreman, D. P., Grant, P. A., Shrestha, B., Moody, S. A., Villiers, F., et al. (2014). In situ metabolic analysis of single plant cells by capillary microsampling and electrospray ionization mass spectrometry with ion mobility separation. *Analyst*, *139*(20), 5079–5085. https://doi.org/10.1039/C4AN01018C.

Zhang, T., Zhu, M. M., Zhu, N., Strul, J. M., Dufresne, C. P., Schneider, J. D., et al. (2016). Identification of thioredoxin targets in guard cell enriched epidermal peels using cysTMT proteomics. *Journal of Proteomics*, *133*, 48–53. https://doi.org/10.1016/j.jprot.2015.12.008.

Zhao, Z. X., Stanley, B. A., Zhang, W., & Assmann, S. M. (2010). ABA-regulated G protein signaling in Arabidopsis guard cells: A proteomic perspective. *Journal of Proteome Research*, *9*(4), 1637–1647. https://doi.org/10.1021/pr901011h.

Zhao, Z. X., Zhang, W., Stanley, B. A., & Assmann, S. M. (2008). Functional proteomics of *Arabidopsis thaliana* guard cells uncovers new stomatal signaling pathways. *The Plant Cell*, *20*(12), 3210–3226. https://doi.org/10.1105/tpc.108.063263.

Zhu, M. M., Dai, S. J., McClung, S., Yan, X. F., & Chen, S. X. (2009). Functional differentiation of *Brassica napus* guard cells and mesophyll cells revealed by comparative proteomics. *Molecular & Cellular Proteomics*, *8*(4), 752–766. https://doi.org/10.1074/mcp.M800343-MCP200.

FSC
www.fsc.org
MIX
Papier | Fördert
gute Waldnutzung
FSC® C083411

Kolibri 360 GmbH
Ferdinand-Jühlke-Straße 7
99095 Erfurt, Deutschland
produktsicherheit@kolibri360.de